AF840578

Biology of Rabbitfishes

Editors

Andrew S. Hoey

College of Science and Engineering
James Cook University
Townsville, Queensland
Australia

Brett M. Taylor

University of Guam Sea Grant and Marine Laboratory
University of Guam
Mangilao, Guam
USA

CRC Press
Taylor & Francis Group
Boca Raton London New York

CRC Press is an imprint of the
Taylor & Francis Group, an **informa** business
A SCIENCE PUBLISHERS BOOK

Cover credit: Images reproduced by permission of Victor Huertas and François Libert.

Clockwise from top left

Siganus corallinus (Victor Huertas)
Siganus lineatus (Victor Huertas)
Siganus spinus (François Libert)
Siganus argenteus (François Libert)

First edition published 2025
by CRC Press
2385 NW Executive Center Drive, Suite 320, Boca Raton FL 33431

and by CRC Press
4 Park Square, Milton Park, Abingdon, Oxon, OX14 4RN

© 2025 Andrew S. Hoey and Brett M. Taylor

CRC Press is an imprint of Taylor & Francis Group, LLC

Reasonable efforts have been made to publish reliable data and information, but the author and publisher cannot assume responsibility for the validity of all materials or the consequences of their use. The authors and publishers have attempted to trace the copyright holders of all material reproduced in this publication and apologize to copyright holders if permission to publish in this form has not been obtained. If any copyright material has not been acknowledged please write and let us know so we may rectify in any future reprint.

Except as permitted under U.S. Copyright Law, no part of this book may be reprinted, reproduced, transmitted, or utilized in any form by any electronic, mechanical, or other means, now known or hereafter invented, including photocopying, microfilming, and recording, or in any information storage or retrieval system, without written permission from the publishers.

For permission to photocopy or use material electronically from this work, access www.copyright.com or contact the Copyright Clearance Center, Inc. (CCC), 222 Rosewood Drive, Danvers, MA 01923, 978-750-8400. For works that are not available on CCC please contact mpkbookspermissions@tandf.co.uk

Trademark notice: Product or corporate names may be trademarks or registered trademarks and are used only for identification and explanation without intent to infringe.

Library of Congress Cataloging-in-Publication Data (applied for)

ISBN: 978-1-032-21649-2 (hbk)
ISBN: 978-1-032-21650-8 (pbk)
ISBN: 978-1-003-26940-3 (ebk)

DOI: 10.1201/9781003269403

Typeset in Times New Roman
by Prime Publishing Services

Foreword

You could certainly be forgiven for questioning why such a small family of fishes, comprised of only 1 genus and 31 species, is deserving of its own monograph. Let me, first and foremost, dispel any such qualms. Rabbitfish are an intriguing family of fishes that punch well above their weight when it comes to fascinating stories. For starters, rabbitfishes are, quite simply, unique among teleost fishes (the group that make up 96% of all extant fish species). In the skull of all other teleosts, the palatine bone connecting the 'cheek' to the jawbones is made up of a single ossification. In rabbitfish, the palatine is made up of two ossifications (as first described by Starks 1907). The presence of this 'pre-palatine' bone, combined with the poorly developed backward spines of the premaxillary and the short process from the back of the maxillary bone mean that the movement of the upper jaw is a swinging motion from a hinge, rather than the outward sliding motion seen in all other teleost fishes, giving rabbitfish their unique and characteristic "nibble".

Beyond their skeletal uniqueness, however, it is the evolutionary, behavioural, demographic, ecological, and economic stories of rabbitfish that are compelling. One of the things I have grown to love most about the family is that they defy generalisations: all rabbitfish were grouped as 'herbivores' until we discovered that some were not; all rabbitfish were classified as 'diurnal', until we discovered that some had the flexibility to alter their circadian rhythm and feed at night. The family holds many other behavioural mysteries: why do reef-based species live in male-female pairs but then migrate to an aggregation to spawn? What are the pairings for, if not for mating success?

They are a family that has secret weapons: the toxin injected by their numerous fin spines is the stuff of fishermens' legends, but yet has not been widely studied in terms of its chemical properties and potential. They are fast-growing, with high-levels of protein in their flesh making them an important food source in many parts of the Indo-Pacific, and giving them an economic story to tell. Their ecological story includes tales of 'migrants', biological invasion, and the discovery of unique ecosystem functions.

And underpinning it all, the family is torn in two by an intriguing series of dichotomies: half are schooling, half form pair-bonds; half live on coral reefs, half in estuaries and seagrass beds; half are colourful and deep-bodied, the other half are drab and fusiform-shaped. Such family division, combined with the many documented cases of hybridisation among rabbitfish species is, quite literally, the stuff of evolutionary mystery and family intrigue that would be worthy of any soap-opera plotline.

I started my PhD in 2008 armed with a copy of, what was then, one of the only reference books on rabbitfishes, David Woodland's seminal, 'Revision of the Fish Family Siganidae with Descriptions of Two New Species and Comments on Distribution and Biology'. This new book, 'The Biology of Rabbitfishes', which I am delighted also has David as a contributor, is a timely update of our scientific knowledge and understanding of this extraordinary family of fishes. My greatest hope is that rabbitfish can finally step out of the shadow of their more famous cousins, the surgeonfish. I hope that this book will cement the position of rabbitfish as a family of significant ecological importance in their own right, so that they can be protected as such.

Rebecca Bathgate (née Fox)
Canberra, Australia

Preface

Rabbitfishes (f. Siganidae) are common members of herbivorous fish assemblages on tropical and subtropical seascapes throughout the Indian and Pacific Oceans, and more recently the Mediterranean Sea, where they are particularly abundant on coastal reefs, seagrass meadows and macroalgal beds. Although there is some uncertainty as to the number of species of extant rabbitfishes with some authors recognising 29 species while others recognise at least 31 species, all species are grouped in a single genus, *Siganus*. Long assumed to be a functionally homogenous group on account of this taxonomic simplicity, the strong within-family morphological similarities and the classification of the vast majority of species as 'herbivores', rabbitfish have tended to languish as a footnote in the story of their more diverse and specious cousins, the surgeonfishes. However, the morphological similarities among rabbitfishes mask a series of intriguing familial dichotomies. All rabbitfishes are not equal. In fact there are marked variations in diet, distribution, socioecology, and life history strategies among rabbitfishes. Unpicking the assumption of homogeneity has opened a veritable Pandora's Box of evolutionary puzzles. What we now know is that this unique group of fishes not only contribute to important ecological processes on tropical reefs, but their 'live fast, die young' strategy means they represent potential highly productive fisheries resources, even on degraded macroalgal-dominated reefs.

Named for their small mouths and rapid bites, rabbitfishes share another characteristic with their terrestrial namesake; the ability to invade, establish and modify novel habitats. Indeed, rabbitfishes have successfully invaded the Mediterranean Sea and several subtropical and temperate environments where they have contributed to the deforestation of macroalgal beds and the functioning of these ecosystems.

This book introduces the reader to this unique and fascinating group of fishes, examining their biogeography, evolution, biology and conservation. It provides important insights into their evolution, ecological versatility, life-history strategies, and associations with non-reef habitats that have been key to their success and capacity to invade novel habitats. The book also highlights the potential resilience of rabbitfish populations to environmental change, their potential as a productive and sustainable fisheries and aquaculture resource, and ends with an overview of current and future research directions.

Andrew Hoey (Townsville, Australia)
Brett Taylor (Guam, USA)

Contents

1

Taxonomy and Biogeography of the Siganidae

David J. Woodland

The Siganidae (rabbitfishes or spinefoots) is a family of marine fishes that live in shallow inshore waters throughout the tropical Indo-West Pacific region from the east coast of Africa to French Polynesia. They are of moderate size; the smallest, *Siganus spinus*, attains at most 20 cm SL, the largest, *Siganus javus*, can reach 45 cm SL, though this is exceptional. A maximum size of 30–35 cm is common even for the largest species. Most of the species frequent coral reefs but some of these species may spend part of their life cycle around mangroves or seagrass beds. One species, *Siganus vermiculatus*, lives exclusively around mangroves, river mouths and in coastal lagoons (Hoey et al. – Chapter 7).

The body is laterally compressed and may be deep or slender, and the caudal fin can be truncate, emarginate or deeply forked. The mouth is small, with a single row of flattened close-set teeth on the premaxilla and on the dentary, creating a serrated "beak" which is not protrusible. Except for two species, the Pacific Ocean *Siganus puellus* and its Indian Ocean sibling *Siganus puelloides*, which mainly feed on monaxonid sponges and sessile colonial tunicates, the bulk of their diet is algae, cyanobacteria or less commonly, seagrasses (Woodland 1990, Hoey et al. 2013, Clements – Chapter 4). They consume the finer filamentous species of algae, especially when young, progressing to turf algae or the coarser fronds of macroalgae with increasing age.

The family Siganidae has several unique anatomical features. Most prominently, there are two spines in each pelvic fin, an inner and an outer spine separated by three soft rays, the inner spine attached to the abdomen by a membrane. Secondly, there are 13 spines and 10 rays in the dorsal fin and seven spines and nine rays in the anal fin. Exceptions to this formula are rare. For example, occasionally the first dorsal fin ray may be replaced by a spine, clearly a development abnormality. Thirdly, the first two dorsal fin spines arise from the first pterygiophore. Projecting

Department of Zoology, University of New England, Armidale, New South Wales, Australia.
Email: davidjwoodland@bigpond.com

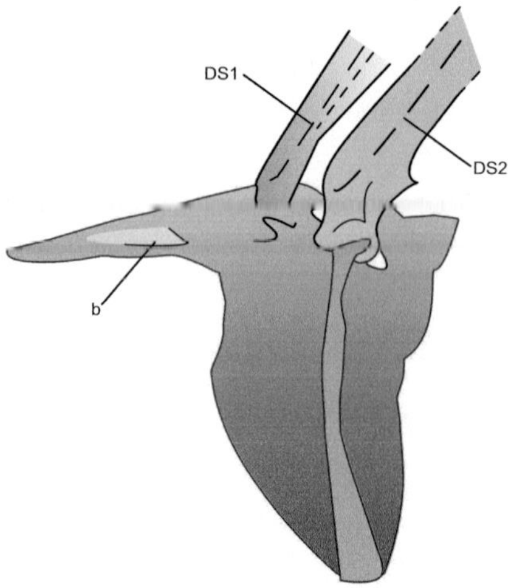

Fig. 1. Barbed procumbent spine (b) of the first dorsal-fin pterygiophore; the first two dorsal-fin spines (DS1 and DS2) are attached to this pterygiophore (modified after Tyler and Bannikov 1997).

from this pterygiophore is a procumbent spine which sits in a pocket that opens via a slit on the nape (Fig. 1). This spine is not part of the dorsal fin, but it is sometimes erroneously treated in this way (Helfman et al. 2009, Swainston 2020, Golani et al. 2021). Because the spine carries a barb it is considered to play a defensive role, but as it barely protrudes, if at all, from the slit and as it is not venomous it must play a minor role compared to the fin spines. All the fin spines are venomous. There is no venom sac, the venom cells lie in a groove along the spine. The venom is released when the tissue surrounding the spine is traumatised as it penetrates the victim. Another unique character is that the palatines are each composed of two separate ossifications joined by a hinge, an arrangement that plays a role in the articulation of the upper jaw (Tyler and Bannikov 1997).

There are at least 31 extant species of rabbitfish. Woodland (1990) recognised 27 species in his revision of the family. Since the publication of that revision Randall and Kulbicki (2005) have described *Siganus woodlandi* from the West Pacific, and Woodland and Anderson (2014) have described *Siganus insomnis* from the Maldives and southern India, while Kuiter and Debelius (2001) and Woodland (2022) have recognised *Siganus stellatus* and *Siganus laqueus* (von Bonde) as full species rather than subspecies of *S. stellatus* (Forsskal). Molecular studies have for the most part validated the classification determined from morphological characters. However, a molecular study of nucleotide sequences by Kuriiwa et al. (2007) identified two incipient species differing not only genetically but morphologically within what Woodland (1990) had treated as a single species, *Siganus corallinus* (Valenciennes). Although this species is yet to be described it, its presence takes the total number of rabbitfish species to 31. Furthermore, based on two sightings and an inconclusive underwater photograph, Stockwell and Carpenter (2012) tentatively recorded *Siganus trispilos* from Sipidan Island off northeastern Sabah. Apart from this single record for the Indo-Malayan Archipelago *S. trispilos* was believed to be endemic to the Northwest Shelf of Western Australia. This curiously disjunct population led the authors to suggest it might be an undescribed species.

There are also some uncertainties regarding the status of the sister taxa *Siganus canaliculatus* (Park) and *Siganus fuscescens* (Houttuyn). Woodland (1990) recognised *S. canaliculatus* and *S. fuscescens* as separate species based on the distribution and number of spots on the sides of the fish, with *S. canaliculatus* having fewer spots on the sides than *S. fuscescens*. The decision to treat these two taxa as separate species was believed to be confirmed by an allozyme study on fish that were caught in the vicinity of the island of Singapore (Adams and Woodland 1994). The two different colour morphs seen in the Singapore market were said to be fished from different habitats, *S. canaliculatus* from inshore sand flats, *S. fuscescens* from offshore reefs. Analysis of the allozyme data indicated that these local populations with their different markings were not interbreeding. However, molecular studies of these putatively different species from specimens obtained from over a much wider area have led to different conclusions. Hsu et al. (2011) found that populations from Indonesia and Taiwan that they identified as *S. canaliculatus* and *S. fuscescens* (*sensu* Woodland) were interbreeding; they concluded they were colour morphs of the one biological species. Kuriiwa et al. (2007) found that specimens from Japan and Indonesia that were identified as these two taxa lacked sufficient genetic differentiation to be considered different species. On the other hand, Borsa et al. (2007) found that two fish from Riau and the Solomon Islands identified as *S. canaliculatus* were genetically different from individuals identified as *S. fuscescens* rom Riau, Solomon Islands, New Caledonia and the Bismarck Archipelago. As Riau and Singapore are essentially the one locality this result supports the conclusion from the allozyme study of Adams and Woodland (1994), namely that there are at least some populations distinguishable from one another by their spot pattern that are good biological species. Clearly, the spot pattern is not a universally reliable taxonomic character for distinguishing between these taxa. Moreover, in another genetic study Ravago-Gotanco et al. (2018) found two cryptic species among fishes from the Philippines and Hong Kong that they identified from their markings as *S. fuscescens*. More studies are needed to resolve the taxonomy of the fishes of this complex, more so because they are important food fishes that are heavily fished in densely populated areas such as the Philippines and Indonesia.

Origins of the Taxonomy of the Siganidae

Fricke et al. (2024 in Eschmeyer's Catalog of Fishes) attribute the genus name *Siganus* to Fabricius (https://www.calacademy.org/scientists/projects/eschmeyers-catalog-of-fishes, accessed 11 June 2024). Previously *Siganus* was attributed to Forsskal (1775). Forsskal was a member of an expedition sponsored by Frederick V of Denmark-Norway to report on the cultures, languages, geography, animals and plants of places in the Middle East. Its primary objective was to gather information which might lead to a clearer interpretation of the Bible. The party consisted of six persons chosen for their expertise in various disciplines. Forsskal was the biologist. He described the species of plants and animals he collected. Some of his fish specimens were sent back to Denmark where they are now in the Zoologisk Museum, Copenhagen.

The party spent 18 months at various locations in the Mediterranean Sea before entering the Red Sea in October 1762. Here Forsskal was confronted by an Indo-Pacific marine fauna, including genera and indeed families of fishes not to be found in the Mediterranean or Atlantic Seas. The party travelled the length of the Red Sea and the adjacent eastern hinterland, encountering numerous difficulties as well as being exposed to malaria. Forsskal died from malaria on 11 July 1763; he was the second member of the expedition to die. The four surviving members decided to leave for Bombay; a further two died while aboard the ship and another at Bombay. The sole survivor was the cartographer Niebuhr who returned to Denmark via the Persian Gulf. It was left to Niebuhr to produce reports on the results of the expedition. Among these was the publication of the descriptions of numerous species of plants and animals which Forsskal had written on slips of paper, the "*Descriptiones animalium, avium, amphibiorum, piscium, insectorum, vermium*" (Forsskal 1775). Niebuhr confessed to having great difficulty in preparing this material for publication, compounded by the fact that some slips appeared to have been lost. He also said that he had intermingled remarks of his own "without distinguishing them as such for the greater part of the whole is Mr Forskal's (*sic*)". On the title page of the "*Descriptiones animalium...*" Niebuhr credits himself only as the Editor. Moreover, he makes no mention in his Foreword of a third party being involved in the preparation of the publication. However, Fricke (2008) found a letter to Niebuhr dated 11 Jan. 1776 from J.C. Fabricius (best known as an insect taxonomist) in which he regrets that Niebuhr had not been happy with an unspecified piece of work that Niebuhr had employed him to do. Fricke concluded that he had been remunerated to help Niebuhr prepare Forsskal's descriptions for publication, and that because of the sum involved his contribution must have been significant (Fricke 2008).

The presentation of the material in the "*Descriptiones animalium...*" can be inconsistent and the entries for *Siganus* are especially confusing. On an unnumbered page preceding the descriptions of the fishes there is a table with the heading "Nova Genera. Piscium" which lists SALARIA, SIGANUS, SCARUS, ACANTHURUS in upper case. Each is followed by a second genus name and a numeral in lower case which refers to the order in which the descriptions of species occur in the text. For example, "SIGANUS (Scarus 9. 10.)" refers to the two species of rabbitfish described later in the text. Similarly, "SCARUS (Scarus 11-18)" refers to the descriptions of eight species Forsskal assigned to the genus *Scarus*. This information is repeated in a subsequent table on a page numbered "X". Confusingly, under the heading "SCARUS: novum genus" species No. 9 is listed as "rivulatus: Djezavi vel Sigian. nov. gen. SIGANUS", where "djezavi" and "sigian" are the common names for this fish transcribed from Arabic. Since part of Forsskal's brief was to record the Arabic names of the animals (and plants) he described, Fricke (2008) contends that Forsskal would have provided the Arabic names for the fish he described as *Scarus rivulatus*. But Fricke goes further, quote "It was Fabricius in Niebuhr who apparently thought this may be another new genus and added the name *Siganus* new genus". Unlike the genus *Scarus*, the genus *Siganus* is not formally defined in the text; its distinctive features are only available from the descriptions of the two included species, notably the two spines in the ventral fins. To summarize Fricke,

Fabricius latinized the Arabic name "Sigian" to create an acceptable genus name for the fish which Forsskal had placed in the genus *Scarus*, but which Fabricius recognized should be in a separate genus. But Niebuhr (1772) appears to contradict this in a passage from his "*Beschreibung von Arabien*" (Travels in Arabia). Quoting from the English edition of this work: "Mr Forskal discovered in the Red Sea several sorts of Ray-fishes which are unknown in Europe. I was told by my friend that in the short passage between Suez and Jidda he observed more than a hundred new species, only part of which he could rank among the known genera. He was obliged to form four new genera, which he named *Salaria*, *Scarus*, *Signanus* (*sic*) and *Acanthurus*".

The billets on which Forsskal described his fishes have not survived so it is virtually impossible to determine any contributions by a third party to the "*Descriptiones animalium...*". Nevertheless, there is a contribution by Niebuhr that can be identified, a footnote to the description of *Siganus rivulatus*: "Pisces Lohajae 1/2 uln. long; Basrae fere ulnae", i.e., "Fishes from Al Luhayyah half an ell long, from Basra about an ell". A Danish ell was 63 cm. This comment must have been added by Niebuhr since he was the only member of the party to reach Basra located at the head of the Persian Gulf. The remark should not be taken literally since the maximum size for this species is about 40 cm TL. The type specimens of *S. rivulatus* in the Zoologisk Museum, Copenhagen are 13 and 16 cm SL. The author previously claimed that Basra might be a second type locality for this species (Woodland 1990, p. 45), but there is no evidence that either of these specimens was collected by Niebuhr. The author believes that the genus *Siganus* and the species *S. rivulatus* should both be attributed to Forsskal (in Niebuhr) 1775, rather than Forsskal and Niebuhr, since that footnote neither contributes to the description of the genus or the species. While Fricke's reasoning is plausible, according to Kottelat (2013a, p. 492), Fabricius does not meet the requirements of the International Commission of Zoological Nomenclature (ICZN) *Code* (art. 50.1.1) to be the author of *Siganus* since there is no *conclusive* evidence that he contributed to its description.

Fowler (1904) created the subgenus *Siganites* with *S. canaliculatus* (Park) as the type species. His defining character was "Cheek naked or only with a small area including a few scattered scales". By contrast Fowler noted that the other siganids he had examined (*S. javus*, *S. vermiculatus*, *S. virgatus*, *S. corallinus*) had "cheeks usually covered with strong overlapping scales". He placed these in the subgenus *Teuthis* Linnaeus, which was at that time frequently used as an alternative to *Siganus*. *S. canaliculatus* is a member of that clade of siganids that are "fusiform" compared to those that are "deep-bodied" (e.g., *S. lineatus*). The molecular classification of the siganids into "deep-bodied" and "fusiform" clades are equivalent to Fowler's subgenera *Teuthis* and *Siganites*, respectively. Fowler's defining character for *Siganites* was necessarily imprecise as the amount of scaling on the cheeks can vary quite widely, even within the one species. A better defining character, apparently unnoticed by Fowler, would have been the presence or absence of scales on the thoracic area. In deep-bodied species this area is scaled. In fusiform species the area from the isthmus through to the base of the pelvic fins is devoid of scales. It seems that the strong overlapping scales on the cheeks and the thoracic area in deep-bodied species serve to protect these areas from abrasion. A deep body is characteristic

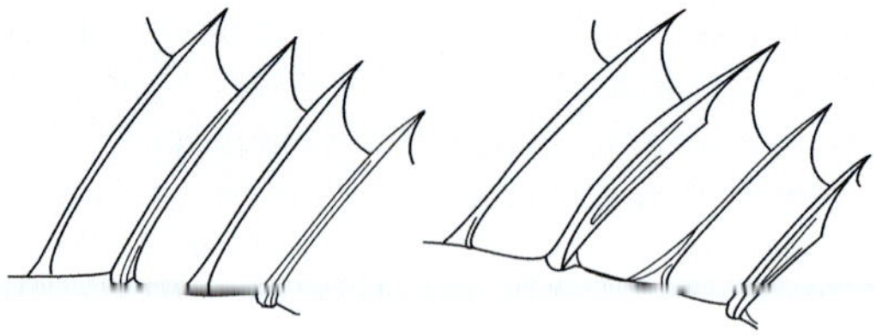

Fig. 2. Spines of the dorsal and anal fins of siganids may be either slender (left) or stout (right). Slender spines occur in the fusiform species that move in schools with the tides to forage over reef flats. Stout spines are typical of the deep-bodied species resident on coral reefs.

of those siganids that manoeuvre their way through a coral reef landscape (*S. vermiculatus* being an exception, it is typically found around estuaries and in coastal lagoons). Except for *S. spinus* the fusiform species have slender fragile fin spines and move with the rising tide from deeper water to feed on shallow reef flats or beach rock zones. On the other hand, small schools of adult *S. spinus* are usually found in shallow water along the reef front, often venturing onto the reef flat to browse on filamentous and macroalgae (Bryan 1975); this would explain why it has robust spines like the deep bodied species that live amongst coral formations (Fig. 2).

The genus *Lo* was created by Seale (1906) for a siganid with a "greatly produced snout", type species, *Lo vulpinus* (Schlegel and Muller 1845). Later in the same year it was redescribed in the publication "Fishes of Samoa" by Jordan and Seale (1906), where it is expressly stated "This is known to us from one specimen brought by Mr Seale from the Solomon Islands". Seale (1906) gave the reason for his choice of "Lo" as it being the name used by the Samoans for the species of Siganidae. Although that is correct there are no species of *Lo* in the islands of Samoa. Jordan (1927) further added to the confusion about the distribution of the species of *Lo* by including *Lo vulpinus* in his "Shore Fishes of Samoa": e.g., Axelrod and Emmens (1973) gave Samoa as a locality record for *Lo vulpinus*. There are five species of *Lo,* and their easterly limit is Tonga (Fig. 3). Their distribution is informative from a biogeography point of view, and it is discussed below.

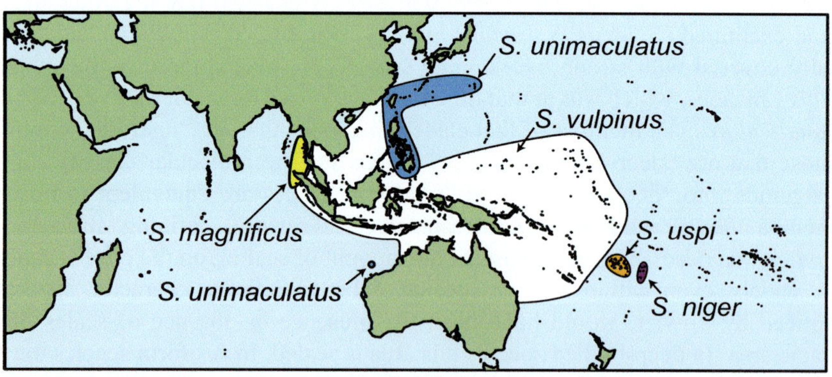

Fig. 3. Distributions of the five species of foxface (subgenus *Lo*): from left to right, *Siganus magnificus* (Andaman Sea), *Siganus vulpinus* (Pacific Ocean), *Siganus unimaculatus* (N.W. Pacific and Australia's N.W. Shelf), *Siganus uspi* (Fiji), *Siganus niger* (Tonga).

Teuthis vs. *Siganus*

At various times rabbitfishes have been assigned to the genus *Siganus* (Siganidae) or alternatively *Teuthis* (Teuthididae). Wanting this situation resolved David Starr Jordan, one of the Commissioners of the ICZN recommended to the Commission that *Teuthis* be used for that family of fishes represented by *Teuthis javus* Linnaeus 1766 (a rabbitfish) and not to the family represented by the acanthurid *Teuthis hepatus* Linnaeus 1766 (ICZN 1926, Opinion 93). While this should have resolved the issue, for an unknown reason the ICZN failed to add *Teuthis* to the Official List of Generic Names in Zoology, and the ruling was ignored by many taxonomists who continued to use *Siganus* for rabbitfishes.

The confusion about the use of these generic names lies with Linnaeus. In the 12th Edition of his "*Systema Naturae*" Linnaeus (1766), convinced of the value of his binomial system of naming species, imposed binomial names on a number of species previously described by earlier authors but identified by polynomial names. Among these were species belonging to families which we now classify in the Acanthuridae and Siganidae but which Linnaeus united in a single genus, *Teuthis*. He described what he said were two species of *Teuthis*: *T. hepatus*, an acanthurid and *T. javus,* a siganid. "Teuthis" had been used previously as a genus by Browne (1756) for an acanthurid, and "Hepatus" had been created as a genus by Gronovius (1763) to accommodate two species that he described from specimens in his collection, one an acanthurid, the other a siganid from Java. Linnaeus did not describe these species from specimens; his descriptions were compiled entirely from the published descriptions of authors that used the polynomial naming system. His description of *Teuthis hepatus* was compiled from five descriptions by different authors of four species of which at least three, and possibly all four, were acanthurids. Furthermore, two of those descriptions were accompanied by excellent colour plates of the fish being described and they are obviously of very different fishes. Considering that more than one species was eligible to be called *Teuthis hepatus* it is not surprising that no acanthurid is now given the genus name *Teuthis*. However, *hepatus* survives in *Paracanthurus hepatus* (Linnaeus), one of the species included by Linnaeus in his genus *Teuthis* and illustrated in colour by Seba (1759). *Teuthis javus* of Linnaeus has fared better. It was based on a description accompanied by an excellent illustration by L.T. Gronovius of a fish in his collection who named it "Hepatus cauda fronteque inermibus" to distinguish it from another specimen in his collection that he described as "Hepatus mucrone reflexo utrinque prope caudam", the latter name referring to the scalpel-like spine that folds into a groove on the caudal peduncle that is not present (*inermibus*) in siganids but is present in some genera of acanthurids (Gronovius 1763).

In 1968, Nielsen and Klausewitz asked the ICZN to validate the generic name *Siganus* Forsskal, 1775. They argued that the posthumous publication of the "*Descriptiones animalium...*", where the genus *Siganus* and its included species *rivulatus* are described, explained the irregularities in the text. Nevertheless, they said the published descriptions met the criteria for *Siganus* to be recognised as a valid genus. Woodland (1972, 1973) supported their proposal on the grounds that

Siganus was being more commonly used than *Teuthis* for the rabbitfishes and went further proposing that the relevant part of Opinion 93 of ICZN be annulled and the genus name *Teuthis* Linnaeus, 1766 be placed on the Official Index of Rejected and Invalid Generic Names in Zoology. Taylor (1970) however argued that the original recommendation of the ICZN (Opinion 93) should be reaffirmed, since there was no point in validating a genus name that would be a junior synonym of *Teuthis* Linnaeus. Kottelat (2013b) has published a comprehensive review of the background to this taxonomic impasse. He concludes that since the ICZN has not responded to any of the proposals dating back to Nielsen and Klausewitz (1968) they now have no alternative but to ratify the current prevailing usage of *Siganus*.

Molecular Taxonomy

In his revision of the family Woodland (1990) treated *Lo* (Seale 1906) as a subgenus to accommodate those species with a tubulate snout referred to in the aquarium trade as "foxfaces" because of a fanciful resemblance of the markings on the head to the markings on the muzzle of the common fox. He left the other species in the subgenus *Siganus*. To quote from Woodland (1990, p. 18): "Even though the subgenus *Siganus* is acting as a repository for species other than those in *Lo*, it would be unwise to subdivide the subgenus without a study of the cladistic relationships of the species involved". Molecular studies on the phylogeny of the extant species have now been conducted; these allocate them to three distinct clades (Borsa et al. 2007, Kuriiwa et al. 2007, Siqueira et al. 2019). One clade includes all the deep bodied species, with the species of *Lo* as a subclade. Another clade contains the seven shallow bodied species, designated as "fusiform" by Borsa et al. (2007). The third clade contains just two species, *S. argenteus* and *S. woodlandi*. *S. argenteus* has a streamlined spindle-shape body, the spines and rays of the median fins are short, the caudal peduncle is slender, and the caudal fin is stiffened and very deeply forked, features associated with being a strong swimmer. This species is often seen swimming in mid water, especially around the wave zone at the reef front or occasionally in surge channels (Tobias 1976). *S. woodlandi* closely resembles *S. argenteus*, but the body is slightly deeper, and the snout is slightly pointed with the cheeks more fully scaled, characters usually associated with being a bottom feeder. Randall and Kulbicki (2005) found its digestive tracts packed with macroalgae but also observed it feeding in mid water, "presumably on zooplankton". Although these two species are obviously related and are united in a separate clade, they are separated from one another by a substantial nucleotide distance (Lemer et al. 2007).

Biogeography

The Siganidae can be traced back in the fossil record to its beginnings in the Early Eocene (*ca.* 50 mya). Ten species in eight genera have been described from skeletal fossils from sites in Iran, Turkmenistan, Russian Caucasus, Switzerland and, best known of all for the exceptional preservation of the specimens, Monte Bolca in northern Italy, two separate sites, a few kilometres apart and slightly different in age - Middle Eocene (Friedman and Carnevale 2018). These deposits were formed in the shallow warm seas of the western Tethys. As the Africa and Arabia Plates moved northward against the Eurasian Plate these deposits were uplifted and displaced onto their continental locations. During this process the western Tethys was constricted to a proto-Mediterranean Sea and finally the Mediterranean Sea of today. The closures of the two existing seaways between the Mediterranean Sea and the Indian Ocean were complete by about 14 mya (Bialik et al. 2019). With the only connection now being with the north Atlantic through the Strait of Gibraltar, conditions in the Mediterranean Sea changed. During the Pleistocene there were successive invasions of cold-water marine fish from the Atlantic into the Mediterranean; these were correlated with the Glacial – Interglacial cycles. Over time there was a progressive loss of Indo-Pacific taxa and an increase of Atlantic taxa. Warm water fish finally disappeared from the Mediterranean Sea during the late Pleistocene (Landini and Sorbini 2005, Agiadi et al. 2018). At the same time a new centre which was to become extremely rich in all kinds of marine organisms, including the Siganidae, was being created in the Indo-Australian Archipelago (IAA) (Fig. 4).

It seems not to have been remarked that almost all the fossilized siganids were extremely small fish. Only two of the specimens that contributed to the descriptions of the eight genera and 10 species had a standard length (SL) greater than 5 cm: an 11.4 cm SL specimen from the Oligocene of Switzerland and a 16.4 cm SL fish from the Middle Eocene of Monte Bolca, Pesciara. The remainder range in size from 1-5 cm SL. Compared to extant siganids fish this small are at most settlement size. They were likely living in quiet backwaters of an inner reef flat or a lagoon where fine sediments and an anoxic event made fossilization possible. Despite their extremely small size their body proportions indicate that with two exceptions they belonged to deep or moderately deep bodied species. The exceptions are the 3.5 cm SL *Caucasiganus eocaenicus* Bannikov, Tyler and Sorbini from the North Caucasus (Bannikov et al. 2010) and the 1-5.5 cm SL specimens of *Ruffoichthys spinosus* Sorbini from the Middle Eocene of Monte Bolca (Tyler and Bannikov 1997). Both these species have oblong slender bodies. Interestingly, *Ruffoichthys bannikovi* Tyler and Sorbini, 4.7 cm SL, also from the Middle Eocene of Monte Bolca, as is its congener, is moderately deep bodied, not elongate. The two large fish are also relatively deep bodied. In extant siganids the slender bodied species migrate in large

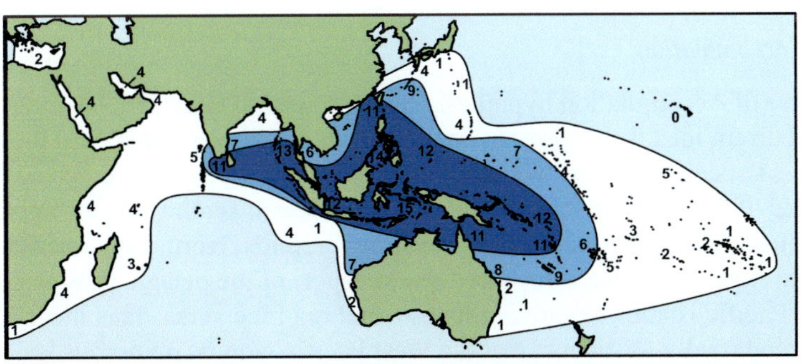

Fig. 4. The number of species of Siganidae found at various localities throughout the range of the family. The two Mediterranean species (*Siganus rivulatus* and *Siganus luridus*) entered that Sea from the Red Sea following the opening of the Suez Canal. White: 1–5 species; light blue: 6–10 species; dark blue: > 10 species

schools with the tides from deeper areas to feed on flooding reef flats. On the other hand, most deep bodied species reside in small schools or in pairs on deeper habitats of a coral reef. It seems that by the Middle Miocene, 16-11.6 mya, siganids may have already evolved to exploit reef habitats in different ways.

High Species Richness in the Indo-Australian Archipelago

Around 20 mya the India Australian plate drifting north contacted the Eurasian plate. Over the next 20 my what had once been a single tropical ocean was divided into two oceans, the Pacific Ocean and the Indian Ocean. This separation between the two oceans became more complete as the movement north generated the Indonesian Archipelago, a process that is continuing today (Hall 2013). Nevertheless, there is always a connection between the two oceans, the Indonesia Through Flow (ITF). The ITF continues to flow, although certainly with reduced intensity at the times of Glacial Maxima, especially when sea levels are at their lowest, around 130 m lower than when at their interglacial maximum level such as exists today.

There are three main hypotheses about the historical factors that have determined the existing pattern of species richness in the Tropical Indo-Pacific Ocean: Centre of Origin, Centre of Accumulation/Survival, and Centre of Overlap. The focus of all three hypotheses has been an explanation for the incredible number of species present in the Indo-Australian Archipelago (IAA), particularly, the centre with most species within that region, the Coral Triangle (CT). Although previously regarded as competing hypotheses, the recent consensus is that the three hypotheses should not be seen as mutually exclusive, rather a number of factors will have contributed to the pattern of species richness we see today. Nevertheless, some factors may still have played a bigger role than others.

Centre of Origin

The Centre of Origin hypothesis asserts that the species richness of the Indo-Australian Archipelago (IAA) is the result of *in situ* speciation and not the result of an accumulation of species by migration from outside the IAA (Briggs 1984). The alternative is the “Centre of Accumulation” hypothesis (a variation of the Centre of Survival hypothesis).

Centre of Accumulation

The Centre of Accumulation hypothesis can be traced to Ladd (1960). Ladd opposed the then current idea that Indonesia, with its rich molluscan fauna, was the centre of dispersal of species into the Western Pacific. Rather, he proposed species arose in the islands of the Western Pacific and drifted west and southwest “to find a suitable niche in hospitable Indonesia”. The prevailing currents, North Equatorial and South Equatorial flowing eastward favoured the transport of the pelagic larvae of molluscs from the Pacific islands towards Indonesia and not vice versa. This theory has been recast more broadly to suggest that the West Pacific with its numerous islands could be the source of many species in groups as diverse as corals, echinoderms and fishes.

Centre of Overlap

Although not called the Centre of Overlap in the original paper, the Centre of Overlap hypothesis was first proposed by the author to explain the presence of sibling pairs of siganid that shared a zone of sympatry within the IAA (Woodland 1983). The classic examples were pairs of sibling species of siganid each with an Indian Ocean and a Pacific Ocean-centred distribution that shared a common border within the IAA. The distribution of one of these pairs of species, *S. doliatus* and *S. virgatus*, is reproduced here from Woodland 1990 (Fig. 5). The sibling species pair, *S. guttatus* and *S. lineatus*, have a similar distribution (Fig. 19 in Woodland 1990). That zone of sympatry/parapatry sits between the Asian continental shelf to the west and the Australian continental shelf to the east, identified as the Sunda Shelf and the Sahul Shelf, respectively. These shelves are submerged at the present time but are emergent at times of sea level lows around the times of Glacial Maxima. At those times of sea level lows, the discharge of freshwater accompanied by sediment into the narrowed seaway between these two shelves would likely lead to the demise of coral reefs in that zone. Lower salinity, sedimentary coastlines, and a loss of coral reefs would for a time reduce gene flow between populations lying to the east and west of that area, leading to the creation of sibling species, one with an Indian Ocean centred distribution and the other with a Pacific Ocean centred distribution. These sibling species are not confined to the IAA, they range eastward to Sri Lanka on the one hand and to islands in the West Pacific Ocean on the other. Then there are other species found in the IAA that are found in the Coral Triangle and islands on the western fringe of the Pacific Ocean: *S. punctatissimus, S. woodlandi, S. randalli.* It can be argued that these species evolved either in the eastern half of the IAA or on islands of the Western Pacific.

Even at present sea levels the Malacca Strait (between Sumatra and Malaya) and the Java Sea (between Borneo, Sumatra and Java) is characterised by low salinity, high turbidity, mud bottoms and poor coral reef development (Defant 1961, Tjia 1966). This means that IAA must act as a partial barrier for corallophilic species between the Indian and Pacific Oceans even when sea levels are high as well as a more effective barrier when sea levels are low. In other words, the Indian and Pacific

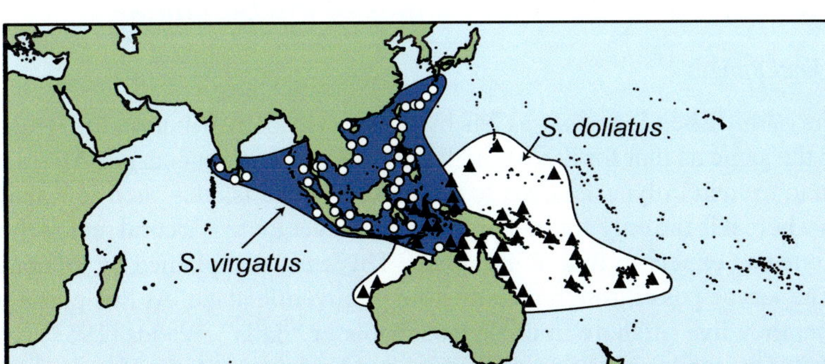

Fig. 5. Locality records and distributions for the sibling species *Siganus doliatus* (black triangles) and *Siganus virgatus* (white circles) showing the overlap in their distributions between the Bird's Head of New Guinea and Sulawesi.

Oceans may be considered as separate biogeographic regions. Many siganid species and other Tropical Indo-West Pacific fishes add support to the Region of Overlap proposition.

Support for the Centre of Overlap proposition comes from a number of studies. The earliest was for anemone fishes, *Amphiprion* spp., by Allen (1975), who also invoked Quaternary lowering in sea level as the likely explanation for the situation. Gaither and Rocha (2013) reviewed the evidence in support of the Centre of Overlap from studies of 36 sibling pairs of fish species which had a Pacific and an Indian Ocean distribution with an overlap in the IAA. The authors of the original studies defined the sibling species on morphological similarities while 17 of those studies also included molecular data that was used to estimate divergence times between the siblings. They found overwhelming support for the importance of sibling species in contributing to the species diversity and species richness of the IAA. Furthermore, the estimates of the times of divergence between the siblings showed that the events leading to their genetic differentiation occurred in the Pleistocene (estimated divergence times range, 0.3–6.6 mya).

Unlike the Centre of Origin and Centre of Accumulation hypotheses the Centre of Overlap proposition offered no suggestions about where the ancestral species of any sibling pair might have originated. It merely stated that a number of sibling pairs of siganid had Indian Ocean and Pacific Ocean-centred distributions that meet within the IAA, their evolutionary divergence arising from the separation of sibling populations during times of sea level lows. However, Matias and Rignios (2018) in their review of the "Centre Hypotheses" interpreted the Centre of Overlap more broadly, "as an amalgamation of divergent groups in the Coral Triangle after the recession of the Indo-Pacific Barrier (Woodland 1983)". They say this highlights the role of peripheral regions and isolated islands in generating biodiversity. This interpretation appears to include aspects of the Centre of Accumulation hypothesis. Nevertheless, since the Siganidae includes many corallophilic species it is likely that some of those species may have evolved in isolation in islands in the Western Pacific Ocean. The peripheral evolution of species of *Lo* in Fiji and Tonga and the Andaman Sea supports this idea.

Longitudinal and Latitudinal Ranges of the Siganidae

Longitudinal Ranges

Of all the rabbitfishes *S. argenteus* has by far the widest distribution. Its longitudinal range is the same as that for the whole family—from the east coast of Africa and the Red Sea to French Polynesia, even to the more isolated islands such as Rapa Iti and Pitcairn where it is the only siganid present. Wherever there is a coral reef in the South Pacific one can expect to find *S. argenteus*. This can be explained by its unique life cycle. The larvae pass through a distinctive prejuvenile stage. At this phase in their life cycle, they live offshore in compact schools or "balls". Woods (1953) collected prejuveniles two miles out to sea off Bikini and Enewetak Atolls. He noted that they had large fatty livers, which he said would improve their buoyancy. The livers might also act as an energy reserve if the prejuveniles were to be transported away from their home reef by adverse currents. After an extended period at sea, they return to the reef flats where they metamorphose into the juvenile form. After spending four to eight weeks on the reef flat they migrate to live in deeper water over the seaward reef slope and the submarine terrace where they school in mid water, diving to the bottom to feed on turf algae. Less frequently they feed on the reef flat or in surge channels (Tobias 1976). Adult fish have all the attributes of fish that live in open water: a spindle shaped streamlined body, a slender caudal peduncle, and a stiffened deeply forked caudal fin.

Siganus spinus is the second most widely distributed species of siganid. It is found on coral reefs throughout most of the Indo-West Pacific Region: from Sri Lanka to the Society Islands, north to southern Honshu and south to New Caledonia (Woodland 1990). It does not pass through a distinctive prejuvenile stage; however, it also spends the first weeks of its life in the open sea living in compact schools. Juveniles were recovered from the guts of skipjack tuna (*Katsuwonis pelamis*) caught 78 km offshore out from Lagonoy Gulf, Philippines. Their age was determined from their otoliths to be three weeks. A few days later juveniles invaded the inshore reef flats of the Gulf in large numbers (Soliman et al. 2010). Unlike *S. argenteus* they have a squat fusiform body and a truncate caudal fin. Adult fish are often seen browsing along the reef front in small schools. *S. argenteus* and *S. spinus* are the only siganids that lead a pelagic life far out to sea before settling as juveniles on coral reefs. Presumably this explains their much broader longitudinal ranges when compared to the other species of siganid.

Latitudinal Ranges

Although typically associated with coral reefs (but see Hoey et al. – Chapter 7) the latitudinal limits of the family extend further north and south than the respective limits of coral reefs in both the Pacific and Indian Ocean. The species involved are all members of the fusiform clade, species that are not obligate coral reef dwellers: *S. fuscescens* north to southern Honshu, Japan and south to Jervis Bay and Port Cockburn on the east and west coasts of Australia (Hutchins 1991), *S. sutor* south to Port Elizabeth, S. Africa (Floros et al. 2012), and lastly *S. luridus* and *S. rivulatus* recent arrivals in the warmer waters of the eastern part of the Mediterranean Sea following the opening of the Suez Canal. The warm waters of the Kuroshio, East Australia, Leeuwin and Agulhas currents, respectively, enable these species to extend their ranges to temperate latitudes. At low latitudes these siganids are found on coral reefs and seagrass habitats, at slightly higher latitudes on rock reefs encrusted with coral, the reefs providing the substrate for the benthic algae on which they feed. At the highest latitudes rock reefs or firm sand bottoms substitute for coral reef flats.

Invasive species in the Mediterranean Sea

Since the opening of the Suez Canal in 1869 more than 100 species of fish have invaded the Mediterranean from the Red Sea. Among these are two of the four siganids present in the Red Sea: *S. rivulatus* was first recorded from the Palestine coast in 1927 and *S. luridus* from the coast of Israel in 1955. Both species have since extended their distributions throughout the warmer southernmost parts of the eastern Mediterranean Sea (Golani et al. 2021). There are sporadic records of these species from the colder western Mediterranean Sea but there are no established populations. These two species have been so successful that they have become important components of commercial catches of inshore fishes. At least part of their success can be attributed to the broad range of algae on which they feed; five green, 10 red and 10 brown species of algae in both seas (Bariche 2006). They feed on the most abundant species when food is plentiful, on a broader range when food is scarce. A comparison of diets of fish from the Libyan coast and the Red Sea showed that although their diets differed at the species level, they were very similar at the genus level. Moreover, the two species occupied different habitats; *S. luridus* lived where algae grew on rock and for food, they targeted macroalgae; *S. rivulatus* was found where algae grew on sand or rock but for food, they targeted the epilithic algae and associated materials (Shakman et al. 2009). The other two siganids that are found throughout the Red Sea (i.e., *S. argenteus* and *S. stellatus*) have not invaded the Mediterranean Sea. Both species live exclusively about coral reefs and there are no coral reefs in the Mediterranean. However, there is a single record of an adult *S. argenteus* caught on the Libyan coast (Abelghani et al. 2020).

There are other records of siganids from the Mediterranean, their presence probably due to being introduced in a ship's ballast water. There are two records for *S. javus*, both based on single adult specimens: one hand-lined off Alexandria, Egypt (Corsini-Foka and Zava 2022) and another taken by trammel net off the Syrian coast (Ibrahim et al. 2010). *S. javus* occurs in the Gulf of Oman but not the Red Sea. An adult *S. fuscescens* was caught by hand line in a shipping harbour on the east coast of Italy; as this is a western Pacific species the authors concluded that it must have been introduced via shipping, e.g., in ballast water or by accompanying a drilling platform being relocated into the area (Azzuro and Tiralongo 2020). The same explanation must apply to the capture by trammel net of an adult siganid off Naples. A genomic study found that this fish was a member of the *fuscescens/canaliculatus* complex, but that it aligned with neither of these species, instead aligning with a clade consisting of specimens from the Western Pacific and East Indian Oceans (Virgili et al. 2023).

Western Australia Pacific Affinities

Although the coast of Western Australia is geographically located in the Indian Ocean its biogeographic affinities are with the Western Pacific. This can be attributed to the influence of the Indonesian Through Flow (ITF). The ITF is the of result of a pressure gradient between the Western Pacific and the eastern Indian Ocean. It transports water through straits and passages between the islands of the Indonesian Archipelago. It fuels the Leeuwin Current which flows southward down the full length of the west coast of Western Australia. The major ITF passageways are between the islands in the eastern part of Indonesia, Makassar Strait, Ombai Passage and Timor Passage. The magnitude of the ITF varies from days to decades and over geological time (Tillinger 2011). Global sea level changes connected to the Glacial-Interglacial cycles affect the strength of the Leeuwin Current. When the sea level sits between 65 m and 120 m below present-day sea level, which occurs throughout the Glacial part of a cycle, the strength of the Leeuwin Current is reduced significantly (Petrick et al.

2019). Therefore, at present day sea levels, the Leeuwin Current carries warm water further south than when sea levels are lower around the times of Glacial Maxima.

Eight species of siganid are found on the coast of Western Australia, all of which have Pacific rather than Indian Ocean affinities. Three of them, *S. unimaculatus* (as *S. vulpinus*), *S. puellus* and *S. punctatissimus* are known only from the coral shoals situated on the fringe of the North-West continental shelf (Allen and Russell 1986). A further four species (*S. punctatus, S. lineatus, S. trispilos, S. doliatus*) are found along the coast in coral reef habitats. *S. punctatus* is the Pacific sibling of the Indian Ocean *S. laqueus*; *S. lineatus* is distributed across northern Australia and the Pacific side of the Indo-Australian Archipelago and other island groups in the Western Pacific such as Palau and New Caledonia, and shares a common border with its sibling species *S. guttatus* around the Bird's Head of New Guinea. *Siganus guttatus* is **not** found in Western Australia; it inhabits northwestern regions of the Western Pacific and the eastern Indian Ocean, such as Okinawa, the Philippines, western Indonesia and the Andaman Sea coasts of Thailand, Burma and Sumatra. *Siganus trispilos* is believed to be an endemic species (although a small disjunct population of what may be this species occurs at Sipidan I., Sabah); it is a sibling of *S. corallinus* that ranges from the Seychelles to New Caledonia. Another endemic is a distinctive colour variant of the Pacific *S. doliatus*; although its taxonomic status needs to be resolved genetically. *S. fuscescens* and *S. canaliculatus* have both been recorded from the coast of Western Australia, but as the taxonomy of these species is unresolved it is not clear whether these represent one or two species. Records are not confined to coral reefs. Fish identified as *S. fuscescens* occur as far south as Port Coburn.

Australia's Indian Ocean Territories

Of special interest is the biogeography of the fishes of the Cocos (Keeling) Islands, isolated islands situated in the Indian Ocean around 1000 km southwest of Java. Of the 602 species present, 78% are widespread Indo-Pacific species, but 9% are Western Pacific species at the western edge of their geographic range, and 6% are Indian Ocean species at the eastern edge of their range (Hobbs et al. 2014a). Two of the four siganids present are examples of this east west divide: sibling species *S. punctatus* (Western Pacific) and *S. laqueus* (Indian Ocean). The others are *S. puellus* at the western edge of its range and the ubiquitous *S. argenteus*.

Christmas Island is the only other remote island in the eastern Indian Ocean. It lies 300 km to the south of Java. It is fringed by a narrow coral reef extending out only 20–100 m from the shore before dropping off steeply at a depth of 15 km. The fish fauna has been thoroughly surveyed: 681 species, of which 89% are found on the shallow fringing reef. Approximately 76% of the species are widespread Indo-West Pacific species, 13% Pacific Ocean, and 4% Indian Ocean species (Hobbs et al. 2014b). Comparable figures for the Cocos (Keeling) Islands are 560 shallow water species but the two share only 48% of their species. Unlike the Cocos (Keeling) Islands there is no lagoon present, and this is believed to be the major determinant of the difference in composition of the two fish communities (Hobbs et al. 2014b). The only siganid present is *S. corallinus*, a coral reef flat inhabitant broadly distributed from East Africa to New Caledonia.

Indian Ocean

The siganids found in the Eastern Indian Ocean are with one exception also found in the IAA. Only *S. insomnis*, found both in the Maldives and southern India has been shown to be significantly different genetically from its siblings in the IAA to be treated as a separate species (Woodland and Anderson 2014). Clearly, the Bay of Bengal is not an effective barrier between the IAA and the southern India area for the Siganidae.

Red Sea/Gulf of Aden

The Red Sea is renowned for its high level of endemism: 14% to 17% for fishes, crabs and echinoderms and 10% for scleractinian corals (DiBattista et al. 2016). This has been attributed to its relative isolation from the Gulf of Aden. The number of endemic fishes is 165; however, there are another 51 fish species that are endemic to the Red Sea and the Gulf of Aden combined (Bogorodsky and Randall 2019). This strongly suggests that there is a second barrier that isolates the Gulf of Aden and the Red Sea combined from the Indian Ocean.

Of the four species of siganid in the Red Sea, *S. stellatus* appears to be endemic to the region, at least there are no records of it from the adjacent Gulf of Aden. Woodland (1990) found that *S. luridus* specimens from the Red Sea were significantly deeper bodied than specimens from both the East Africa region and the Persian Gulf. Whether the two are different species needs to be investigated by genetic analysis, but if they are different species there would be a second siganid endemic to the area. *S. rivulatus* is known from the Red Sea, but also from the Gulf of Aden and the Persian Gulf. The ubiquitous *S. argenteus* is the fourth Red Sea resident.

The Red Sea has been periodically isolated, although perhaps not completely, from the Gulf of Aden at times of sea level lows during periods around Glacial Maxima. There is a sill at the southern end of the Red Sea at the Bab-el-Mandab strait at a depth of around 137 m below current sea levels. It has been estimated that during the last glacial maximum (LGM) around 18,000 years ago the sill depth would have been around 17 m. This means that a moderate exchange of water with the Gulf of Aden was still possible. A continuous connection has been supported by the study of Hemleben et al. (1996) who analysed stable isotope and faunal records for the past 380,000 yr from a core taken from the floor of the central Red Sea. At the time of the last glacial maximum all planktonic foraminifera disappeared but the benthic foraminifera survived. Salinities as high as 55% were inferred from the stable isotope record. They remained high till between 8000 to 9000 years ago when the missing foraminifera started to reappear in the core. From this study the authors concluded that the very high salinities exceeded the tolerance of the planktonic foraminifera but that they were able to reinvade the Red Sea from the Gulf of Aden as salinities returned to interglacial levels. Extrapolating from those results, fishes living in the upper zone of the Red Sea such as the siganids might also have taken refuge in the Sea of Aden during times of hypersalinity.

Isolation of the Gulf of Aden from the surrounding seas probably also plays a role in the evolution of the Red Sea/Gulf of Aden endemism. Strong upwellings of cold water that occur on the south coast of Oman and the east coast of Somalia must contribute to the faunal separation between the Africa/Arabia Region and areas to its east. These upwellings are seasonal. They are generated by the south-west monsoon that occurs from April to October. Sea surface temperatures fall to as low as 16ºC on the coast of Oman and even lower, 14ºC, on the Somali coast (Bruce 1974, Priest et al. 2016). These extreme temperatures lead to coastlines largely devoid of coral reefs and mangroves from the Gulf of Oman through to southern Somalia (Stoddart 1971). The absence of these important habitats for fishes may be contributing to the faunal break between the Africa/Arabia region and the eastern side of the Indian Ocean. Additionally, tropical fishes cannot tolerate temperatures this low. During an oceanographic survey of the east coast of Somalia Foxton (1965) witnessed a mass mortality of tropical inshore fishes, e.g., Lutjanidae and Serranidae, which coincided with an area of exceptionally cold surface water.

Western Indian Ocean Province

There is a fall in the number of species between the southern India area and the Western Indian Province, here defined as the east coast of Africa and its offshore islands such as the coral reef rich Seychelles, Mauritius and Madagascar with its great variety of habitats, the Red Sea, Gulf of Aden through to the Gulf of Oman and its extension, the Persian Gulf. There are only seven species of siganid inhabiting this large area but four of them are endemic: *S. stellatus*, *S. luridus*, *S. rivulatus* and *S. sutor* (*S. sutor* is reported from the east coast of Malaysia, but this needs to be confirmed as there are no records of it from the coasts of India). The Red Sea *S. stellatus* is found only on coral reefs, but the other three are not obligatory coral reef dwellers. *S. luridus* and *S. rivulatus* are successful invaders of the Mediterranean Sea. The three non-endemic species found in the region are all coral reef dwellers that range to the east of the IAA (*S. laqueus*) and beyond to the West Pacific Ocean (*S. corallinus* and *S. argenteus*).

With both *S. laqueus* and its Red Sea sibling *S. stellatus* recognised as full species there are seven species of siganid inhabiting the east coast of Africa through to the Gulf of Oman and its extension, the Persian Gulf. Excluding *S. stellatus,* three of these species are endemic to the region: *S. luridus, S. rivulatus* and *S. sutor*. All three are not obligatory coral reef dwellers, e.g., *S. luridus* and *S. rivulatus* are successful invaders of the Mediterranean. The other three species in the region are all coral reef dwellers that range to the east to the IAA (*S. laqueus*) and beyond: *S. corallinus* and *S. argenteus*.

Biogeography of the species of Foxface

The foxfaces (i.e., species in the subclade "Lo") are distinguished from all other siganids by an excessively pointed snout. As juveniles they occur in large schools, and as adults in pairs. The adults feed on macroalgae and cyanobacteria (Hoey et al. 2013, Clements – Chapter 4), with the elongate snout being suggested to be adaptation to feeding in crevices. The ancestral species possibly evolved from a species with similar body proportions that lived in pairs when adult. A species like *S. corallinus* was conjectured to be a likely candidate, because as well as possessing these features it has a semi protuberant snout. This is not supported by phylogenies generated from mitochondrial and nuclear DNA analyses (Borsa et al. 2007, Kuriiwa et al. 2007, Siqueira et al. 2019). In these studies the deep bodied species resolve into two subclades with the foxfaces in one clade and *S. corallinus* in the other.

The geographic distribution of these five species clearly shows that the creation of distinct species has occurred in peripheral populations, *S. magnificus* in the Andaman Sea, *S. uspi* in Fiji and *S. niger* in the Tonga Group. Siqueira et al. (2019) have produced a molecular phylogeny for the Siganidae based on two mitochondrial genes and one nuclear gene which includes four of the five species of foxface; *S. uspi, S. magnificus, S. vulpinus* and *S. unimaculatus.* Their maximum clade credibility tree (Fig. S1 in Siqueira et al. 2019) suggests that *S. uspi* is ancestral to the other three species examined, but the resolution is probably not sufficient to accept this with any degree of confidence since the radiation of all the species is constrained within the last one to two million years. Specimens of *S. niger* are occasionally reported in New Caledonia, presumably carried there from Tonga in the prevailing current. Therefore *S. niger* may be ancestral to *S. uspi* and even the most widespread of the species, *S. vulpinus. Siganus unimaculatus*, which differs from *S. vulpinus* only in having a black spot on the side and considered by some to be simply a colour variant, occurs peripherally to *S. vulpinus* in the Ryukyu and Ogasawara Islands, and in the Philippines where it occurs along the east coasts of the chain of islands. Curiously, there is also a population of *S. unimaculatus* on the Rowley Shoals and Scott and Seringapatam reefs on the fringe of the continental shelf of the North-West coast of Western Australia presumably dispersed there some time in the past by the Indonesian Through Flow.

Ecological Biogeography

Biogeography is the study of the distribution of organisms in geographic space and through geological time, known as ecological and historical biogeography, respectively. Much has been written about the historical biogeography of those organisms that inhabit the shallow seas of the Tropical Indo-Pacific Ocean. Much less has been written about how these organisms are able to coexist, how they apportion the resources available to make coexistence possible. Taxonomists classify species by differences in their morphology, with these differences often reflecting the different niches they occupy (e.g., habitat niche: Bellwood et al. 2002; feeding niche: Barnett et al. 2006). Competition between species is reduced if they occupy different niches. For example, competition for dietary resources can be reduced by having a different diet in the same habitat or in some circumstances by having the same diet but feeding in a different habitat. It follows that habitat diversity will play a predominant role in determining the number of species that coexist within a defined area.

Indeed, areas of high fish species richness often coincide with a high levels of habitat diversity. For example, the Bird's Head Peninsula of West Papua supports *ca.* 1,500 fish species in an area of approximately 50,000 km^2, and this high species richness has been related to "the amazing level of habitat diversity" (Allen and Erdman 2009), among other factors (e.g., location at the crossroads of faunal elements from Indonesia, the Melanesian Archipelago and the Timor/Arafura Sea, geologic history, and widely fluctuating sea levels). Moreover, the importance of habitat richness to the Siganidae may be illustrated by comparisons of fish assemblages between the islands of the main Palauan Archipelago and the outlying seamounts of the Southwest Palau Islands, some 270–600 km south of the main archipelago. The main Palauan Archipelago consists of 340 islands with a wide diversity of habitats including fringing coral reefs, seagrass meadows, macroalgal beds, mangroves and estuaries. In contrast, the Southwest Palau Islands consist of six emergent coral reefs sitting atop seamounts on the Palau Ridge, five of which are narrow reefs with steep drop offs., and the sixth (Helen Reef) is an atoll with considerable habitat diversity (Donaldson 1996). Twelve siganid species occur on the main Palauan Archipelago, six species on Helen Reef, and one species at one of the other Southwest Palauan Islands, while siganids are absent from the remaining four Southwest Palau reefs (Donaldson 1996). Although this comparison is confounded by habitat area (Donaldson 2002) it shows a clear link between the richness of the siganids and habitat availability. All the species of siganid found at the Southwest Palau Islands are corallophilic, while species that rely on estuaries, seagrass meadows, macroalgal beds and/or mangroves during their juvenile and/or adult life stages (e.g., *S. vermiculatus, S. canaliculatus, S. lineatus*; Sambrook et al. 2019, Hoey et al. – Chapter 7) were absent.

Similarly, comparisons of the stomach content and distribution of 11 species on siganid across inner-, mid- and outer-shelf reefs of the northern Great Barrier Reef show some differentiation among species (Hoey et al. 2013). For example, analysis of the stomach contents identified four distinct groups: browsers of foliose and leathery algae (*S. canaliculatus, S. javus*), croppers of red and green macroalgae (*S. argenteus, S. corallinus, S. doliatus, S. spinus*), mixed feeders on diverse algal material, cyanobacteria, detritus and sediment (*S. lineatus, S. puntatissimus, S. punctatus, S. vulpinus*), and a sponge feeder, *S. puellus.* Similarly, there was variation in the composition of siganid assemblages among reef zones (slope, crest, flat and back reef) and reefs (inner-, mid-, outer-shelf) suggesting there is some partitioning of resources.

While some studies have provided some initial insights into the ecological biogeography of rabbitfishes, it is generally a neglected area. Our current understanding of the life cycles of siganids is largely based on a few studies species of potential aquaculture interest (Camus et al. – Chapter 9). There are an increasing number of studies that have reported the habitat associations of recently-settled and juvenile rabbitfishes (e.g., Tang et al. 2020, Webber et al. 2024), however our understanding of ontogenetic shifts in habitat and diet of siganids is limited and is an area for future research.

Conclusion

Despite being a relatively small family of predominantly tropical marine fishes, the Siganidae are an interesting group in terms of their history, taxonomy and biogeography. There are 31 species of extant siganid currently recognised, however this number may need to be revised once the genetic relationships of disjunct populations of *S. trispilos,* unique colour variants of *S. doliatus*, and the uncertainties of the sister taxa *S. canaliculatus* and *S. fuscescens* are resolved. The biogeography of the Siganidae is also unique and shaped by the combined influence of historical events and their ecology. This chapter has provided an overview of the current knowledge of the history, taxonomy and biogeography of the Siganidae, and it is hoped that it stimulates future research on the taxonomy, biogeography and ecology of this group.

References

Abdelghani, A., S.A.A. Al Mabruk, F. Crocetta and D. Golani. 2020. The streamlined rabbitfish *Siganus argenteus* (Quoy & Gaimard, 1825) in the Mediterranean Sea. Thalassas. 37: 287–290.

Adams, M. and D.J. Woodland. 1994. Molecular systematics of the rabbitfishes of the Oramin complex: Towards a resolution of the *Siganus fuscescens/S. canaliculatus* species problem using allozyme electrophoresis (Siganidae: Pisces). Proc. Fourth Indo-Pacific Fish Conf. 28: 373–485.

Agiadi, K., A. Girone, E. Koskeridou, P. Moisette, J.-J. Coanée and F. Quillévéré. 2018. Pleistocene marine fish invasions and paleoenvironmenal reconstructions of the eastern Mediterranean. Quat. Sci. Revs. 196: 80–99.

Allen, G.R. 1975. The Anemone Fishes. 2nd Edition. T.F.H. Publications, Neptune City, New Jersey, USA.

Allen, G.R. and B.C. Russell. 1986. Part VII: Fishes. Faunal surveys of the Rowley Shoals, Scott Reef and Seringapatam Reef, North-Western Australia. Rec. West Aust. Mus. Suppl. 25: 75–103.

Axelrod, H.R. and C.W. Emmens. 1973. Exotic Marine Fishes. Looseleaf. Supplement, Book 3, p. 435.00. T.F.H. Publications, Neptune City, New Jersey, USA.

Azzurro, E. and F. Tiralongo. 2020. First record of the mottled spinefoot *Siganus fuscescens* (Houttuyn, 1782) in Mediterranean waters: a Facebook based detection. Medit. Mar. Sci. 21: 448–451.

Bannikov, A.F., J.C. Tyler and C. Sorbini. 2010. Two new taxa of Eocene rabbitfishes (Perciformes, Siganidae) from the North Caucasus (Russia), with redescription of *Acanthopygaeus agassizi* (Eastman) from Monte Bolca (Italy) and a phylogenetic analysis of the family. Bollettino del Museo Civico di Storia Naturale di Verona, Geologia Paleontologia Preistoria 34: 3–21.

Bariche, M. 2006. Diet of the Lessepsian fishes, *Siganus rivulatus* and *S. luridus* (Siganidae) in the eastern Mediterranean: A bibliographic analysis. Cybium. 30: 41–49.

Barnett, A., D.R. Bellwood and A.S. Hoey. 2006. Trophic ecomorphology of cardinalfish. Marine Ecology Progress Series 322: 249–257.

Bellwood, D.R., P.C. Wainwright, C.J. Fulton and A. Hoey. 2002. Assembly rules and functional groups at global biogeographical scales. Functional Ecology 16: 557–562.

Bialik, O.M., M. Frank, C. Betzler, R. Zammit and N.D. Waldeman. 2019. Two step closure of the Miocene Indian Ocean gateway to the Mediterranean. Sci. Rep. 9: 8842.

Bogorodsky, S.V. and J.E. Randall. 2019. Endemic fishes of the red sea. pp. 239–265. *In*: Rasul, N.M.A. and I.C.F. Stewart [eds.]. Oceanographic and Biological Aspects of the Red Sea. Springer International, Cham, Switzerland.

Borsa, P., S. Lemer and D. Aurelle. 2007. Patterns of lineage diversification in rabbitfishes. Mol. Phyl. Evol. 44: 427–435.

Briggs, J.C. 1984. Centres of origin in biogeography. Biogeogr. Monogr. 1: 1–105.

Browne, P. 1756. The Civil and Natural History of Jamaica. B. White and Son.

Bruce, J.G. 1974. Some details of upwelling off the Somali and Arabian coasts. J. Mar. Res. 32: 419–423.

Bryan, P.G. 1975. Food habits, functional digestive morphology, and assimilation efficiency of the rabbitfish *Siganus spinus* (Pisces, Siganidae) on Guam. Pacific Sci. 29: 269–277.

Corsini-Foka, M. and B. Zava. 2022. Second occurrence of *Siganus javus* (Siganidae) in the Mediterranean waters. Annales: Ser. Hist. Nat. 32: 287–292.

DiBattista, J.D., J.H. Choat, M.R. Gaither, J.P.A. Hobbs, D.F. Lozano-Cortes, R.F. Myers et al. 2016. On the origin of endemic species in the Red Sea. J. Biogeogr. 43: 13–30.

Donaldson, T.J. 1996. Fishes of the remote Southwest Palau Islands, a zoogeographic perspective. Pacific Sci. 50: 285–308.

Donaldson, T.J. 2002. High islands versus low islands: a comparison of fish faunal composition of the Palau Islands. Environ. Biol. Fish. 65: 241–248.

Drew, A.E. 1973. The comparative ecology of *Siganus lineatus* and *Siganus canaliculatus*; Two Fishes of the Palau Islands (Pisces: Siganidae). M.S. Thesis, Northern Illinois University, De Kalb, Ill. 46 pp.

Floros, C., M.H. Schleyer, J.Q. Maggs and L. Celliers. 2012. Baseline assessment of high-latitude coral reef fish communities in southern Africa. Afr. J. Mar. Sci. 34: 55–69.

Forsskal, P. 1775. Descriptiones animalium avium, amphibiorum, piscium, insectorum, vermium; quae in itinere orientali observavit Petrus Forskal [....]. Post mortem auctoris edidit Carsten Niebuhr. Adjuncta est materia medica kahirina atque tabula Maris Rubra geographica. Moller, Hauniae.

Fowler, H.W. 1904. A collection of fishes from Sumatra. J. Acad. Nat. Sci. Phila. 12: 495–560, pls. VII-XXVII.

Foxton, P. 1965. A mass fish mortality on the Somali coast. Deep-Sea Res. 12: 17–19.

Fricke, R. 2008. Authorship, availability and validity of fish names described by Peter (Pehr) Simon Forsskål and Johann Christian Fabricius in the 'Descriptiones animalium' by Carsten Niebuhr in 1775 (Pisces). Stuttgarter Beitrage zur Naturkunde A. Neue Serie 1: 1–76.

Friedman, M. and G. Carnevale. 2018. The Bolca Lagerstätten: shallow marine life in the Eocene. J. Geol. Soc. 175: 569–579.

Gaither, M.R. and L. Rocha. 2013. Origins of species richness in the Indo-Malay-Philippine biodiversity hotspot: evidence for the centre of overlap hypothesis. J. Biogeogr. 40: 1638–1648.

Golani, D., E. Azzurro, J. Dulcic, E. Massuti and L. Orsi-Relini. 2021. Atlas of Exotic Fishes in the Mediterranean Sea. CIESM, Paris, Monaco.

Gronovius, L.T. 1763. Zoophylacii Gronoviani fasciculus primus exhibens animalia quadrupeda, amphibia atque pisces, quae in museo suo adservat, rite examinavit, systematice disposuit, descripsit atque iconibus illustravit Laur. Theod. Gronovius. 1: 1–140, pls. 1–17. Sumptibus auctoris [published by author], Lugduni Batavorum [Leiden].

Hall, R. 2013. The palaeogeography of Sundaland and Wallacea since the late Jurassic. J. Limnol. 72: 1–17.

Helfman, G.S., B.B. Collette, D.E. Facey and B.W. Bowen. 2009. The Diversity of Fishes. Biology, Evolution and Ecology. 2nd Edition. Wiley-Blackwell, Oxford.

Hemleben, C., D. Meischner, R. Zahn, A. Almogi-Labin, H. Erlenkeuser and B. Hiller. 1996. Three hundred eighty thousand year long stable isotope and faunal records from the Red Sea: Influence of global sea level change on hydrography. Paleoceanography 11: 147–156.

Hobbs, J.P.A., S.J. Newman, G.E.A. Mitsopoulos, M.J. Travers, C.L. Skepper, J.J. Gilligan et al. 2014a. Fishes of the Cocos (Keeling) Islands: new records, community composition and biogeographic significance. Raffles Bull. Zool., Suppl. 30: 203–219.

Hobbs, J.P.A., S.J. Newman, G.E.A. Misopoulos, M.J. Travers, C.L. Skepper, J.J. Gilligan et al. 2014b. Checklist and new records of Christmas Island fishes: the influence of isolation, biogeography and habitat availability on species abundance and community composition. Raffles Bull. Zool., Suppl. 30: 184–202.

Hoey, A.S., S.J. Brandl and D.R. Bellwood. 2013. Diet and cross shelf distribution of rabbitfishes (f. Siganidae) on the northern Great Barrier Reef: implications for ecosystem function. Coral Reefs 32: 973-984.

Hsu, T.H., Y.T. Adiputra, C.P. Burridge and J.C. Gwo. 2011. Two spinefoot colour morphs: mottled spinefoot *Siganus fuscescens* and white-spotted spinefoot *Siganus canaliculatus* are synonyms. J. Fish Biol. 79: 1350–1355.

Hutchins, J.B. 1991. Dispersal of tropical fishes to temperate seas in the southern hemisphere. J. Roy. Soc. West. Aust. 74: 79–84.

Ibrahim, A., M. Lahlah, M.Y. Kassab, W. Ghanem and S. Ogaily. 2010. *Siganus javus*, a new record from the Syrian waters, with a reference to growth and feeding of two lessepsian fish. Rapp. Comm. Int. Mer. Medit. 39: 544.

ICZN. 1926. Opinion 93. Twelve generic names of fishes placed in the Official List, by suspension of the rules. Smithsonian Miscellaneous Collections 73: 5–11.

Jordan, D.S. 1927. Shore fishes of Samoa. J. Pan-Pacif. Res. Inst. 2: 3–11.

Jordan, D.S. and A. Seale. 1906. The fishes of Samoa: description of the species found in the archipelago, with a provisional check-list of the fishes of Oceania. Bull. U.S. Bur. Fish. 25: 173–455.

Kottelat, M. 2013a. The fishes of the inland waters of Southeast Asia: A catalogue and core bibliography of the fishes known to occur in freshwaters, mangroves and estuaries. Raffles Bull. Zool. Suppl. 27: 1–663.

Kottelat, M. 2013b. Comment on the proposed 'validation' of *Siganus* Forskal, 1775 (Pisces, Siganidae): awaiting a ruling since 1968 (Case 1721). Bull. Zool. Nomencl. 70: 114–120.

Kuiter, R.H. and H. Debelius. 2001. Surgeonfishes, Rabbitfishes and their Relatives: A Comprehensive Guide to Acanthuroidea. Chorleywood Publishing, Herts, UK.

Kuriiwa, K., N. Hanzawa, T. Yoshino, S. Kimura and M. Nishida. 2007. Phylogenetic relationship and natural hybridization in rabbitfishes (Teleostei: Siganidae) inferred from mitochondrial and nuclear DNA analyses. Mol. Phyl. Ecol. 45: 69–80.

Ladd, H.S. 1960. Origin of the Pacific Island molluscan fauna. Amer. J. Sci. 258-A: 137–150.

Landini, W. and C. Sorbini. 2005. Evolutionary trends in the Plio-Pleistocene ichthyofauna of the Mediterranean Basin: nature, timing and magnitude of the extinction events. Quat. Internat. 131: 101–107.

Lemer, S., D. Aurelle, L. Vigliola, J.D. Durand and P. Borsa. 2007. Cytochrome *b* barcoding, molecular systematics and geographic differentiation in rabbitfishes (Siganidae). C. R. Biologies 330: 86–94.

Linnaeus, C. 1766. Systema Naturae per Regna Tria Naturae secundum Classes, Ordines, Genera, Species, cum Characteribus, Differentiis Synonymis, Locis. 12th Edition. Vol. 1. L. Salvii, Holmiae.

Matias, A.M.A. and C. Rignios. 2018. Revisiting the "Centre Hypotheses" of the Indo-West Pacific: Idiosyncratic genetic diversity of nine reef species offers weak support for the Coral Triangle as a centre of genetic biodiversity. J. Biogeogr. 45: 1–12.

Niebuhr, M. 1772. Travels through Arabia and Other Countries in the East. Vol. II. English language edition, translated by R. Heron. (1792). Edinburgh.

Nielsen, J.G. and W. Klausewitz. 1968. *Siganus* Forsskal, 1775 (Pisces: Siganidae): proposed validation under the plenary powers. Z.N.(S.) 1721. Bull. Zool. Nomencl. 25: 26–28.

Petrick, B., A. Martinez-Garcia, G. Auer, L. Reuning, A. Auderset and H. Deik. 2019. Glacial Indonesian throughflow weakening across the Mid-Pleistocene climatic transition. Sci. Rep. 9: 16995.

Priest, M.A., J.D. Battista, J.L. McIlwain, B.M. Taylor, N.E. Hussey and M.L. Berumen. 2016. A bridge too far: dispersal barriers and cryptic speciation in an Arabian Peninsula grouper (*Cephalopholis hemistiktos*). J. Biogeogr. 43: 820–832.

Randall, J.E. and M. Kulbicki. 2005. *Siganus woodlandi*, new species of rabbitfish (Siganidae) from New Caledonia. Cybium. 29: 185–189.

Ravago-Gotanco, R.G., T.L. de la Cruz, M.J. Pante and P. Borsa. 2018. Cryptic genetic diversity in the mottled rabbitfish *Siganus fusce*scens with mitochondrial introgression at a contact zone in the South China Sea. PLoS One 13: e0193220.

Sambrook, K., A.S. Hoey, S. Andréfouët, G.S. Cumming, S. Duce and M.C. Bonin. 2019. Beyond the reef: The widespread use of non-reef habitats by coral reef fishes. Fish and Fisheries 20: 903–920.

Seale, A. 1906. Fishes of the South Pacific. Occ. Pap. B.P. Bishop. Mus. 1: 61–128.

Seba, A. 1759. Locupletissimi rerum naturalium thesauri accurata descriptio et iconibus artificiosissimis expressio per universam physices historiam, Opus, cui, in hoc rerum genere, nullum par exstitit. Ex toto terrarum orbe collegit, digessit, descripsit, et depingendum curavit. Tomus III. Van Waesberge, Amsterdam.

Shakman, E., C. Boedeker, M. Bariche and R. Rinzebach. 2009. Food and feeding habits of the Lessepsian migrants *Siganus luridus* Rȕppell, 1828 and *Siganus rivulatus* Forsskal, 1775 (Teleostei:Siganidae) in the southern Mediterranean (Libyan coast). J. Biol. Res. Thessaloniki 12: 115–124.

Siqueira, A.S., D.R. Bellwood and P.F. Cowman. 2019. Historical biogeography of herbivorous coral reef fishes: the formation of an Atlantic fauna. Proc. R. Soc. B. 286: 20182672.

Soliman, V.S., H. Yamada and K. Yamaoka. 2010. Early life-history of the spiny siganid *Siganus spinus* (Linnaeus 1758) inferred from otolith microstructure. J. Appl. Ichthyol. 26: 540–545.

Stockwell, B. and K. Carpenter. 2012. Fish fauna. pp. 79–100. *In*: Kassem, K., B. Hoeksema and Y.A. Affendi [eds.]. Semporna Marine Ecological Expedition. WWF-Malaysia, NCB Naturalis, Universiti Malaysia Sabah.

Stoddart, D.R. 1971. Environment and history in Indian Ocean reef morphology. pp. 3–38. *In:* Stoddart, D.R. and M. Yonge [eds.]. Regional Variation in Indian Ocean Coral Reefs. Symp. Zool. Soc. London, No. 28.

Swainston, R. 2020. Swainston's Fishes of Australia. Viking, Penguin Random House, Australia.

Tang, S., A. Graba-Landry and A.S. and Hoey. 2020. Density and height of *Sargassum* influence rabbitfish (f. Siganidae) settlement on inshore reef flats of the Great Barrier Reef. Coral Reefs 39: 467–473.

Taylor, W.R. 1970. Comment on the proposed validation of *Siganus* Forskal, 1775, and request that the name *Teuthis* Linnaeus (Pisces) be placed on the Official List of Generic Names in Zoology. Z. N.(S.) 1721. Bull. Zool. Nomencl. 26: 178–179.

Tillinger, D. 2011. Physical oceanography of the present day Indonesian Throughflow. pp. 267–281. *In*: Hall, R., M.A. Cottam and M.E.J. Wilson [eds.]. The SE Asian Gateway: History and Tectonics of the Australia-Asia Collision. Geol. Soc., London.

Tjia, H.D. 1966. Java Sea. pp. 424–429. *In*: Fairbridge, R.W. [ed.]. The Encyclopedia of Oceanography. Vol. I. Reinhold, New York.

Tobias, W.J. 1976. Ecology of *Siganus argenteus* (Pisces: Siganidae) in relation to its mariculture potential on Guam. Univ. Guam Mar. Lab. Tech. Rept. 29: 58–93.

Tyler, J.C. and A.F. Bannikov. 1997. Relationships of the fossil and Recent genera of rabbitfishes (Acanthuroidei: Siganidae). Smithsonian Contrib. Paleobiol. 84: 1–35.

Virgili, R., P. Fasciglione, F. Tiralongo, R. Pannone, M. Palomba, M. Santoro et al. 2024. Interdisciplinary approaches at early invasion stages maximise the evidence on human-induced and climate-driven dispersal. Hydrobiologia http://dx.doi.org/10.1007/s10750-024-05537-y.

Webber, K., K. Fabricius, S.K. Wilson and A.S. Hoey. 2024. The influence of *Sargassum* biomass and thallus density on the recruitment of coral reef fishes. Coral Reefs 43: 687–699.

Woodland, D.J. 1972. Proposal that the genus name *Teuthis* Linnaeus (Pisces) be suppressed. Bull. Zool. Nomencl. 29: 190–193.

Woodland, D.J. 1973. Addendum to the proposal that genus name *Teuthis* Linnaeus (Pisces) be suppressed. Bull. Zool. Nomencl. 30: 6–7.

Woodland, D.J. 1983. Zoogeography of the Siganidae (Pisces): An interpretation of distribution and richness patterns. Bull. Mar. Sci. 33: 713–717.

Woodland, D.J. 1990. Revision of the fish family Siganidae with descriptions of two new species and comments on distribution and biology. Indo-Pacific Fishes No. 19, 136 pp. B.P. Bishop Mus., Hawaii.

Woodland, D.J. and R.C. Anderson. 2014. Description of a new species of rabbitfish (Perciformes: Siganidae) from southern India, Sri Lanka and the Maldives. Zootaxa 3811: 129–136.

Woodland, D.J. 2022. Family Siganidae. pp. 209–217, 654–656. *In*: Heemstra, P.C., E. Heemstra, D.A. Ebert, W. Holleman and J.E. Randall [eds.]. Coastal Fishes of the Western Indian Ocean. Vol. 5. NRF-SAIAB, Makhanda, South Africa.

Woods, L.P. 1953. Family Siganidae. pp. 649–658, pls. 73–74. *In*: Schultz, L.P., E.S Herald, E.A. Lachner, A.D. Welander and L.P. Woods [eds.]. Fishes of the Marshall and Marianas Islands. Bull. U.S. Nat. Mus. 202, 1: 1–685.

2

The Morphology and Colouration of Rabbitfishes

Victor Huertas[1,*] and *Christopher R. Hemingson*[2]

Introduction

Two of the most recognizable features of tropical fishes are their diversity of shapes and their striking colour patterns. If one was to describe a rabbitfish to someone unfamiliar with coral reefs it would make sense to start by pointing out that they belong to a widespread family of reef fishes. However, this may evoke images of the more iconic butterflyfishes, angelfishes, anemonefishes, and parrotfishes. While research on rabbitfishes has received less attention than the aforementioned groups, they are an ecologically important and widespread component of the fish fauna in Indo-Pacific seascapes. In this chapter, we provide an overview of the main morphological axes of variation and colour patterns found across rabbitfishes.

The study of the biological morphology of fishes is important for several reasons. First, it helps us understand the diversity and complexity of fish forms and structures. Morphology can be used as a tool towards identifying and classifying species, which is fundamental for fields like taxonomy and evolutionary biology. The anatomical differences and similarities between species provide indications of their potential evolutionary relationships and shared ancestry (i.e., it helps trace the evolutionary history of organisms and understand how they have evolved over time). Secondly, morphology provides clues into the function and adaptation of fishes. By examining the physical characteristics of a fish, researchers can generate hypotheses about its functional performance. These hypotheses can then be tested through ecomorphological trials to infer how the fish interacts with its environment—how it moves, feeds, reproduces, and survives. This knowledge, therefore, constitutes

[1] College of Science and Engineering, James Cook University, Townsville, Queensland 4811, Australia.

[2] Department of Marine Science, The University of Texas at Austin, Marine Science Institute, Port Aransas, TX 78373, United States of America.
Email: christopher.hemingson@austin.utexas.edu

* Corresponding author: victor.huertas@my.jcu.edu.au

the foundation of our understanding of the ecological relationships of fishes with other organisms and their environment, and provides hints into their evolutionary history. Due to the interwoven nature of evolution, morphology, and ecology, some of the information presented in this chapter overlaps with other chapters in this volume that delve into the evolution and the functional ecology of rabbitfishes with more detail (Siqueira and Cowman – Chapter 3, Brandl and Tebbett – Chapter 6, respectively).

The Morphology of Rabbitfishes

With 31 recognised species within a single genus (*Siganus*; Woodland – Chapter 1), Siganidae is a relatively small family when compared to other common reef fish families such as the Acanthuridae (84 spp), Chaetodontidae (136 spp), Labridae (661 spp), and Pomacentridae (419 spp). Rabbitfishes (or siganids) generally share a common set of morphological traits, although certain species may display slight variations. This limited morphological variation, particularly in features relevant to traditional taxonomy (e.g., meristic characters), hindered the description of rabbitfish species early on and lead to the misidentification of some of the members of this group (Woodland 1990).

To evaluate variation of the external morphology across rabbitfishes, we conducted a quantitative analysis of morphological traits aimed at capturing the disparity in the shape of the body, head, and pectoral and caudal fins. We focused on these traits because they are involved in vital functions such as swimming and feeding. We measured 11 morphological traits (Fig. 1) from 99 lateral images of pinned dead fishes representing 22 of the 31 extant rabbitfish species (~ 70% of the family). With the exception of three images obtained from FishBase (Froese and Pauly 2023), rabbitfish images were obtained from the Division of Fishes Collections

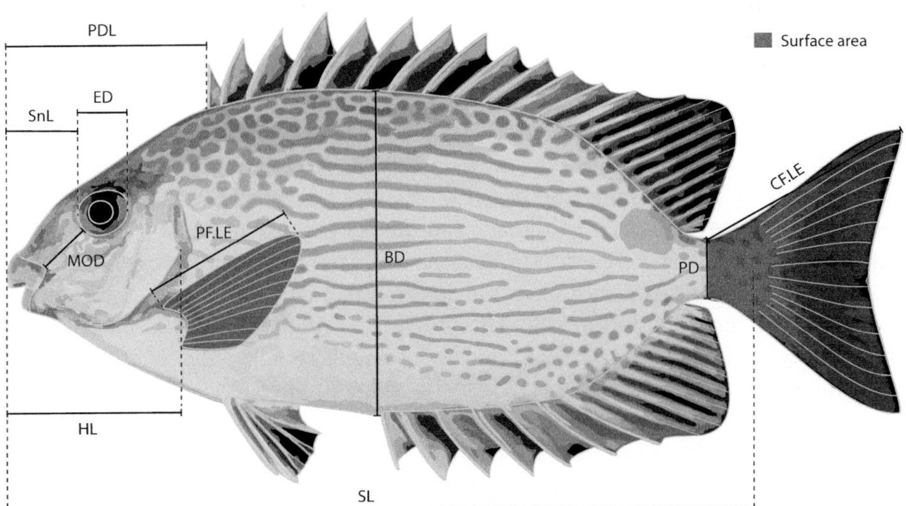

Fig. 1. Morphometric data collected to describe and analyse morphological traits of rabbitfishes. Vectorized image of a lined rabbitfish (*Siganus lineatus*) sourced from the Division of Fishes Collections at the Smithsonian Institute's National Museum of Natural History.

Fig. 2. Rabbitfishes, as many other animals, display allometric growth. These images show the juvenile (left) and adult (right) life stages of *Siganus doliatus*. Note the proportionally larger eye and longer spines in the juvenile when compared to the adults. Photos: Victor Huertas.

at the Smithsonian Institute's National Museum of Natural History (https://collections.nmnh.si.edu/). Importantly, each image includes the standard length of the fish photographed, allowing for the size-standardization of measurements. Measurements were collected using the open-source software FIJI (Schindelin et al. 2012). All the fishes measured were adults to avoid introducing biases due to allometric growth. Certain morphological features are known to vary among life stages in rabbitfishes. For example, juveniles have proportionally larger eyes and longer spines than adult conspecifics (Fig. 2), and the caudal fin of several species of rabbitfishes such as *Siganus punctatus*, *S. punctatissimus*, and *S. stellatus*, becomes incrementally more forked as the fish grows (Woodland 1990). In fact, ontogenetic changes in rabbitfishes extend beyond mere morphology. Many species also undergo a shift in the habitat they occupy as they grow older (Sambrook et al. 2019, Hoey et al. – Chapter 7) and their sociality (Mirbach and Brandl 2016).

The eye diameter (ED) was measured from the anterior to the posterior rims of the eye. The maxilla-orbit distance (MOD) was measured as the shortest distance between the maxilla and the orbit. We used this distance as an indicator of the ability of the fish to access concealed reef surfaces (i.e., crevice feeding; Brandl et al. 2014, Fox and Bellwood 2013). Maximum body depth (BD) was measured perpendicular to the anteroposterior axis of the fish at the point of greatest distance between the dorsal and ventral surfaces. Conversely, the caudal peduncle depth (PD), also measured perpendicular to the anteroposterior axis, was the shortest distance between the dorsal and ventral surfaces at the caudal peduncle region. Several measurements were taken starting at the anteriormost part of the body and running parallel to the anteroposterior axis: The pre-dorsal length (PDL) was measured as the distance to the first dorsal spine; the snout length (SnL) to the anterior edge of the orbit; the head length (HL) to the posterior edge of the operculum. The standard length (SL) was measured from the anteriormost point of the fish to the base of the hypural joint. The length of the leading edge of the pectoral and caudal fins (PF.LE and CF.LE, respectively) were also measured. Two informative traits of swimming performance—pectoral and caudal fin aspect ratios—were calculated as the length of the respective fin's leading edge squared divided by the fin area. For each species, the mean of every variable was calculated.

Table 1. Loadings for 11 morphological traits for the first two axes of a phylogenetic principal components analysis based on standard length-standardized measurements. Highest loadings for principal components 1 and 2 are shown in bold with values above the cut-off of ± 0.8.

Morphological trait	**PC1**	**PC2**
Maximum body depth (BD)	0.586	–0.187
Caudal peduncle depth (PD)	–0.463	0.517
Pre-dorsal length (PDL)	**0.818**	0.134
Snout length (SnL)	**0.910**	0.180
Eye diameter (ED)	–0.303	–0.313
Maxilla-orbit distance (MOD)	0.604	–0.146
Head length (HL)	**0.890**	–0.091
Pectoral fin length of leading edge (PF.LE)	0.784	0.365
Pectoral fin aspect ratio (PF.AR)	0.773	0.124
Caudal fin length of leading edge (CF.LE)	–0.159	**0.843**
Caudal fin aspect ratio (CF.AR)	–0.309	**0.888**

To visualize general patterns in rabbitfish morphology, we analyzed this data set using a phylogenetic Principal Components Analysis (pPCA). This approach accounts for statistical non-independence due to phylogenetic relatedness (Revell et al. 2008). Combined, the first two principal components (PC1 and PC2) accounted for 61.5% of the total variance. The traits with the highest loadings on the first two principal components (i.e., the traits with the highest correlation with PC1 or PC2) were SnL, HL, and PDL for PC1; and CF.AR and CF.LE for PC2, respectively (Table 1).

The Body Shape of Rabbitfishes

The body of rabbitfishes is laterally compressed with adults reaching a maximum total length between 20 and 55 cm depending on the species. Across the family, the body shape ranges from oblong (e.g., *Siganus guttatus*, *S. lineatus*, *S. vermiculatus*) to fusiform (e.g., *Siganus argenteus*, *S. fuscescens*, *S. spinus*). Although some rabbitfish species are known to utilise different habitats (Sambrook et al. 2019, 2020, Hoey et al. – Chapter 7), the body shape of rabbitfishes appears to be influenced by the primary habitat they occupy, as shown for other fishes (Larouche et al. 2020, Rincon-Sandoval et al. 2020). For example, rabbitfishes that primarily live in non-reefal habitats such as shallow algal flats and seagrass meadows (e.g., *Siganus argenteus*, *S. fuscescens*, and *S. spinus*) tend to be more streamlined with lower body depths and have a shorter, rounded snout (Fig. 3). By contrast, rabbitfishes inhabiting heterogeneous habitats with high structural complexity such as coral reefs (e.g., *Siganus doliatus*, *S. vulpinus*) usually have deeper bodies than those in less intricate environments like sand aprons and seagrass meadows. Deeper bodies have been related to greater manoeuvrability and facilitate sharp turns and sudden speed boosts (Larouche et al. 2020) which can be advantageous in an environment where predators may lurk at every corner. A deep body, however, may not be enough to enhance fast-starts if the fish does not possess a high percentage of muscle mass relative to body size (Webb 1978).

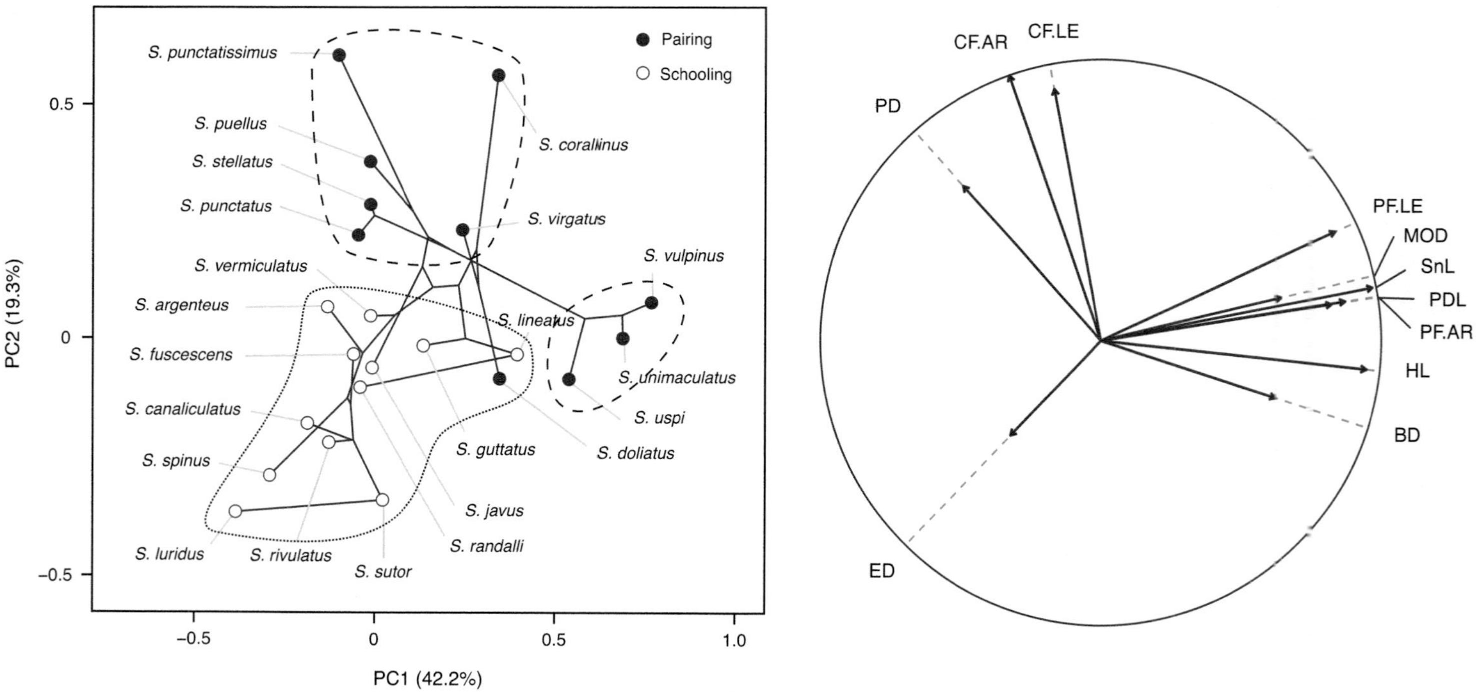

Fig. 3. Phylomorphospace of 22 species of rabbitfishes based on mean values of 11 morphological traits. Filled circles represent pairing species and empty circles represent schooling species. Dashed lines group clusters of pair-forming species and dotted line groups a cluster of schooling species. Measurements displayed in Fig. 1. BD: maximum body depth, PD: caudal peduncle depth, PDL: predorsal length, SnL: snout length, ED: eye diameter, MOD: maxilla-orbit distance, HL: head length, PF.LE: length of pectoral fin's leading edge, PF.AR: pectoral fin's aspect ratio, CF.LE: length of caudal fin's leading edge, CF.AR: caudal fin's aspect ratio. A pruned rabbitfish tree from Rabosky et al. (2018) accessed through https://fishtreeoflife.org/ (Chang et al. 2019) is superimposed. Data collected from images provided with the permission of the National Museum of Natural History, Smithsonian Institution (https://collections.nmnh.si.edu/) and FishBase (Froese and Pauly 2023).

Siganids have homocercal (i.e., symmetrical) caudal fins that vary in shape from truncate (e.g., *Siganus lineatus*, *S. spinus*, *S. sutor*) to deeply forked (e.g., *Siganus argenteus*, *S. woodlandi*). Truncate caudal fins have a low aspect ratio and are generally more prevalent in fishes inhabiting complex habitats (Langerhans and Reznick 2010). Conversely, the high aspect ratio of forked caudal fins reduces drag and enhances thrust, improving swimming efficiency (Vogel 1994). Thus, reef fishes with forked caudal fins are usually fast swimmers, or cruisers, that are predominantly found in the water column (Rincon-Sandoval et al. 2020). The shape of the caudal fin in rabbitfishes does not appear to be tightly associated with either of the two main habitats they occupy (i.e., coral reefs and shallow algal flats). As mentioned previously, it is also worth noting that some rabbitfishes experience ontogenetic changes in the shape of the caudal fin. For example, the caudal fins of *Siganus punctatus* and *S. stellatus* are emarginate in individuals smaller than 125 mm and 100 mm of standard length, respectively, but they develop into deeply forked caudal fins as they grow larger (Woodland 1990).

Siganids also exhibit relatively low pectoral fin polymorphism compared to other families such as labrids (Wainwright et al. 2002) and pomacentrids (Aguilar-Medrano et al. 2013). The pectoral fin shape (a high pectoral fin aspect ratio, in particular) is important for the locomotion of reef fishes with a labriform swimming mode in high-flow and/or wave-swept environments (Bellwood and Wainwright 2001, Fulton et al. 2001, 2017, Binning and Roche 2015, Bejarano et al. 2017). Rabbitfishes use these habitats (Hoey et al. 2013) but they generally employ a swimming mode that is characterized by a combination of body and caudal fin (BCF) undulation (Webb 1994, Fulton 2010), and as such the pectoral fins likely perform a minor role in locomotion which may explain the relatively uniform pectoral fin shape across the family.

The dorsal and anal fins of rabbitfishes are long and extend almost the entire length of their body. The dorsal fin has between 13 to 17 spines that can be locked in an erect position to deter would-be predators. When erect, the first dorsal spine is pointed forward. Anterior to the first dorsal spine, siganids have an additional small spine directed forward that can be embedded but it sometimes protrudes from a groove (Woodland 1990). Anal fins have seven spines, compared to three or fewer in most reef fishes (Kuiter and Debelius 2001). An important trait that is characteristic of siganids is the presence of two pelvic fin spines (Woodland 1997). Due to the unusually large amount of fin spines, rabbitfishes are also known popularly in Australia as "spinefeet". In addition to the physical damage such as punctures and cuts that spines in dorsal, pelvic and anal fins can inflict, these injuries are aggravated by venom secreted in glands at the base of each spine. The venom, a stonefish-like toxin with lower toxicity (Yoshinaga-Kiriake et al. 2020), is not life-threatening to humans but it can cause severe localized pain, numbness, swelling, and allergic reactions in some cases.

The Morphology of the Mouth

The mouth of rabbitfishes is terminal, small, and faces slightly downward. Unlike the majority of teleosts, their upper jaw is not protrusible (Woodland 1990), a trait that increases bite force and facilitates the removal of food items attached to the substratum. Furthermore, their lower jaw is short, resulting in a reduced out-lever distance (Bellwood 2003, Bellwood and Hoey 2004). Thanks to the upper jaw's lack of protrusion and the lower jaw's high closing ratio, siganids are able to deliver a slow and forceful bite. These traits, along with the interlocking bicuspid teeth of siganids (discussed below), are ideally suited for cropping and tearing food items growing on the benthos. These include algae, seagrass, sponges, and cyanobacteria (Bryan 1975, Mantyka and Bellwood 2007, Hoey et al. 2013, Loffler et al. 2015, Ebrahim et al. 2020).

At least one species, the barred rabbitfish *Siganus doliatus*, possesses an intra-mandibular joint (IMJ) (Konow et al. 2008). An IMJ provides an additional point of inflexion between the dentary and articular bones in the lower jaw. By allowing the lower jaw to bend outward, the IMJ increases contact of the jaws with the substratum by expanding the maximum gape (Gibb et al. 2015). Intra-mandibular joints are present in multiple families of teleosts, mainly in fishes that bite on food items attached to the substratum (Konow et al. 2008). Aside from the reported IMJ in *S. doliatus* and possibly *S. canaliculatus* (Yeomans and Beech 2021), the presence of an IMJ in other siganids has yet to be investigated and as such its prevalence within the family is unclear. However, since the IMJ is a basal symplesiomorphic trait of siganids and acanthurids (Konow et al. 2008), it is possible that IMJs are widespread across siganids. If the IMJ facilitates cropping on open surfaces, another anatomical modification—a tubulate snout—enables a few rabbitfish species to feed in concealed areas that are impenetrable to other herbivores.

The shape, number, and arrangement of teeth in the mouths of reef fishes has been shown to be an informative trait delineating various feeding modes (Motta 1989, Purcell and Bellwood 1993, Bellwood et al. 2014, Mihalitsis and Bellwood 2019). In rabbitfishes, teeth in both the upper and lower jaws are arranged in a single row with no gaps (Woodland 1997). Teeth are incisiform and asymmetrically bicuspid with a primary cusp and a smaller lateral cusp (Woodland 1990, Streit et al. 2015). Sayed et al. (2019) reported *Siganus rivulatus* as having tricuspid teeth along the premaxilla. However, upon careful examination of the scanned electron micrographs in this publication, it seems there might have been a slight misinterpretation by the authors regarding a chip in one of the three teeth depicted, which could be mistaken for a gap between the primary cusp and an additional 'third' cusp. When the mouth is closed, the teeth in both jaws interlock (Streit et al. 2015). The shape and arrangement of the teeth in rabbitfishes have been suggested to be particularly well-suited for cropping the leaves of macroalgae (Streit et al. 2015) and turf algae. No teeth are found inside the oral cavity.

Little information exists on the structure of the oral cavity of rabbitfishes. A hardened surface posterior to the teeth has been documented in *Siganus luridus* and *S. rivulatus* (Fishelson and Delarea 2014). A very similar structure present in several surgeonfishes (family Acanthuridae) is thought to act as a retention plate (Fishelson and Delarea 2014). A histological analysis of the oral cavity of the detritivorous surgeonfish *Ctenochaetus striatus*, revealed that the anterior surface of the palate is keratinized, which suggests that this region may be exposed to abrasion resulting from detrital particles brushed inside the mouth by their bristle-like teeth (Tebbett et al. 2018). However, no keratinization has been documented in the oral cavity of rabbitfishes to our knowledge. Fishelson and Delarea (2014) hypothesized that the presence of the hardened surface in the oral cavity of *S. luridus* and *S. rivulatus* may be the result of a shared ancestry of rabbitfishes and surgeonfishes rather than a structure serving a functional purpose in rabbitfishes. Whether this hardened surface is keratinized or not is unknown. The oral cavity of several species of fairy wrasses (f. Labridae) is lined with abundant mucus-secreting cells that produce a coat of mucus that may be involved in the retention of food particles and the protection of the oral mucosa from stinging prey ingested by these planktivorous reef fishes (Huertas and Bellwood 2020). To our knowledge, there is no evidence of mucus secretion in the mouth of siganids. Since rabbitfishes feed on non-stinging prey such as macroalgae, sponges, and cyanobacteria, it is likely that only a limited number of mucous cells are present in their oral cavity, if any.

Of the two subgenera of the genus *Siganus* (*Siganus* and *Lo*), the latter is characterised by a pronounced tube-like snout (Woodland 1990, Brandl and Bellwood 2013). It is therefore not surprising that the maxilla-orbit distance (MOD) and snout length (SnL) separate rabbitfishes with a tubulate snout (*Siganus unimaculatus*, *S. uspi*, and *S. vulpinus*) from the remaining species in the pPCA space (Fig. 3). The prominent snout of the five species in the subgenus *Lo* (*Siganus magnificus*, *S. niger*, *S. unimaculatus*, *S. uspi*, and *S. vulpinus*) is similar to that of some acanthurids (e.g., *Zebrasoma*), which further facilitates foraging on algae growing on concave substratum (Brandl and Bellwood 2013, Nanami 2018). It is worth noting, however, that other siganids are also capable of accessing reef crevices. In addition to *S. vulpinus*, a species with one of the longest snouts across siganids (Fig. 3), Fox and Bellwood (2013) noted that *S. corallinus*, and *S. puellus* were also able to routinely feed in reef gaps and fissures. They proposed that this may be due to a more elongated and narrower head and snout when compared to other nominally herbivorous reef fishes such as those within the genera *Scarus* and *Acanthurus*.

The pPCA of the morphometric data set we collected suggests that there are morphological differences between schooling and pairing siganids (Fig. 3). Pair-forming species appeared grouped in two clusters. The first of these encompasses predominantly reef-dwelling pairing siganids of the subgenus *Siganus* (but see Hoey et al. – Chapter 7) and their position is primarily driven by a higher aspect ratio and longer leading edge of the caudal fin, and a deeper caudal peduncle. Although *Siganus doliatus* appears separated from the rest of the pairing reef-dwelling species, this may be an artifact due to lack of replication for this species because we were only able to include measurements from a single individual. Moreover, the relatively small size of this fish (90.0 mm SL) could have placed the position of *S. doliatus* in the plot lower than expected due to the effect of allometric growth. Specifically, due to the comparatively low aspect ratio of the caudal fin, a trait that is known to increase in full-grown adults of this species (Woodland 1990). Therefore, it is likely that the inclusion of additional individuals of *S. doliatus* would place it near its sister species *S. virgatus*. The second cluster includes species of the subgenus *Lo*. In this case, their position is driven by a longer head and traits associated with having a long, pointed snout such as longer distances between the orbit and the maxilla and between the first dorsal spine and the tip of the snout. In addition, these siganids also had a deeper body, and a higher aspect ratio and longer leading edge of the pectoral fins, and may reflect their close association with structurally complex coral reef habitats. Finally, a third cluster consists of schooling species, whose position is mainly influenced by a larger eye diameter.

In a study evaluating potential associations between morphology, pairing behaviour, and ecology, Brandl and Bellwood (2013) found similar results but with a notable difference. In addition to a pointed snout, a deep body, a concave forehead, and a comparatively shorter maximum length, the authors also identified large eyes as a key morphological trait typically exhibited by pair-forming reef fishes. Our analysis, on the other hand, suggests that it is schooling species that have the largest eyes amongst the rabbitfishes we examined (Fig. 3). Although this discrepancy in the relationship of the eye diameter with pairing behaviour may seem conflicting, several factors may explain it. First, our analysis focused solely on siganids with the 22 species included belonging to this family. By contrast, the analysis in Brandl and Bellwood (2013) encompassed a broader taxonomic diversity with 47 species from six families (i.e., Chaetodontidae, Acanthuridae, Pomacanthidae, Kyphosidae, Labridae, and Siganidae), and a lower number of siganid species (n=9). Thus, the taxonomic diversity of their analysis reflects a broader pattern across reef fishes while our findings reveal a family-specific pattern. Furthermore, differences in the measurements collected in both studies might have also contributed to this apparent discrepancy.

While the series of morphological features described above hint at a potential role of morphology underpinning pairing behaviour in siganids, we warn that this relationship may be circumstantial. Instead, several studies suggest that the habitat in which siganids live may have a stronger influence than morphology on whether they pair or not (Brandl and Bellwood 2013; Mirbach and Brandl 2016). In addition to the shared morphological features described above, pairing siganids such *as S. corallinus* (Fig. 4) often display brighter colours than schooling species. These differences in colouration are discussed in greater detail in the sections below.

Fig. 4. Anterior portion of the body of the coral rabbitfish *Siganus corallinus*. Photo: Victor Huertas.

External and Internal Eye Morphology

The external morphology of the eye of rabbitfishes is highly conserved across the family, likely driven by similar ecological constraints (Schmitz and Wainwright 2011; Schweikert et al. 2019). A spherical, crystalline lens is located centrally within a circular pupil. The eyes of rabbitfishes lack an aphakic gap—an open space in the pupil not occupied by the lens—which prevents lens accommodation and greatly reduces their ability for variable focus (Sivak 1978, Schmitz and Wainwright 2011). However, small adaptive changes to their lenses can occur over larger timespans, giving them the ability to tailor their vision to suit different environmental needs (Gagnon et al. 2011). Their lenses possess ocular filtering media which are pigments and materials that absorb certain wavelengths of light before they stimulate photoreceptor cells in the retina (Siebeck and Marshall 2001, Gagnon et al. 2011). This filtering media functions by absorbing shorter and UV wavelengths, essentially act as sunglasses protecting the eye from UV damage (Siebeck and Marshall 2001). While published data on the spectral sensitivity of rabbitfish photoreceptors is limited, measurements from *S. rivulatus* indicate three photoreceptors with maximum absorbances peaking at 440, 450, and 512 nm (Karpestam et al. 2007, Gagnon et al. 2011). If indicative of the broader spectral sensitivities within the family, it appears rabbitfishes lack UV vision and have a rather narrow spectral range suggesting limited colour vision. The specific combination of lens filtering media and spectral sensitivities create a unique visual system that is similar to that of other herbivorous fish species on coral reefs (Karpestam et al. 2007) suggesting that their ecology (most notably, their diet) likely shapes their visual capabilities (Siebeck and Marshall 2007).

Colours and Patterns in Rabbitfishes

In many ways, colouration is simply an extension of an organism's morphology; forming an important component of an organism's phenotype. Although it does not appear to be governed by structure or shape to the naked eye, organismal colours and patterns are in fact a product of the density and distribution of specific cells, pigments, and compounds within the tissues of an organism (Sun et al. 2013, Shawkey and D'Alba 2017). These cells and corresponding pigments and microstructures (discussed in detail in the following section) absorb and reflect various wavelengths of light; generating the unique appearances that we and other organisms recognize as a colour pattern. As with most morphological features (e.g., teeth), colouration confers various fitness benefits that are often tailored to a specific life history strategy and ecology (Cuthill et al. 2017). For instance, cryptic and camouflaged colouration helps individuals blend in better to avoid predators or capture prey more successfully (Troscianko et al. 2016, Merilaita et al. 2017). Conspicuous colour patterns help relay information to viewers, like the fitness of an individual (Hill 1990, Trigo et al. 2020) or that a potential prey item is chemically defended (e.g., poison dart frogs; Summers and Clough 2001, Maan and Cummings 2012).

The colouration of rabbitfishes uses a mixture of these strategies, with some species being brightly coloured and displaying ornate patterns, while others having a drabber appearance. Some of the most dominant colours found on the 31 species are yellow hues and white/silvers, followed by black and brown shades. While their colour palette is relatively limited compared to other common reef fish families (e.g., Labridae, Pomacanthidae), specific combinations of these colours are highly contrasted, maximizing their signalling potential (Marshall et al. 2003a). White and black are the colour stimuli/spectra with the largest perceptive difference as they represent absolute changes in luminance lacking a chromatic (i.e., colour or hue) component. On coral reefs, yellow is also a highly salient colour as it contrasts strongly against the saturated blue colour of clear water reef environments (Marshall 2000). Yet, this colour can still provide some level of camouflage when viewed against the "average" colour of a healthy coral reef (Marshall et al. 2018). Indeed, some of the more colourful species (e.g., *S. vulpinus*) are highly sought after and frequently kept in aquaria by hobbyists.

The Development of Colour Patterns and Colour Change

Chromatophores are the pigment-bearing cells responsible for creating colouration in most teleost fishes (Sköld et al. 2016, Salis et al. 2019, Goda and Kuriyama 2021). The main chromatophores contributing to the colouration of rabbitfishes are likely xanthophores (yellow), melanophores (black), iridiophores/cyanophores (blue/cyan), and leucophores (silver/reflective). The specific pigment compounds within these cells are a combination of those that are synthesized within the individual (e.g., melanin) as well as those that are acquired strictly though their diet (e.g., carotenoids). The presence of certain carotenoids in the algal component of their diet contributes to their ability to maintain these vivid colours (most notably yellow), as this class of pigments are not naturally synthesized by animals. Indeed, rabbitfish reared and fed prescribed diets varying in their algal component showed changes in the vibrancy of their yellow hues (Sulaeman et al. 2022). As a consequence, their appearance and bright colouration is a factor often considered when these fishes are purchased at markets for consumption (Sulaeman et al. 2022).

The specific cellular formation of patterns in rabbitfishes appears to be highly conserved within the family. Typically, rabbitfishes show patterns that originate following a reaction-diffusion process; a process most frequently used to model how two chemicals react within a spatial context. Alan Turing demonstrated how this process can be used to describe how intricate colour patterns are formed which are nearly identical to those seen in numerous organisms (Turing 1952). Subsequent research has refined these models to more accurately mimic pattern formation in animals (Kondo 2002, Morales et al. 2015). Patterns created by this process are defined by a recurring and reticulate nature, ranging from repeated small spots and dashes, to meandering, labyrinthine lines (Fig. 5A). These types of colorations are widespread on rabbitfishes with some species displaying an entire spectrum of these patterns across their body. *S. puellus* displays larger blotches on the head that transition to near complete vertical lines, which gradually become more labyrinthine as the pattern moves ventrally along the body. As the pattern progresses caudally, it becomes less punctuated and chaotic in nature, ultimately terminating into complete horizontal lines that cease at the caudal peduncle. These types of patterns are phylogenetically widespread and found on many rabbitfish species, as well as many other teleosts (Morales et al. 2015, Miyazawa 2020).

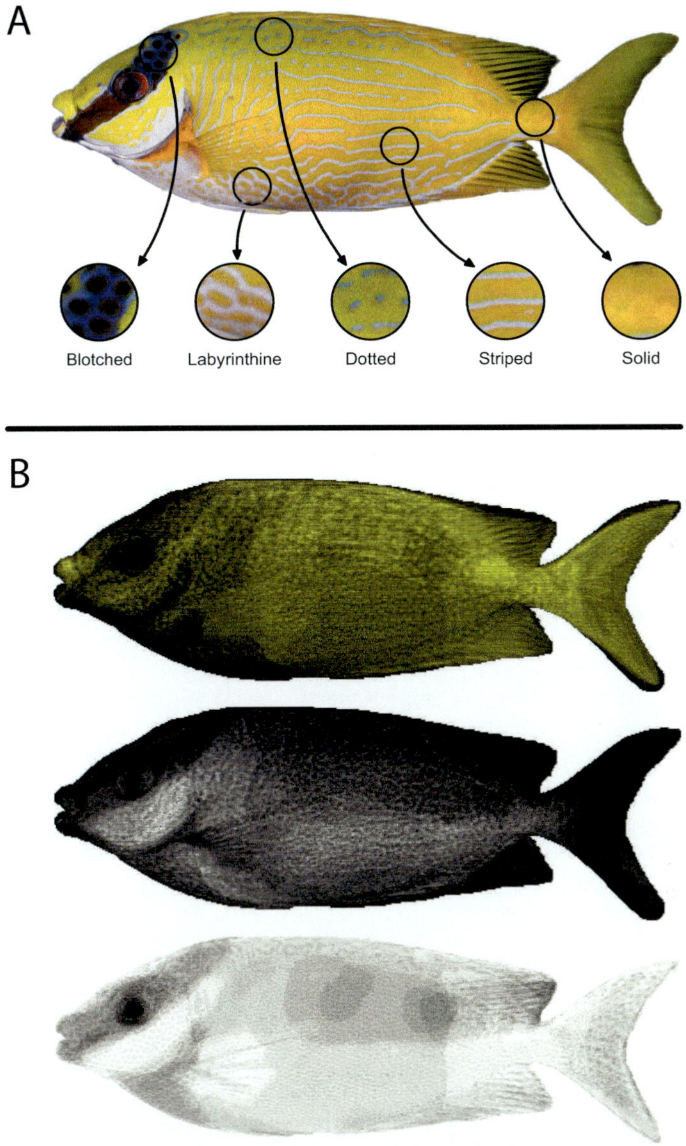

Fig. 5. (A) The diversity of patterns created by reaction-diffusion processes. *Siganus puellus* shows a variety of the unique patterns that are created by this process. Photo with permission from François Libert. (B) Heat maps of yellow, white, and black across the bodies of 24 rabbitfish species. The heat intensity has been rescaled so that the highest density location for a given colour is close to 100% saturated. Please note: *S. puellus* was used as the reference image required for the colour detection technique, hence the similar appearance of the general shape to this species.

Apart from displaying differences in colour patterns among species, rabbitfishes exhibit temporal changes in their appearance. Many rabbitfish species are capable of undergoing both morphological and physiological colour change, two processes that differ in both the mechanisms underpinning the colour change and the timescale over which the changes occur (Figon and Casas 2018). Morphological colour change occurs as a result of changes in the density and type of chromatophores within their skin and/or the total amount of pigmentary material within the cells (Sugimoto 2002, Goda and Kuriyama 2021). Thus, the timeframe for morphological colour change is constrained by the production of new cells, new cell types, and intracellular components (Sköld et al. 2016). This form of colour change most prominently occurs as juvenile rabbitfish develop into adults (Seale and Ellis 2019). Morphological colour change is widespread in teleost fishes and, like rabbitfishes, is primarily associated with ontogeny (e.g., Cortesi et al. 2016). By contrast, physiological colour change occurs within the timeframe from seconds to minutes, making it more obvious to the human eye. This type of change is caused by the expansion and contraction of pigments within the chromatophores, which alters the concentration, allowing the colours to change rapidly (Amiri 2008, Sköld et al. 2013). This type of colour change is frequently used by rabbitfishes when they sleep. Reef-dwelling rabbitfishes will often find a concealed shelter at night to avoid nocturnal predators (most often a hole, crevice, or overhang in the reef). While sleeping, some species will adopt a highly cryptic and mottled pattern that differs greatly from their daytime appearance (Fig. 6A day time, Fig. 6B night time). This camouflaged colouration likely aids

Fig. 6. Rabbitfishes are capable of profound physiological colour changes which are most frequently observed at night. Above is *Siganus luridus* displaying typical daytime colouration. Below is the same species at night showing a mottled and highly cryptic appearance. Please note that these are different individuals. Photos with permission from François Libert.

in reducing detection by hunting nocturnal predators. When stressed or frightened, rabbitfishes will also adopt this mottled pattern (Seale and Ellis 2019). Aquarium hobbyist frequently observe this appearance when an individual is acclimating to a new aquarium/environment.

Broadscale Patterns of Colouration and Ecological Correlates

When assessing the full colourspace of all rabbitfishes (by using multivariate techniques that measure and compare the complete colour distribution of each species), distinct axes of variation arise (Fig. 7A). Most notably, the first principal coordinate axis primarily reflects a distinction between highly conspicuous and brightly coloured species dominated by yellow hues (e.g., *S. vulpinus*) to species that are relatively drab in appearance characterized mostly by silvers and browns (e.g., *S. rivulatus*). The second principal component mainly reflects differences along a dark to light axis with *S. niger* representing the most extreme end of the spectrum defined by an almost entirely black appearance.

Indeed, when we explore the specific distribution of common colours across the body of rabbitfishes, some distinct motifs emerge, giving clues as to the possible function of each respective colour. Yellow hues show the most widespread presence across the body; however, this colour occurs with greatest frequency towards the posterior section of the body; peaking around the caudal peduncle (Fig. 5B). Many reef fish species possess yellow colouration on the posterior half of their body possibly suggesting that this colour functions as an important signal, however, one that has been selected to occur in a location that is far from vital body organs. White occurs with less frequency and at more specific locations on rabbitfishes (Fig. 5B). Interestingly, white is most commonly present on the ventral half of the body as well as a diagonal streak below the eye along the cheek/operculum. White present on the lower half of the body is indicative of countershading which aids in concealment when the individual is viewed from below (Hamilton III and Peterman 1971, Kelley et al. 2017). The function of this white streak along the cheek is uncertain when considered in isolation. However, it may help boost contrast when the distribution of black is considered. The highest density of black occurs on the head as a diagonal stripe (immediately above the white stripe) that completely surrounds the eye, called an "eye stripe" (Fig. 5B). This pattern is phylogenetically widespread in fishes (Kelley et al. 2013, Zapfe 2021) and helps conceal the pupils, making it more difficult for predators to locate and orient their prey. Eye stripes in rabbitfishes appear to be phylogenetically labile as this trait has originated independently in at least three lineages within the family. Oftentimes, this stripe is paired with other patterns to confer further fitness benefits (Kelley et al. 2013, Miyazawa 2020).

A rabbitfish's colour pattern appears to be linked to its ecology. Species that are typically found exclusively in coral dominated environments tend to be more colourful in appearance; harbouring a greater coverage of yellow hues (e.g., *S. corallinus* and *S. vulpinus*)—as evident by a significant phylogenetic linear regression of PCoA 1 and reef association (Fig. 7B, $p < 0.001$, adjusted $R^2 = 0.61$). The clear waters found on coral reefs permit the use of more distinct and vibrant colours and patterns as they transmit best in transparent environments (Marshall et al. 2003b). However, the

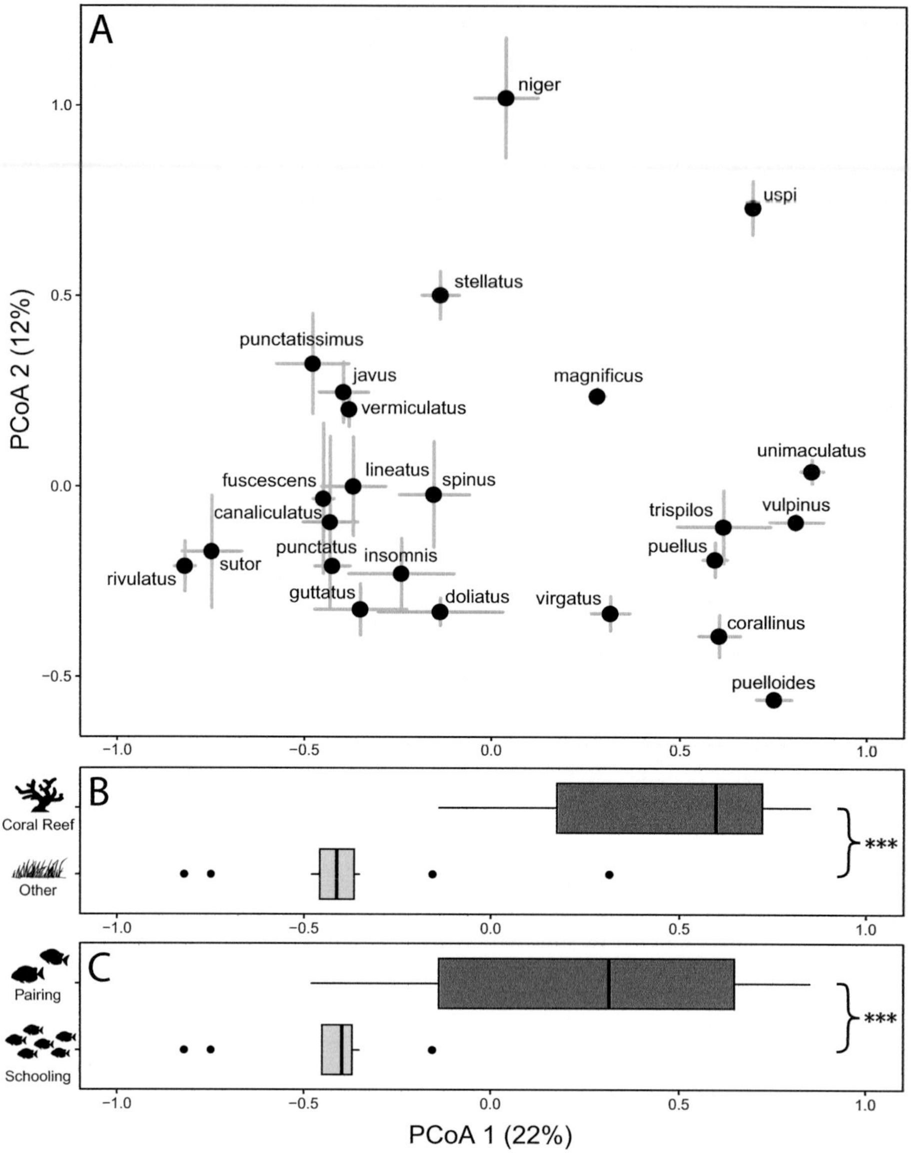

Fig. 7. A: The complete multivariate colourspace of 24 rabbitfish species that have both images and phylogenetic data available. The colourspace was created by measuring the amount of colour on each fish that falls within 27 colour bins using the *colordistance* R package. The multivariate colour data was analysed using dissimilarity measures (specifically χ^2 distance) and visualized using a PCoA. The black circles represent the mean centroid based on three separate images analysed for each species. The grey lines represent the Standard Error along both PCoA1 and PCoA2. In the two lower panels, the values along principal coordinate axes 1 have been grouped by whether the species is more associated with coral reefs or other habitats (B) and if the species is typically found in pairs or schools (C). In both instances, groupings were significantly different using a phylogenetic generalized least squares regression which accommodates shared ancestry between species.

unique nature of the benthos on coral reefs likely also plays a crucial role in selecting vibrant colours and patterns (Hemingson et al. 2022) as many non-coral reef species in the same taxonomic families tend to be less colourful (Hodge et al. 2020). There is also a similar pattern when assessing how colouration is correlated to social behaviour (schooling vs. pairing, Fig. 7C). The most colourful species also tend to be those that form monogamous pairs ($p < 0.001$, adjusted $R^2 = 0.46$). As pairing can confer unique fitness benefits, like allowing for extended foraging bouts and increased partner vigilance (Fox and Donelson 2014, Brandl and Bellwood 2015) it has been argued that reciprocity may be restricted to animals that can meet these demands. Here, we provide evidence for the potential presence of direct reciprocity in teleost fishes. We demonstrate that in pairs of coral reef rabbitfishes (f. Siganidae, the brighter and more salient colours may make it easier to identify the partner over greater distances (Marshall et al. 2003a) or in more turbid waters (Woodland 1990). However, these hypotheses need to be formally tested and validated with behavioural experiments.

Conversely, non-reefal, and schooling rabbitfish species (e.g., *S. canaliculatus*) tend to be more uniform and drabber in colouration than reef-associated rabbitfishes. In general, schooling fishes often have more uniform colour patterns, like single-coloured reflective flanks or bars/stripes to make target detection more difficult from predators (Hobson 1979). This drabber, more uniform appearance likely helps these species evade predators more effectively by breaking up individual outlines and matching backgrounds (Cott 1940, Merilaita et al. 2017). Furthermore, the less colourful nature of non-coral dominated ecosystems paired with reduced water clarity may select for similar drab appearances in their resident organisms (Kelley et al. 2012).

Conclusion and Future Directions

Rabbitfishes have unique morphological and phenotypic adaptations that confer specific fitness benefits, ultimately shaping their behaviour and ecology. Despite the relatively small number of species when compared to other common coral reef fish families, rabbitfishes form an important functional component of Indo-Pacific coastal ecosystems (Fox et al. 2009, Cheal et al. 2012). The narrow taxonomic breadth makes future morphological and colouration studies possible with near complete taxonomic sampling. Below, we propose avenues for future research which aim to better understand the diversity and significance of morphological traits and colouration in rabbitfishes:

1. **Kinematic and ecomorphological studies in rabbitfishes.** A reasonable understanding of the interplay between the morphology and biomechanics underpinning feeding and locomotion performance exists for several families of reef fishes, including acanthurids (Friedman et al. 2016, Mihalitsis and Wainwright 2024, Tebbett et al. 2017), labrids (Westneat 1990, Clifton and Motta 1998, Wainwright et al. 2002, 2004), pomacentrids (Cooper and Westneat 2009, Aguilar-Medrano et al. 2011) chaetodontids (Motta 1988, Ferry-Graham et al. 2001, Konow and Ferry-Graham 2013), and pomacanthids (Konow and Bellwood 2011). An in-depth assessment of the biomechanics of feeding and locomotion in rabbitfishes, however, is lacking. Advances in our understanding

of the functional anatomy of rabbitfishes may help explain the observed variation in feeding preferences across the family in different habitats despite similar external morphology (Fox et al. 2009, Hoey et al. 2013) and clarify how fins and body shapes influence their swimming ability.

2. **Complete phenotypic comparisons across the family.** The limited phylogenetic diversity within this family means conducting phylogenetic comparative analyses with 100% species representation (or very close to) is an attainable goal. For many common reef fish families, this requires extensive sampling effort. This allows for the inquiry of intriguing questions about character evolution and their respective selective pressures to be conducted at the family level.
3. **Characterizing the visual capabilities of rabbitfishes.** Our current understanding of the visual capabilities of rabbitfishes is extremely limited. To our knowledge, there is only one published record of the spectral sensitivities of rabbitfish (*S. rivulatus*) and very limited knowledge of their lens filtering media (which provides clues as to what they likely can and cannot see). Any interpretation of colour and appearance is contingent on knowing the visual capabilities within the family. Thus, identifying spectral sensitivities and visual acuity within the family will fill this knowledge gap and provide further identification of how rabbitfishes see the world. This would then allow further work investigating how vision impacts behaviour and appearance.
4. **Understanding the genetic mechanisms underpinning pattern formation and expression.** The high prevalence of reaction-diffusion driven elements in rabbitfishes and their ability to be reared easily in captivity make them an excellent group to research pigment development and the genetic mechanism underpinning pigment expression.

Further work investigating these lines of research will yield valuable insights as to what shapes the phenotypic evolution and functional morphology of this coral reef fish family.

References

Aguilar-Medrano, R., B. Frédérich, E. De Luna and E.F. Balart. 2011. Patterns of morphological evolution of the cephalic region in damselfishes (Perciformes: Pomacentridae) of the Eastern Pacific. Biol. J. Linn. Soc. 102: 593–613. https://doi.org/10.1111/j.1095-8312.2010.01586.x.

Aguilar-Medrano, R., B. Frédérich, E.F. Balart and E. De Luna. 2013. Diversification of the pectoral fin shape in damselfishes (Perciformes, Pomacentridae) of the Eastern Pacific. Zoomorphology 132: 197–213. https://doi.org/10.1007/s00435-012-0178-8.

Aguilar-Medrano, R., B. Frédérich, E. De Luna and E.F. Balart. 2011. Patterns of morphological evolution of the cephalic region in damselfishes (Perciformes: Pomacentridae) of the Eastern Pacific. Biol. J. Linn. Soc. 102: 593–613. https://doi.org/10.1111/j.1095-8312.2010.01586.x.

Amiri, M.H. 2008. Postsynaptic Alpha 2-adrenoceptors mediate melanosome aggregation in melanophores of the white-spotted rabbitfish (*Siganus canaliculatus*). Pak. J. Biol. Sci. 12: 1–10. https://doi.org/10.3923/pjbs.2009.1.10.

Bejarano, S., J.-B. Jouffray, I. Chollett, R. Allen, G. Roff, A. Marshell et al. 2017. The shape of success in a turbulent world: wave exposure filtering of coral reef herbivory. Funct. Ecol. 31: 1312–1324. https://doi.org/10.1111/1365-2435.12828.

Bellwood, D. and P. Wainwright. 2001. Locomotion in labrid fishes: implications for habitat use and cross-shelf biogeography on the Great Barrier Reef. Coral Reefs 20: 139–150. https://doi.org/10.1007/s003380100156.

Bellwood, D.R. 2003. Origins and escalation of herbivory in fishes: a functional perspective. Paleobiology 29: 71–83. https://doi.org/10.1666/0094-8373(2003)029<0071:OAEOHI>2.0.CO;2.

Bellwood, D.R. and A.S. Hoey. 2004. Feeding in Mesozoic fishes: a functional perspective. pp. 639–649. *In*: Arratia, G. and A. Tintori [eds.]. Mesozoic Fishes 3—Systematics, Paleoenvironments and Biodiversity. Verlag Dr. Friedrich Pfeil, München.

Bellwood, D.R., A.S. Hoey, O. Bellwood and C.H.R. Goatley. 2014. Evolution of long-toothed fishes and the changing nature of fish-benthos interactions on coral reefs. Nat. Commun. 5: 3144. https://doi.org/10.1038/ncomms4144.

Binning, S.A. and D.G. Roche. 2015. Water flow and fin shape polymorphism in coral reef fishes. Ecology 96: 828–839. https://doi.org/10.1890/14-0426.1.

Brandl, S.J. and D.R. Bellwood. 2015. Coordinated vigilance provides evidence for direct reciprocity in coral reef fishes. Sci. Rep. 5: 14556. https://doi.org/10.1038/srep14556.

Brandl, S.J. and D.R. Bellwood. 2013. Morphology, sociality, and ecology: Can morphology predict pairing behavior in coral reef fishes? Coral Reefs 32: 835–846. https://doi.org/10.1007/s00338-013-1042-0.

Brandl, S.J., A.S. Hoey and D.R. Bellwood. 2014. Micro-topography mediates interactions between corals, algae, and herbivorous fishes on coral reefs. Coral Reefs 33: 421–430. https://doi.org/10.1007/s00338-013-1110-5.

Bryan, P.G. 1975. Food habits, functional digestive morphology, and assimilation efficiency of the Rabbitfish Siganus spinus (Pisces, Siganidae) on Guam. Pac. Sci. 29: 269–277.

Chang, J., D.L. Rabosky, S.A. Smith and M.E. Alfaro. 2019. An R package and online resource for macroevolutionary studies using the ray-finned fish tree of life. Methods Ecol. Evol. 10: 1118–1124. https://doi.org/10.1111/2041-210X.13182.

Cheal, A., M. Emslie, I. Miller and H. Sweatman. 2012. The distribution of herbivorous fishes on the Great Barrier Reef. Mar. Biol. 159: 1143–1154. https://doi.org/10.1007/s00227-012-1893-x.

Clifton, K.B. and P.J. Motta. 1998. Feeding morphology, diet, and ecomorphological relationships among five Caribbean labrids (Teleostei, Labridae). Copeia 4: 953–966. https://doi.org/10.2307/1447342.

Cooper, W.J. and M.W. Westneat. 2009. Form and function of damselfish skulls: Rapid and repeated evolution into a limited number of trophic niches. BMC Evol. Biol. 9: 1–17. https://doi.org/10.1186/1471-2148-9-24.

Cortesi, F., Z. Musilová, S.M. Stieb, N.S. Hart, U.E. Siebeck, K.L. Cheney et al. 2016. From crypsis to mimicry: changes in colour and the configuration of the visual system during ontogenetic habitat transitions in a coral reef fish. J. Exp. Biol. 219: 2545–2558. https://doi.org/10.1242/jeb.139501.

Cott, H.B. 1940. Adaptive coloration in animals. Methuen & Co. Ltd., London.

Cuthill, I.C., W.L. Allen, K. Arbuckle, B. Caspers, G. Chaplin, M.E. Hauber et al. 2017. The biology of color. Science 357: 1–7. https://doi.org/10.1126/science.aan0221.

Ebrahim, A., T.S.H. Martin, P.J. Mumby, A.D. Olds and I.R. Tibbetts. 2020. Differences in diet and foraging behaviour of commercially important rabbitfish species on coral reefs in the Indian Ocean. Coral Reefs 39: 977–988. https://doi.org/10.1007/s00338-020-01918-6.

Ferry-Graham, L.A., P.C. Wainwright, C.D. Hulsey and D.R. Bellwood. 2001. Evolution and mechanics of long jaws in butterflyfishes (Family Chaetodontidae). J. Morphol. 248: 120–143. https://doi.org/10.1002/jmor.1024.

Figon, F. and J. Casas. 2018. Morphological and physiological colour changes in the animal kingdom. pp. 1–11. *In*: John Wiley and Sons, Ltd [ed.]. eLS. Wiley, https://doi.org/10.1002/9780470015902.a0028065.

Fishelson, L. and Y. Delarea. 2014. Comparison of the oral cavity architecture in surgeonfishes (Acanthuridae, Teleostei), with emphasis on the taste buds and jaw "retention plates". Environ. Biol. Fishes 97: 173–185. https://doi.org/10.1007/s10641-013-0139-1.

Fox, R., T. Sunderland, A. Hoey and D. Bellwood. 2009a. Estimating ecosystem function: contrasting roles of closely related herbivorous rabbitfishes (Siganidae) on coral reefs. Mar. Ecol. Prog. Ser. 385: 261–269. https://doi.org/10.3354/meps08059.

Fox, R., T. Sunderland, A. Hoey and D. Bellwood. 2009b. Estimating ecosystem function: contrasting roles of closely related herbivorous rabbitfishes (Siganidae) on coral reefs. Mar. Ecol. Prog. Ser. 385: 261–269. https://doi.org/10.3354/meps08059.

Fox, R.J. and D.R. Bellwood. 2013. Niche partitioning of feeding microhabitats produces a unique function for herbivorous rabbitfishes (Perciformes, Siganidae) on coral reefs. Coral Reefs 32: 13–23. https://doi.org/10.1007/s00338-012-0945-5.

Fox, R.J. and J.M. Donelson. 2014. Rabbitfish sentinels: first report of coordinated vigilance in conspecific marine fishes. Coral Reefs 33: 253–253. https://doi.org/10.1007/s00338-013-1108-z.

Friedman, S.T., S.A. Price, A.S. Hoey and P.C. Wainwright. 2016. Ecomorphological convergence in planktivorous surgeonfishes. J. Evol. Biol. 29: 965–978. https://doi.org/10.1111/jeb.12837.

Froese, R. and D. Pauly. Editors. 2024. FishBase. World Wide Web electronic publication. www.fishbase.org, (06/2024).

Fulton, C.J. 2010. The role of swimming in reef fish ecology. pp. 374–406. *In*: Domenici, P. and B.G. Kapoor [eds.]. Fish Locomotion. CRC Press, https://doi.org/10.1201/b10190-12.

Fulton, C.J., D.R. Bellwood and P.C. Wainwright. 2001. The relationship between swimming ability and habitat use in wrasses (Labridae). Mar. Biol. 139: 25–33. https://doi.org/10.1007/s002270100565.

Fulton, C.J., P.C. Wainwright, A.S. Hoey and D.R. Bellwood. 2017. Global ecological success of Thalassoma fishes in extreme coral reef habitats. Ecol. Evol. 466–472. https://doi.org/10.1002/ece3.2624.

Gagnon, Y.L., N. Shashar and R.H.H. Kröger. 2011. Adaptation in the optical properties of the crystalline lens in the eyes of the Lessepsian migrant *Siganus rivulatus*. J. Exp. Biol. 214: 2724–2729. https://doi.org/10.1242/jeb.048066.

Gibb, A.C., K.L. Staab, C.J. Moran and L.A. Ferry-Graham. 2015. The teleost intramandibular joint: A mechanism that allows fish to obtain prey unavailable to suction feeders. Integr. Comp. Biol. 1–12. https://doi.org/10.1093/icb/icv042.

Goda, M. and T. Kuriyama. 2021. Physiological and morphological color changes in teleosts and in reptiles. pp. 387–423. *In*: Hashimoto, H., M. Goda, R. Futahashi, R. Kelsh and T. Akiyama [eds.]. Pigments, Pigment Cells and Pigment Patterns. Singapore.

Hamilton, III, W.J. and R.M. Peterman. 1971. Countershading in the colourful reef fish Chaetodon lunula: conceament, communication or both? Anim. Behav. 19: 357–364.

Hemingson, C.R., M. Mihalitsis and D.R. Bellwood. 2022. Are fish communities on coral reefs becoming less colourful? Glob. Change Biol. 28: 3321–3332. https://doi.org/10.1111/gcb.16095.

Hill, G.E. 1990. Female house finches prefer colourful males: sexual selection for a condition-dependent trait. Anim. Behav. 40: 563–572. https://doi.org/10.1016/S0003-3472(05)80537-8.

Hobson E.S. 1979. Interaction between piscivorous fishes and their prey. pp. 231–242. *In*: Clepper, H, [ed]. Predator–Prey Systems in Fisheries Management. Washington, DC: Sport Fishing Institute.

Hodge, J.R., F. Santini and P.C. Wainwright. 2020. Colour dimorphism in labrid fishes as an adaptation to life on coral reefs. Proc. R. Soc. B 287: 1–8. https://doi.org/10.1098/rspb.2020.0167.

Hoey, A.S., S.J. Brandl and D.R. Bellwood. 2013. Diet and cross-shelf distribution of rabbitfishes (f. Siganidae) on the northern Great Barrier Reef: Implications for ecosystem function. Coral Reefs 32: 973–984. https://doi.org/10.1007/s00338-013-1043-z.

Huertas, V. and D.R. Bellwood. 2020. Trophic separation in planktivorous reef fishes: a new role for mucus? Oecologia 192: 813–822. https://doi.org/10.1007/s00442-020-04608-w.

Karpestam, B., J. Gustafsson, N. Shashar, G. Katzir and R.H.H. Kroger. 2007. Multifocal lenses in coral reef fishes. J. Exp. Biol. 210: 2923–2931. https://doi.org/10.1242/jeb.002956.

Kelley, J.L., J.L. Fitzpatrick and S. Merilaita. 2013. Spots and stripes: ecology and colour pattern evolution in butterflyfishes. Proc. R. Soc. B 280: 20122730. https://doi.org/10.1098/rspb.2012.2730.

Kelley, J.L., B. Phillips, G.H. Cummins and J. Shand. 2012. Changes in the visual environment affect colour signal brightness and shoaling behaviour in a freshwater fish. Anim. Behav. 83: 783–791. https://doi.org/10.1016/j.anbehav.2011.12.028.

Kelley, J.L., I. Taylor, N.S. Hart and J.C. Partridge. 2017. Aquatic prey use countershading camouflage to match the visual background. Behav. Ecol. 28: 1314–1322. https://doi.org/10.1093/beheco/arx093.

Kondo, S. 2002. The reaction-diffusion system: a mechanism for autonomous pattern formation in the animal skin: Pattern formation in the animal skin. Genes Cells 7: 535–541. https://doi.org/10.1046/j.1365-2443.2002.00543.x.

Konow, N. and D.R. Bellwood. 2011. Evolution of high trophic diversity based on limited functional disparity in the feeding apparatus of marine angelfishes (f. Pomacanthidae). PLoS ONE 6. https://doi.org/10.1371/journal.pone.0024113.

Konow, N., D.R. Bellwood, P.C. Wainwright and A.M. Kerr. 2008. Evolution of novel jaw joints promote trophic diveristy in coral reef fishes. Biol. J. Linn. Soc. 93: 545–555. https://doi.org/10.1111/j.1095-8312.2007.00893.x.

Konow, N. and L.A. Ferry-Graham. 2013. Functional morphology of butterflyfishes. pp. 19–47. *In*: Pratchett, M.S., M.L. Berumen and B.G. Kapoor [eds.]. Biology of Butterflyfishes. CRC Press, Boca Raton, FL.

Kuiter, R.H. and H. Debelius. 2001. Surgeonfishes, rabbitfishes and their relatives. TMC Publishing, Chorleywood, UK.

Langerhans, R.B. and D.N. Reznick. 2010. Ecology and evolution of swimming performance in fishes: predicting evolution with biomechanics. pp. 200–248. *In*: Domenici, P. and B.G. Kapoor [eds.]. Fish Locomotion. CRC Press, https://doi.org/10.1201/b10190-7.

Larouche, O., B. Benton, K.A. Corn, S.T. Friedman, D. Gross, M. Iwan et al. 2020. Reef-associated fishes have more maneuverable body shapes at a macroevolutionary scale. Coral Reefs. https://doi.org/10.1007/s00338-020-01976-w.

Loffler, Z., D.R. Bellwood and A.S. Hoey. 2015. Among-habitat algal selectivity by browsing herbivores on an inshore coral reef. Coral Reefs. https://doi.org/10.1007/s00338-015-1265-3.

Maan, M.E. and M.E. Cummings. 2012. Poison frog colors are honest signals of toxicity, particularly for bird predators. Am. Nat. 179. https://doi.org/10.1086/663197.

Mantyka, C.S. and D.R. Bellwood. 2007. Direct evaluation of macroalgal removal by herbivorous coral reef fishes. Coral Reefs 26: 435–442. https://doi.org/10.1007/s00338-007-0214-1.

Maoka, T. 2020. Carotenoids as natural functional pigments. J. Nat. Med. 74: 1–16. https://doi.org/10.1007/s11418-019-01364-x.

Marshall, N.J. 2000. Communication and camouflage with the same "bright" colours in reef fishes. Philos. Trans. R. Soc. Lond. B. Biol. Sci. 355: 1243–1248. https://doi.org/10.1098/rstb.2000.0676.

Marshall, N.J., F. Cortesi, F. de Busserolles, U.E. Siebeck and K.L. Cheney. 2018. Colours and colour vision in reef fishes: Past, present and future research directions. J. Fish Biol. 95: 5–38. https://doi.org/10.1111/jfb.13849.

Marshall, N.J., K. Jennings, W.N. McFarland, E.R. Loew and G.S. Losey. 2003a. Visual biology of Hawaiian coral reef fishes. II. Colors of Hawaiian coral reef fish. Copeia 2003, 455–466. https://doi.org/10.1643/01-055.

Marshall, N.J., K. Jennings, W.N. McFarland, E.R. Loew and G.S. Losey. 2003b. Visual biology of hawaiian coral reef fishes. III. Environmental Light and an Integrated Approach to the Ecology of Reef Fish Vision. Copeia 2003, 467–480. https://doi.org/10.1643/01-056.

Merilaita, S., N.E. Scott-Samuel and I.C. Cuthill. 2017. How camouflage works. Philos. Trans. R. Soc. B Biol. Sci. 372. https://doi.org/10.1098/rstb.2016.0341.

Mihalitsis, M. and D.R. Bellwood. 2019. Functional implications of dentition-based morphotypes in piscivorous fishes. R. Soc. Open Sci. 6: 190040.

Mihalitsis, M. and P.C. Wainwright. 2024. Feeding kinematics of a surgeonfish reveal novel functions and relationships to reef substrata. Commun. Biol. 7: 13. https://doi.org/10.1038/s42003-023-05696-z.

Mirbach, C.E. and S.J. Brandl. 2016. Ontogenetic shifts in the social behaviour of pairing coral reef rabbitfishes (Siganidae). Mar. Biol. Res. 12: 874–880. https://doi.org/10.1080/17451000.2016.1203952.

Miyazawa, S. 2020. Pattern blending enriches the diversity of animal colorations. Sci. Adv. 6: 1–10. https://doi.org/10.1126/sciadv.abb9107.

Morales, M.A., J.F. Rojas, J. Oliveros and A.A. Hernández S. 2015. A new mechanochemical model: Coupled Ginzburg-Landau and Swift-Hohenberg equations in biological patterns of marine animals. J. Theor. Biol. 368: 37–54. https://doi.org/10.1016/j.jtbi.2014.12.005.

Motta, P.J. 1989. Dentition patterns among Pacific and Western Atlantic butterflyfishes (Perciformes, Chaetodontidae): relationship to feeding ecology and evolutionary history. Environ. Biol. Fishes 25: 159–170. https://doi.org/10.1007/BF00002209.

Motta, P.J. 1988. Functional morphology of the feeding apparatus of ten species of Pacific butterflyfishes (Perciformes, Chaetodontidae): an ecomorphological approach. Environ. Biol. Fishes 22: 39–67. https://doi.org/10.1007/BF00000543.

Nanami, A. 2018. Spatial distributions, feeding ecologies, and behavioral interactions of four rabbitfish species (*Siganus unimaculatus*, *S. virgatus*, *S. corallinus*, and *S. puellus*). PeerJ 6: e6145. https://doi.org/10.7717/peerj.6145.

Purcell, S.W. and D.R. Bellwood. 1993. A functional analysis of food procurement in two surgeonfish species, *Acanthurus nigrofuscus* and *Ctenochaetus striatus* (Acanthuridae). Environ. Biol. Fishes 37: 139–159. https://doi.org/10.1007/BF00000589.

Rabosky, D.L., J. Chang, P.O. Title, P.F. Cowman, L. Sallan, M. Friedman et al. 2018. An inverse latitudinal gradient in speciation rate for marine fishes. Nature 559: 392–395. https://doi.org/10.1038/s41586-018-0273-1.

Revell, L.J., L.J. Harmon and D.C. Collar. 2008. Phylogenetic signal, evolutionary process, and rate. Syst. Biol. 57: 591–601. https://doi.org/10.1080/10635150802302427.

Rincon-Sandoval, M., E. Duarte-Ribeiro, A.M. Davis, A. Santaquiteria, L.C. Hughes, C.C. Baldwin et al. 2020. Evolutionary determinism and convergence associated with water-column transitions in marine fishes. Proc. Natl. Acad. Sci. 117: 33396–33403. https://doi.org/10.1073/pnas.2006511117.

Salis, P., T. Lorin, V. Laudet and B. Frédérich. 2019. Magic traits in magic fish: understanding color pattern evolution using reef fish. Trends Genet. 35: 265–278. https://doi.org/10.1016/j.tig.2019.01.006.

Sambrook, K., M.C. Bonin, M. Bradley, G.S. Cumming, S. Duce, S. Andréfouët et al. 2020. Broadening our horizons: seascape use by coral reef-associated fishes in Kavieng, Papua New Guinea, is common and diverse. Coral Reefs 39: 1187–1197. https://doi.org/10.1007/s00338-020-01954-2.

Sambrook, K., A.S. Hoey, S. Andréfouët, G.S. Cumming, S. Duce and M.C. Bonin. 2019. Beyond the reef: The widespread use of non-reef habitats by coral reef fishes. Fish Fish. 20: 903–920. https://doi.org/10.1111/faf.12383.

Sayed, A.E.H., U.M. Mahmoud and F. Essa. 2019. The microstructure of buccal cavity and alimentary canal of *Siganus rivulatus*: Scanning electron microscope study. Microsc. Res. Tech. 82: 443–451. https://doi.org/10.1002/jemt.23185.

Schindelin, J., I. Arganda-Carreras, E. Frise, V. Kaynig, M. Longair, T. Pietzsch et al. 2012. Fiji: an open-source platform for biological-image analysis. Nat. Methods 9: 676–682. https://doi.org/10.1038/nmeth.2019.

Schmitz, L. and P.C. Wainwright. 2011. Nocturnality constrains morphological and functional diversity in the eyes of reef fishes. BMC Evol. Biol. 11: 338. https://doi.org/10.1186/1471-2148-11-338.

Schweikert, L.E., E.M. Caves, S.E. Solie, T.T. Sutton and S. Johnsen. 2019. Variation in rod spectral sensitivity of fishes is best predicted by habitat and depth. J. Fish Biol. 95: 179–185. https://doi.org/10.1111/jfb.13859.

Seale, A.P. and S. Ellis. 2019. Sustainable Capture-Based Aquaculture of Rabbitfish in Pacific Island Lagoons. Aquac. Aquaponics.

Shawkey, M.D. and L. D'Alba. 2017. Interactions between colour-producing mechanisms and their effects on the integumentary colour palette. Philos. Trans. R. Soc. B Biol. Sci. 372: 20160536. https://doi.org/10.1098/rstb.2016.0536.

Siebeck, U.E. and N.J. Marshall. 2007. Potential ultraviolet vision in pre-settlement larvae and settled reef fish—A comparison across 23 families. Vision Res. 47: 2337–2352. https://doi.org/10.1016/j.visres.2007.05.014.

Siebeck, U.E. and N.J. Marshall. 2001. Ocular media transmission of coral reef fish--can coral reef fish see ultraviolet light? Vision Res. 41: 133–149. https://doi.org/10.1016/S0042-6989(00)00240-6.

Sivak, J.G. 1978. The functional significance of the aphakic space of the fish eye. Can. J. Zool. 56: 513–516. https://doi.org/10.1139/z78-072.

Sköld, H.N., S. Aspengren, K.L. Cheney and M. Wallin. 2016. Fish chromatophores—from molecular motors to animal behavior. International Review of Cell and Molecular Biology 321: 171–219. https://doi.org/10.1016/bs.ircmb.2015.09.005.

Sköld, H.N., S. Aspengren and M. Wallin. 2013. Rapid color change in fish and amphibians - function, regulation, and emerging applications. Pigment Cell Melanoma Res. 26: 29–38. https://doi.org/10.1111/pcmr.12040.

Streit, R.P., A.S. Hoey and D.R. Bellwood. 2015. Feeding characteristics reveal functional distinctions among browsing herbivorous fishes on coral reefs. Coral Reefs 34: 1037–1047. https://doi.org/10.1007/s00338-015-1322-y.

Sugimoto, M. 2002. Morphological color changes in fish: Regulation of pigment cell density and morphology. Microsc. Res. Tech. 58: 496–503. https://doi.org/10.1002/jemt.10168.

Sulaeman, H.A. and A. Zainuddin, Laining. 2022. Evaluation of seaweed, Ulva lactuca as a fresh diet for nursery stage of golden rabbitfish, *Siganus guttatus*. IOP Conf. Ser. Earth Environ. Sci. 1119: 012057. https://doi.org/10.1088/1755-1315/1119/1/012057.

Summers, K. and M.E. Clough. 2001. The evolution of coloration and toxicity in the poison frog family (Dendrobatidae). Proc. Natl. Acad. Sci. U. S. A. 98: 6227–6232. https://doi.org/10.1073/pnas.101134898.

Sun, J., B. Bhushan and J. Tong. 2013. Structural coloration in nature. RSC Adv. 3: 14862. https://doi.org/10.1039/c3ra41096j.

Tebbett, S.B., C.H.R. Goatley and D.R. Bellwood. 2017. Clarifying functional roles: algal removal by the surgeonfishes *Ctenochaetus striatus* and *Acanthurus nigrofuscus*. Coral Reefs 36: 803–813. https://doi.org/10.1007/s00338-017-1571-z.

Tebbett, S.B., C.H.R. Goatley, V. Huertas, M. Mihalitsis and D.R. Bellwood. 2018. A functional evaluation of feeding in the surgeonfish Ctenochaetus striatus: the role of soft tissues. R. Soc. Open Sci. 5: 171111. https://doi.org/10.1098/rsos.171111.

Toomey, M.B., C.I. Marques, P.M. Araújo, D. Huang, S. Zhong, Y. Liu et al. 2022. A mechanism for red coloration in vertebrates. Curr. Biol. 32: 4201–4214.e12. https://doi.org/10.1016/j.cub.2022.08.013.

Trigo, S., A.C.R. Gomes, S.C. Cardoso, M. Teixeira, G.C. Cardoso and M.C. Soares. 2020. Cleaner blues: Condition-dependent colour and cleaner fish service quality. Behav. Processes 181. https://doi.org/10.1016/j.beproc.2020.104246.

Troscianko, J., J. Wilson-Aggarwal, M. Stevens and C.N. Spottiswoode. 2016. Camouflage predicts survival in ground-nesting birds. Sci. Rep. 6: 1–8. https://doi.org/10.1038/srep19966.

Turing, A.M. 1952. The Chemical Basis of Morphogensis 37–72.

Vogel, S. 1994. Life in moving fluids: the physical biology of flow., Second Edition. ed. Princeton University Press, Princeton, New Jersey, USA.

Wainwright, P.C., D.R. Bellwood and M.W. Westneat. 2002. Ecomorphology of locomotion in labrid fishes. Environ. Biol. Fishes 65: 47–62. https://doi.org/10.1023/A:1019671131001.

Wainwright, P.C., D.R. Bellwood, M.W. Westneat, J.R. Grubich and A.S. Hoey. 2004. A functional morphospace for the skull of labrid fishes: Patterns of diversity in a complex biomechanical system. Biol. J. Linn. Soc. 82: 1–25. https://doi.org/10.1111/j.1095-8312.2004.00313.x.

Webb, P.W. 1994. The biology of fish swimming. pp. 45–62. *In*: Maddock, L., Q. Bone and J.M.V. Rayner [eds.]. The Mechanics and Physiology of Animal Swimming. Cambridge University Press, https://doi.org/10.1017/CBO9780511983641.005.

Webb, P.W. 1978. Fast-start performance and body form in seven species of teleost fish. J. Exp. Biol. 74: 211–226. https://doi.org/10.1242/jeb.74.1.211.

Westneat, M.W. 1990. Feeding mechanics of teleost fishes (Labridae; Perciformes): a test of four-bar linkage models. J. Morphol. 295: 205–269.

Woodland, D.J. 1997. Siganidae. Spinefoots, rabbitfishes. pp. 3627–3650. *In*: Carpenter, K.E. and V. Nien [eds.]. FAO Identification Guide for Fishery Purposes. The Western Central Pacific. FAO.

Woodland, D.J. 1990. Revision of the fish family Siganidae with descriptions of two new species and comments on distribution and biology. Indo-Pac. Fishes 19.

Yeomans, L. and M.J. Beech. 2021. An aid to the identification of fish bones from southeast Arabia: The influence of reference collections on taxonomic diversity. Int. J. Osteoarchaeol. 31: 3–17. https://doi.org/10.1002/oa.2920.

Yoshinaga-Kiriake, A., Y. Nagashima, S. Ishizaki and K. Shiomi. 2020. Primary structures and conformations of stonefish toxin-like toxins from three species of rabbitfish, *Siganus puellus*, *Siganus unimaculatus*, and *Siganus virgatus*. Fish. Sci. 86: 889–901. https://doi.org/10.1007/s12562-020-01455-9.

Zapfe, K. 2021. Macroevolution of Bold Color Patterns Across Teleostean Fishes (Master). Clemson University.

3

The Origins and Diversification of Rabbitfishes

Alexandre C. Siqueira[1,*] and *Peter F. Cowman*[2]

Introduction

The complexity of coral reef systems is reflected in the diversity of fishes that inhabit them. While a formal definition of what constitutes a "coral reef fish" has been debated (Bellwood 1996, Robertson 1998), there is now broad consensus based on systematics, biogeography, ecology, and paleontology about what the typical coral reef fish groups are (Bellwood and Wainwright 2002). The evolutionary success of these coral reef fish lineages, regardless of taxonomic rank, is often attributed to the extent and speed at which they have diversified on reef systems over evolutionary time (e.g., Alfaro et al. 2007, Cowman and Bellwood 2011, Siqueira et al. 2020, 2023), or their functional importance on coral reefs (usually associated with their trophic status) (e.g., Cowman et al. 2009, Bellwood et al. 2017, Siqueira et al. 2019a). For example, the family of wrasses and parrotfishes, the Labridae, is a "heavy-hitter" when it comes to its status as a coral reef fish lineage. This family contains over 670 species (although many are not directly associated with coral reefs; Bonaldo et al. 2014, Siqueira et al. 2023), a paleontological and phylogenetic history that spans over 60 million years (Cowman et al. 2009, Bellwood et al. 2019), and an astonishing diversity of trophic modes (Floeter et al. 2018). Similarly, the family of butterflyfishes, the Chaetodontidae, also shares an iconic status as common inhabitants of coral reefs globally, being the one group that contains over 50% of all fishes that feed directly on corals (Cole et al. 2008).

[1] Centre for Marine Ecosystems Research, School of Science, Edith Cowan University, Perth, WA 6027, Australia; And Research Hub for Coral Reef Ecosystem Functions, College of Science and Engineering, James Cook University, Townsville, QLD 4811, Australia.

[2] Biodiversity and Geosciences Program, Queensland Museum, Tropics Queensland Museum Network, Townsville, QLD 4810, Australia; And Centre for Tropical Bioinformatics and Molecular Biology, James Cook University, Townsville, QLD 4811, Australia.
Email: peter.cowman@jcu.edu.au

* Corresponding author: alexandre.siqueira@ecu.edu.au

The rabbitfishes of the family Siganidae, on the other hand, are not nearly as diverse as the Labridae, nor are they considered as iconic as the butterflyfishes. With only 31 extant species restricted to the Indo-Pacific realm (Woodland – Chapter 1), the family is not generally considered a prime example of evolutionary success on coral reefs, with many species also occurring in non-coral reef environments (Hoey et al. – Chapter 7). Yet, this low diversity on present-day reefs, in combination with their rich paleontological record (Bannikov et al. 2010), makes rabbitfishes one of the most interesting reef fish groups to study in terms of evolutionary history, particularly when compared with other roving piscine herbivores. For instance, despite sharing similar early habitats and presenting an analogous expansion in the number of fossil genera to surgeonfishes (Acanthuridae) (Bellwood et al. 2014), rabbitfishes seem to have followed very different evolutionary and biogeographic trajectories compared to surgeonfishes (Siqueira et al. 2019b). As a result of these distinct evolutionary pathways, rabbitfishes today are generally less abundant and largely restricted to coastal inner to mid shelf reefs of the Indo-Pacific (Russ 1984, Hoey et al. 2013), while surgeonfishes are globally widespread and more abundant in mid to outer shelf and oceanic reef locations (Cheal et al. 2012, Tebbett et al. 2022). The extant diversity of siganids also represents less than one third of parrotfish (Scarini) species, even though parrotfishes are part of a much younger evolutionary radiation (Choat et al. 2012). This suggests that patterns of lineage diversification among these nominally herbivorous coral reef fish groups have been extremely heterogeneous.

Despite these interesting aspects, rabbitfish evolution has received less attention in the literature when compared to other typical coral reef fish families. Early work on molecular phylogenetics has sought to clarify the higher taxonomic groupings within the Siganidae (Lacson and Nelson 1993, Borsa et al. 2007, Kuriiwa et al. 2007, Lemer et al. 2007). But only more recently have rabbitfishes been analysed in terms of the timing of divergence between lineages, historical biogeography, and trait reconstructions (Siqueira et al. 2019a, b). In this chapter, we expand on these recent works to explore the evolutionary and biogeographic history of rabbitfishes from paleontological and phylogenetic perspectives. We chart their evolutionary origins by combining fossil and phylogenetic evidence; and explore patterns of lineage diversification and biogeography. We then describe the evolution of novel traits that allowed rabbitfishes to perform distinct functional roles on Indo-Pacific corals reefs, and finish with some recommendations for future evolutionary studies with this family of fishes.

Origins in Time

As with many coral reef fish families, our understanding of the evolutionary origins of Siganidae comes from the extensive fossil record of the Eocene (~ 50 Million years ago [Ma]) of Monte Bolca in northern Italy (Bellwood 1996). However, the richness of the paleontological record in rabbitfishes is not limited to Bolca, and siganid fossils are also known from other localities (Bannikov et al. 2010). In contrast to the relatively moderate number of extant rabbitfish species, the diversity of siganid fossils is comparatively high when other typical coral reef fish families are considered (Friedman and Carnevale 2018). In total, there are ten described fossil species placed within the family Siganidae (based on osteological characters shared with extant species; see Tyler and Bannikov 1997), although no synapomorphies have been identified to justify their placement within the only extant genus *Siganus* (Bannikov et al. 2010; Fig. 1). In fact, the fossil representation of rabbitfishes comprises eight unique fossil genera with varied morphologies (Tyler and Bannikov 1997, Bannikov and Tyler 2002, Bannikov et al. 2010)—all considered extinct—spanning approximately 30 million years of rabbitfish evolution (Fig. 1). The paleontological record of Siganidae extends from the late Paleocene (~ 57 Ma; *Siganopygaeus rarus*) to the Lower Oligocene (~ 30 Ma; *Caucasiganus* sp. and *Protosiganus glaronensis*) (Tyler and Bannikov 1997, Bannikov et al. 2010), while

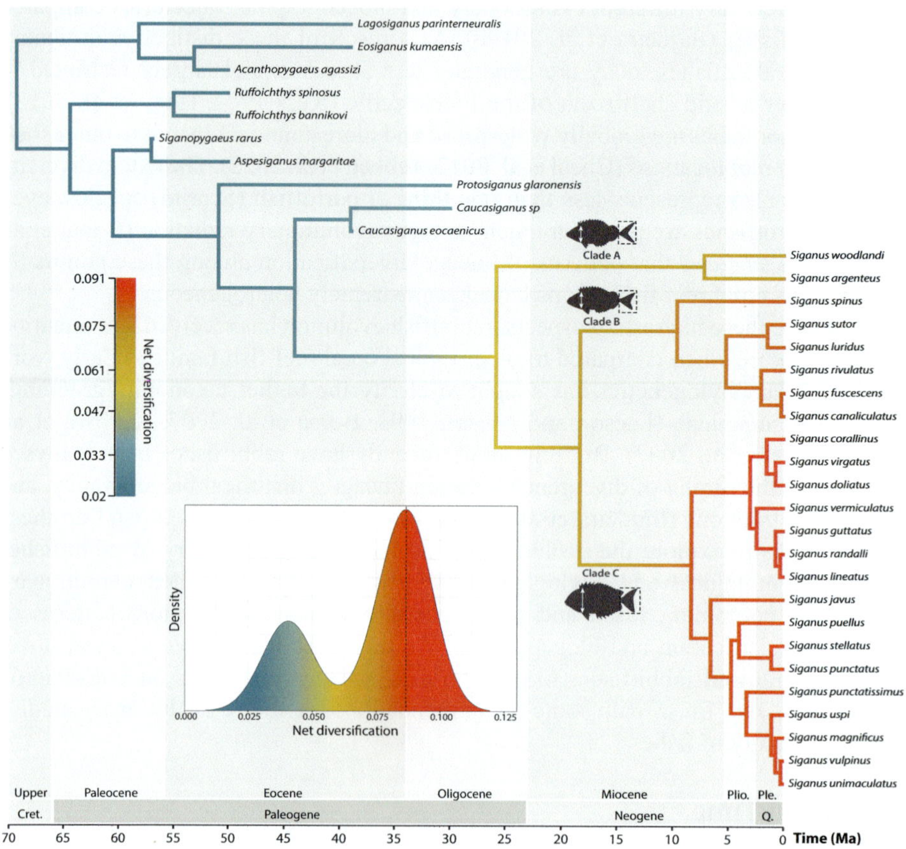

Fig. 1. Time-calibrated rabbitfish phylogeny including fossil and extant species. Maximum clade credibility phylogeny of Siganidae built with the Fossilized Birth-Death model (see Siqueira et al. 2019b), including most described living species and all recognized fossils based on the cladogram produced by the unordered analysis of Bannikov et al. (2010). The branches are coloured according to per-lineage net diversification rates estimated through BAMM (Rabosky 2014), varying from low (blue) to high (red) rates. The inset shows the density distribution of diversification rate values for all tips in the phylogeny, also coloured according to the same scale as the branches. The fish silhouettes highlight some morphological features that generally separate the three main clades in Siganidae. Plio. – Pliocene; Ple. – Pleistocene; Cret. – Cretaceous; Q. – Quaternary.

the origin of *Siganus* has been estimated to have occurred around the Late Oligocene/ Early Miocene (25.7 Ma; 16.2–37.6 [95% highest posterior density, HPD]; Fig. 1) using modern phylogenetic methods (i.e., the Fossilized Birth-Death (FBD) model; Heath et al. 2014).

The same analysis of divergence times in Siganidae, combining fossil and molecular evidence, revealed a crown age estimate at around 69 Ma for the family (Siqueira et al. 2019b). This origin date, however, is at odds with the paleontological record given the lack of fossils belonging to most extant reef fish families prior to the Cretaceous/Paleogene (K/Pg) boundary at 66 Ma (reviewed in Friedman et al. 2023, Schwarzhans et al. 2024). As discussed in Siqueira et al. (2019b), the accumulation of fossils from Monte Bolca at 50 Ma might have exerted a strong influence in these dating results. Under the FBD model, the diversification of rabbitfish fossils is required to conform to a similar pattern to the extant species. This would inevitably push the estimated crown age of Siganidae back to a time when the paleontological record suggests a very different composition of fish lineages. Recent fossil discoveries point to the origin of some extant reef fish groups straight after the K/Pg boundary (Cantalice et al. 2022). The presence of a damselfish (Pomacentridae), a seabass (Serranidae) and a trumpetfish (Aulostomoidea) at 63 Ma suggests a rapid rise in modern reef fish lineages post K/Pg at 66 Ma (Cantalice and Alvarado-Ortega 2016, Cantalice et al. 2018, 2020, 2022), which could also potentially be the case in rabbitfishes. This scenario is supported by the large proportion of the estimated crown age HPD that falls within the Cenozoic (see supplementary material in Siqueira et al. 2019b). Therefore, the crown age estimates shown here (Fig. 1) should be viewed with considerable caution. It is more likely that the origin of Siganidae was associated with the wider expansion of previously unexplored trophic niches post K/Pg, leading to the rise and escalation of herbivory in marine fishes (Bellwood 2003).

Lineage Diversification

Based on an analysis of per-lineage rates of evolution (Rabosky 2014), rabbitfishes displayed very modest diversification rates (i.e., speciation minus extinction) at the start of their evolutionary history (Fig. 1). Although there was a substantial generic expansion at that time, likely associated with morphological disparity (Marramà et al. 2016), the rate of lineage diversification was estimated to have been slow close to the origin of the family in the Paleocene/Eocene (66-34 Ma; Fig. 1). Diversification rates only started accelerating at the crown of the extant genus *Siganus*, during the Oligocene/Miocene (~ 25–20 Ma; Fig. 1). This date largely coincides with the historical moment of development of coral-dominated reefs (Kiessling 2009, Bellwood et al. 2017), and the shift in the global marine biodiversity hotspot towards its modern location in the Indo-Australian Archipelago (IAA) (Renema et al. 2008). Hence, the initial acceleration in rabbitfish diversification rates is likely associated with the development of novel reef habitat structures in the Oligocene/Miocene and their major biogeographic rearrangement at that time, as previously proposed for other coral reef fish families (Cowman and Bellwood 2011, Bellwood et al. 2017).

Towards the end of the Miocene, the pace of rabbitfish diversification increased even further, reaching its peak in the last 5 million years on the Pliocene/Pleistocene (Fig. 1). As a result, the distribution of net diversification rates across the tips of the phylogeny is bimodal, with extant species having high to intermediate net diversification, and fossil lineages having generally low rates of evolution (Fig. 1; inset). Interestingly, the fastest diversifying lineages are represented by species that today display a significant variation in diets, ranging from croppers of red and green macroalgae (e.g., *S. corallinus* and *S. doliatus*) to mixed feeders of diverse algal material, cyanobacteria, detritus (e.g., *S. lineatus*, *S. punctatissimus*, *S. punctatus*, *S. vulpinus*), and even sponges (*S. puellus*) (Hoey et al. 2013, Clements – Chapter 4). This tends to suggest that the recent habitat changes associated with the rise of *Acropora*-dominated reefs (Renema et al. 2016, Siqueira et al. 2022) might have promoted opportunities for lineage diversification in rabbitfishes through the exploitation of diverse trophic pathways (Fox et al. 2009, Nanami 2018, Semmler et al. 2021). Although none of the extant rabbitfish species are known to feed directly on corals, it is likely that the capacity for feeding in concealed microhabitats by reef-associated species (Brandl and Bellwood 2013, 2014, Fox and Bellwood 2013, Brandl and Tebbett – Chapter 6) may have conferred evolutionary advantages in the tridimensional complexity created by branching acroporid corals. The timing of these trophic innovations in rabbitfishes also seems to align with the diversification of parrotfishes and their specialisation on microscopic autotrophs (Clements et al. 2016, Nicholson and Clements 2021), highlighting that the recent reorganisation of coral reef benthic biota has indeed promoted novel ecological strategies. Therefore, the rapid diversification of some extant rabbitfish lineages might be, at least in part, associated with their ecological success in the reef environments that developed during the Pliocene/Pleistocene.

Temporal patterns of lineage expansion and contraction in rabbitfishes become even more evident when analysing their lineage through time plot (Fig. 2), derived from the same phylogenetic trees (Fig. 1). Lineages started accumulating at a regular pace close to the origin of the family in the Paleocene/Eocene, but subsequently stalled from the mid-Eocene onwards (Fig. 2). The Eocene/Oligocene transition was then a crucial moment in rabbitfish evolution, with the likely extinction of many ancient lineages (Fig. 2). This period is associated with a major cooling of the earth (Hutchinson et al. 2021), which is known to have impacted both reef-building corals (Siqueira et al. 2022), and reef-associated fish lineages (Cowman and Bellwood 2011). Rabbitfishes seem to be no exception, with the number of lineages dropping substantially after the transition (Fig. 2), which suggests a major impact from this cryptic extinction event (Cowman and Bellwood 2011) in the family's diversity. It was only more recently, from the late Miocene onwards (last 10 Myr), that lineage numbers started to recover, returning to pre-Oligocene values (Fig. 2), and subsequently surpassing those values with the rapid diversification of extant lineages in the Pliocene/Pleistocene.

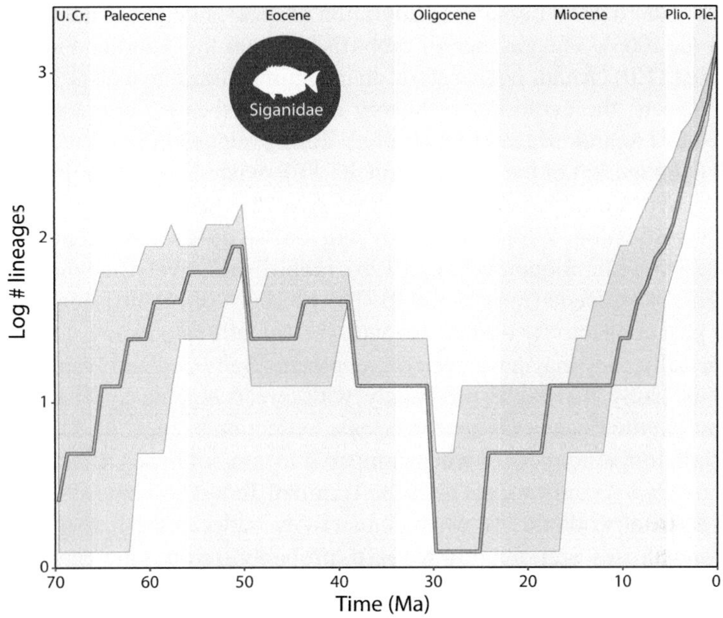

Fig. 2. Lineage through time plot based on the rabbitfish phylogeny. Number of rabbitfish lineages (log scale) through time based on the molecular phylogenies built with the Fossilized Birth-Death model. The solid line represents the median value across 100 phylogenies derived from the posterior distribution of trees from Siqueira et al. (2019b), and the shading represents the 95% credibility interval. Plio. – Pliocene; Ple. – Pleistocene; U. Cr – Upper Cretaceous.

Origins in Space and Biogeography

With the oldest rabbitfish fossil being described from Turkmenistan (*Siganopygaeus rarus*; Bannikov et al. 2010), the western Tethys region gets highlighted as the likely biogeographic place of origin of the Siganidae (Siqueira et al. 2019b). Even though new fossil discoveries can potentially lead to major reinterpretations of the biogeographic origins of fish lineages (e.g., Cantalice et al. 2022), the richness of rabbitfish fossils from the western Tethys is remarkable and suggests that it was an important location for the initial diversification of the family. In addition to the fossil from Turkmenistan, there are four rabbitfish fossil species described from Italy (Monte Bolca), three from Russia (Caucasus), one from Iran and one from Switzerland (Tyler and Bannikov 1997, Bannikov and Tyler 2002, Bannikov et al. 2010). All these different locations were once part of a larger shallow-water marine system in the western Tethys and Paratethys regions during the Eocene/Oligocene (Steininger and Rögl 1979).

Interestingly, despite this richness of rabbitfish fossils from the western Tethys, the geographic range of extant siganids is limited to the Indo-Pacific realm (Woodland 1983, 1990). Apart from Hawaii and Easter Island, rabbitfishes occur naturally from the Indian Ocean and Red Sea, all the way to Pitcairn Island in the Central Pacific (i.e., not considering the Lessepsian migration of species into the Mediterranean Sea; Hassan et al. 2003). The absence of rabbitfishes from the Atlantic and the Tropical East Pacific (TEP) Oceans is, therefore, an intriguing puzzle in marine biogeography. In many aspects, the evolutionary history of rabbitfishes is very similar to that of surgeonfishes (Acanthuridae; Tebbett et al. 2022), suggesting that the potential for historical colonization of the Atlantic and the TEP existed. However, one of the main differences between the evolution of these two families is perhaps the timing of lineage diversification. While the bulk of surgeonfish diversification happened in the Miocene (23-5.3 Ma; Siqueira et al. 2019b), rabbitfishes diversified more recently in the Pliocene/Pleistocene (Figs. 1 and 2). The presence of rabbitfishes in Tethys and the likely past connectivity of that area with the Atlantic (Hou and Li 2018) suggests that, historically, they may have been present there (Bellwood and Wainwright 2002). Hence, it is likely that these potentially widespread siganid species from Tethys went extinct in the Eocene/Oligocene cryptic extinction event (Fig. 2) and rabbitfish lineages took longer to recover when compared to surgeonfishes. Lineage expansion in the genus *Siganus* only started after the Terminal Tethyan Event (18-12 Ma; Fig. 2; Cowman and Bellwood 2013), when connectivity between the Indo-Pacific and the Atlantic was already significantly reduced, probably limiting the chances of cross-oceanic colonization by extant rabbitfishes.

Within the Indo-Pacific, rabbitfish species vary substantially in their geographic ranges. While some species have limited ranges, being restricted to island chains (e.g., *Siganus niger*; endemic to Tonga – Fig. 3), others are quite widespread (e.g., *S. argenteus*; occurs from Africa to Pitcairn and is the only species known to possess a pelagic, pre-juvenile stage) (Woodland 1990). Yet, the family follows the classic bullseye pattern of marine diversity (Cowman et al. 2017) with species richness peaking in the IAA (Woodland 1983, 1990). Among the most interesting patterns of geographic distribution within the IAA is the one from the subgenus *Lo* (Woodland 1990), that contains the species *S. vulpinus*, *S. unimaculatus*, *S. magnificus*, *S. uspi*, and *S. niger*. These species are morphologically unique for having pronounced tubulate snouts (see *Trait evolution* section) and display very shallow genetic (Lacson and Nelson 1993, Kuriiwa et al. 2007) and temporal patterns of divergence, with most lineages arising in the Pleistocene (Fig. 1). Despite being very similar morphologically, these species present markedly distinct colour patterns across their peripatric distributions (Fig. 3). Four out of five species within the subgenus have restricted geographic ranges, while one (*S. vulpinus*) has a wider distribution encompassing the IAA (Woodland 1983; Fig. 3). This pattern suggests that areas of endemism and isolation within the IAA (Cowman et al. 2017) may have largely underpinned speciation within the subgenus. The relatively recent origin of these species is highly congruent with the timing of marked sea-level changes in the Pleistocene (2.58-0.01 Ma), which has created barriers for dispersal and connectivity during low sea-level stands (Ludt and Rocha 2015). Therefore, in combination with the ecological hypothesis of trophic diversification explained above, peripatric processes also appear to have influenced speciation in *Siganus* through isolation during recent low sea-levels.

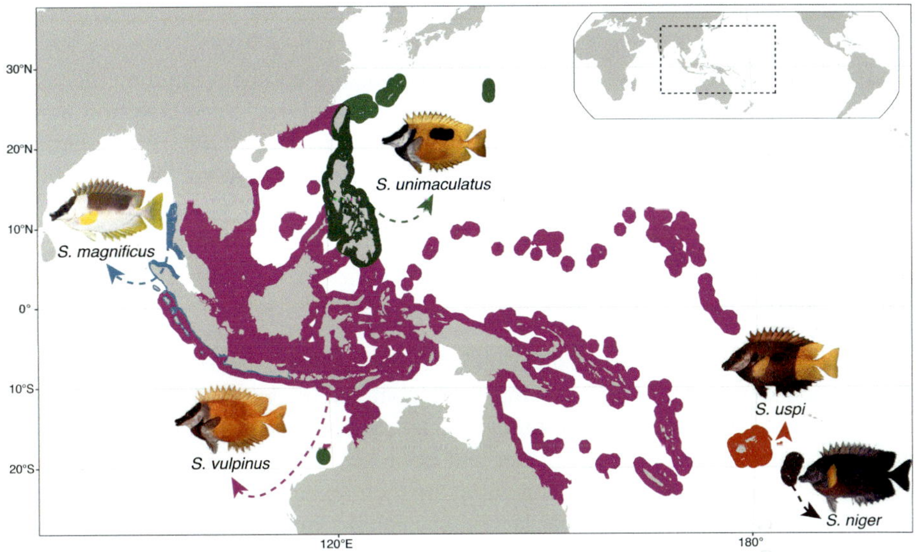

Fig. 3. Geographic distribution of *Siganus* species within the subgenus *Lo*. Geographic distributions for the species *S. vulpinus* (purple), *S. unimaculatus* (green), *S. magnificus* (blue), *S. uspi* (red) and *S. niger* (dark purple) derived from the IUCN database. The area of overlap between *S. vulpinus* and *S. unimaculatus* is shown in dark green. Fish photos by J. E. Randall.

Another notable example of geographic distributions associated with body colouration and vicariance is the one from the species *Siganus doliatus* and *S. virgatus*. These sister species have generally very similar colour patterns but differ slightly in their geographic range. Although their ranges overlap in the centre of the IAA, *S. doliatus* predominantly occurs on the eastern side of the archipelago and some Western Pacific islands, while *S. virgatus* is distributed throughout the west of the IAA and parts of the Eastern Indian Ocean (Woodland 1990). This same east-west IAA pattern of distribution is found in other species complexes (e.g., *S. guttatus* and *S. lineatus*; Kuriiwa et al. 2007), which highlights potentially congruent allopatric events driving rabbitfish evolution. With the emergence of the Sunda Shelf during low sea-level stands (Bellwood and Wainwright 2002), slight body coloration differences might have emerged that were sufficient to maintain species boundaries after their secondary contact within the IAA. Interestingly, despite also occurring in the IAA and having very similar colour patterns to *S. doliatus* and *S. virgatus*, *S. puellus* is part of a different clade in the phylogenetic tree (Fig. 1). This might represent a case of convergent evolution in colour patterns, or it could suggest the retention of an ancient colouration within the lineages. In combination with the patterns of divergence in the subgenus *Lo*, these examples show that slight modifications of body colouration are prevalent across the temporally shallow divergences found across the rabbitfish phylogeny.

The temporally shallow divergences among rabbitfish species are also reflected in the high prevalence of interspecific hybridization reported for the family

(Kuriiwa et al. 2007, Lemer et al. 2007). Closely related species, such as *S. doliatus* and *S. virgatus*, for example, were found to form hybrids where their geographic ranges overlap in the IAA. However, hybridization was also found between more distantly related species (e.g., *S. corallinus* and *S. puellus*) and was very common across the rabbitfish phylogeny (Kuriiwa et al. 2007), occurring in 10 out of 31 described species (Ng et al. 2022). This suggests that the level of genetic incompatibility between rabbitfish species is generally low (Kuriiwa et al. 2007) and reinforces the relatively recent divergence times found using molecular data (Fig. 1).

Trait Evolution

Rabbitfishes from the western Tethys during the Paleocene/Eocene had morphological features associated with herbivory that were completely absent in Mesozoic fishes (Bellwood 2003). However, despite having synapomorphies that justify their placement within the Siganidae (Tyler and Bannikov 1997), rabbitfish fossils were morphologically distinct when compared to their extant counterparts (Bellwood et al. 2014). The extinct siganids lacked specialised morphologies seen in extant species—such as the characteristic long snouts found in the subgenus *Lo*—and had generally smaller distances between their eyes (Bellwood 2003, Bellwood et al. 2014). Some fossil species, such as *Protosiganus glaronensis*, *Eosiganus kumaensis*, and *Acanthopygaeus agassizi*, also had conical teeth that differ from the extant *Siganus* (Tyler and Bannikov 1997, Bannikov et al. 2010). As a consequence, extant species seem to occupy new areas of morphospace in relation to fossil rabbitfishes (Bellwood et al. 2014), potentially indicating major habitat and/or dietary shifts throughout their evolution.

While there is some indication that fossil rabbitfishes were croppers of low algae or browsers of macroalgae like some extant species (Bellwood et al. 2014), they lacked the recent specialisations associated with crevice-feeding (Brandl and Bellwood 2013, 2014, Fox and Bellwood 2013, Semmler et al. 2021). This might have allowed modern species to feed on more specific items, such as cyanobacteria and sponges (Hoey et al. 2013), in a parallel episode of piscine herbivore diversification with an increasing focus on smaller autotrophs and reef detritus (Clements et al. 2016, Nicholson and Clements 2021). Despite propositions that early siganids might have been non-reef-associated (Kuriiwa et al. 2007), morphological evidence suggests that some fossil species were already potentially pair-forming (Bellwood et al. 2014), which is a common behaviour among extant reef-dwelling rabbitfishes. Therefore, by reconstructing some of these general ecological traits (e.g., schooling behaviour, feeding habitat, diet, feeding mode, and maximum size) through time (see Siqueira et al. 2019a), we show that some of the innovations that differentiate fossil and extant rabbitfishes only occurred within the last 10 million years (Fig. 4). It is important to note that these ecological traits differ from the morphological measurements used to describe the patterns above (Bellwood et al. 2014), where fossils and extant species do not overlap. Our ecological trait reconstructions in the siganid phylogeny indicate that the few lineages that survived the Eocene/Oligocene extinction event were likely general reef algae croppers, overlapping with the ecological trait space occupied by fossil species (Fig. 4). This general ecology, characteristic of the Paleocene/

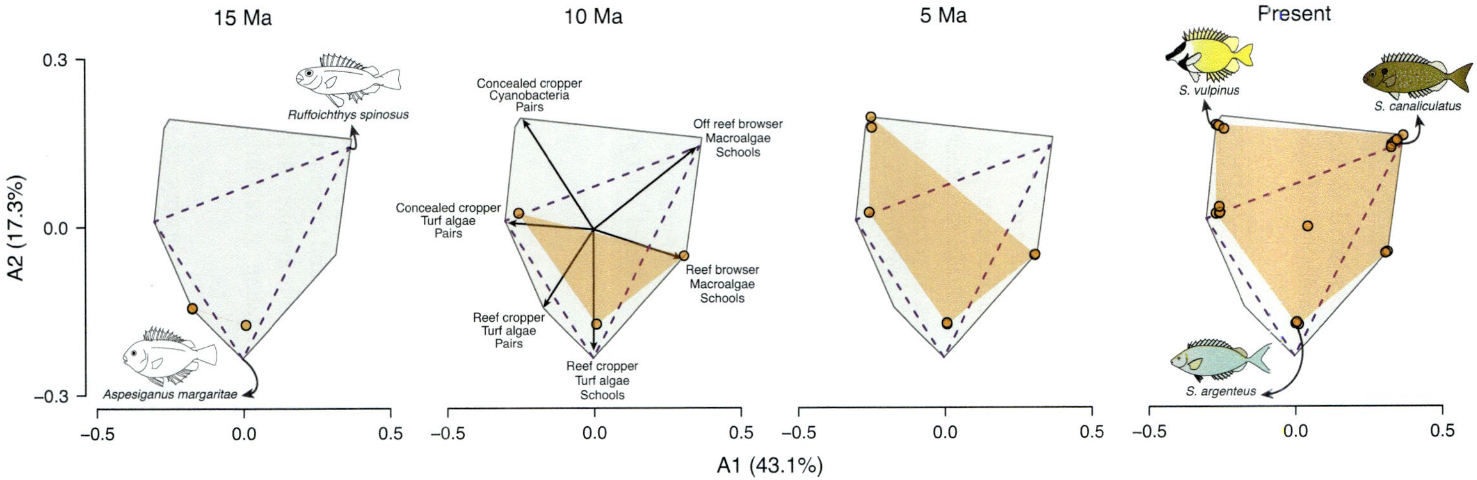

Fig. 4. Ecological trait space occupied by rabbitfishes through time. Two axes (A1-A2) derived from a principal coordinate analysis (PCoA) performed on ecological traits related to feeding (see main text). Each column represents a time-slice (15 Ma to present) with respective ancestral trait reconstructions (data and analysis sourced from Siqueira et al. 2019a). Background grey area shows the total space occupied combining fossils, time-slices, and extant species. Orange convex hulls represent the trait space at each time slice reconstructed through the phylogeny, while the fossil trait space (Eocene species only) is represented in the purple dashed lines. Points represent lineages present in each time-slice. The arrows on the 10 Ma plot represent examples of trait combinations (feeding habitat and mode, diet, and schooling behaviour) on the extremes of the polygon. Percentages on the labels represent the amount of variance explained by each axis in the multivariate space. Two fossil species are illustrated on the leftmost plot (illustrations by C. Goatley), and three extant species are illustrated on the rightmost plot (illustrations by S. B. Tebbett).

Eocene fossils, was prevalent until 10 Ma (Fig. 4). However, between 10 and 5 Ma, specialized crevice feeding lineages had already evolved, which then expanded in response to the major habitat modifications of the last 5 million years (Renema et al. 2016, Siqueira et al. 2022) to form the present-day rabbitfish assemblage (Fig. 4).

Morphological and ecological traits have also been used as key diagnostic features to distinguish the three major clades of extant rabbitfishes (Woodland 1990, Borsa et al. 2007, Kuriiwa et al. 2007, Zolkaply et al. 2021). Species in the earliest-diverging clade (*S. woodlandi* and *S. argenteus*; clade A – Fig. 1) have deeply forked caudal fins, slender fin spines, a scaleless thorax (Woodland 1990, Kuriiwa et al. 2007). In addition, these species lay pelagic eggs, as opposed to demersal eggs laid by the representatives of the other clades (Kuriiwa et al. 2007). Species in the clade B (*S. spinus*, *S. sutor*, *S. luridus*, *S. rivulatus*, *S. fuscescens*, *S. canaliculatus*; Fig. 1) also share the slender fin spines and scaleless thorax with clade A, but they are characterised by truncated caudal fins and more fusiform bodies (Woodland 1990, Borsa et al. 2007). Some species in clade B (e.g., *S. canaliculatus*, *S. fuscescens*, and *S. sutor*), tend to occur in more inshore environments such as estuaries and seagrass beds (e.g., Henderson et al. 2019, Ebrahim et al. 2020). On the other hand, most species in clade C (*S. corallinus*, *S. virgatus*, *S. doliatus*, *S. vermiculatus*, *S. puellus*, *S. stellatus*, *S. punctatus*, *S. punctatissimus*, *S. uspi*, *S. magnificus*, *S. vulpinus*, and *S. unimaculatus*; Fig. 1) are characteristic of coral reefs (although not all species are exclusively associated with coral reefs; e.g., *S. lineatus, S. javus*, *S. guttatus*, *S. randalli, S. vermiculatus*), with many displaying pairing behaviour as adults (Brandl and Bellwood 2013, 2014). Species in clade C have stout fin spines and a scaled thorax (Kuriiwa et al. 2007), besides having deeper bodies and longer snouts when compared with the other clades (Woodland 1990, Borsa et al. 2007, Zolkaply et al. 2021). In conjunction, these diagnostic features highlight that, for rabbitfishes, there is an overall agreement between the relationships based on morphological, ecological, and phylogenetic data.

Future Directions

Despite extant rabbitfishes being largely uniform in morphological and meristic characters traditionally used in fish systematics (Borsa et al. 2007), their taxonomy is relatively well resolved (Woodland 1990). However, recent species descriptions (Randall and Kulbicki 2005, Woodland and Anderson 2014) and geographic range expansions (Moleana et al. 2016, Zolkaply et al. 2021), highlight that we are still adding important pieces of information to the evolutionary and biogeographic puzzle of the Siganidae. Therefore, we highlight a few research avenues that we believe would represent significant advances in the understanding of rabbitfish evolution moving forward:

1. Most of our knowledge about the early evolutionary history of rabbitfishes stems from important fossil discoveries (Tyler and Bannikov 1997, Bannikov et al. 2010) that can now be added to the phylogenetic framework of the family (Siqueira et al. 2019b). Hence, the work of paleontologists in describing new fossil species is fundamental for calibrating our estimates of the time and place of origin of siganid lineages.

2. Although we have a very good coverage of extant rabbitfish species in the molecular phylogeny (24 out of 31 described species are sampled), evolutionary studies would certainly benefit from having a fully sampled phylogeny. Thus, obtaining genetic data for the few species missing from the tree should be a priority, given that it would provide a more complete understanding of the phylogenetic relationships within the Siganidae.
3. Not only should we sequence the missing species using traditional molecular markers, but we should also aim to use more recent phylogenomic approaches (e.g., ultraconserved elements; DiBattista et al. 2018, Tea et al. 2022) by sequencing genomes across all rabbitfish species. This would help to better delineate the relationships and divergence times in Siganidae and would allow a detailed investigation of the genetic basis of recent species divergences within the family.
4. With a comprehensive phylogenetic framework, it will then be possible to apply modern phylogenetic comparative methods to disentangle the role of ecology and biogeography in driving patterns of speciation and extinction in rabbitfishes. These methods will also help to resolve the history of trait evolution through well-informed ancestral state reconstructions.
5. Finally, studies at the population level are warranted. To the best of our knowledge, there have been very few examinations of the phylogeography of rabbitfishes to date (e.g., Ravago-Gotanco and Juinio-Meñez 2010). More studies using this approach would shed some light into the processes driving the intriguing patterns of allopatric and peripatric distribution displayed by some siganid species and add genetic insight into the more recent Lessepsian expansions of species ranges.

Conclusions

By analysing the paleontological record in combination with phylogenetic data, we show that rabbitfishes have been a component of reef environments since the Paleocene/Ecocene (66-34 Ma). While it is still hard to precisely pinpoint their time and place of origin, available fossil evidence suggests that the family Siganidae likely arose sometime right after the K/Pg boundary with many genera originating in the shallow waters of the western Tethys region. Most of this early diversity, however, was probably wiped out during the Eocene/Oligocene cryptic extinction event. After this event, rabbitfish lineages took some time to recover, and diversification rates only started picking up pace in the late Miocene. The last 10 million years (late Miocene/Pliocene/Pleistocene) were, therefore, the most important phase for the evolution of modern species in Siganidae. This period saw substantial variations in sea-levels, which likely shaped the allopatric and peripatric patterns of extant species distributions. Furthermore, it marked the rise and diversification of specialised long-snouted species that today play unique ecosystem roles in Indo-Pacific coral reefs.

Acknowledgements

We would like to thank Prof. David R. Bellwood for the invaluable lessons provided about rabbitfish ecology and evolution to both authors of this chapter. We are also grateful to two anonymous reviewers whose comments improved this chapter and S. B. Tebbett for helpful comments on earlier drafts and for providing the fish illustrations used in Fig. 4. This research was supported by an ECU Vice Chancellor's Research Fellowship to ACS, and an ARC DECRA Fellowship to PFC (DE170100516).

References

Alfaro, M.E., F. Santini and C.D. Brock. 2007. Do reefs drive diversification in marine teleosts? Evidence from the pufferfish and their allies (Order Tetraodontiformes). Evolution 61: 2104–2126.

Bannikov, A.F. and J.C. Tyler. 2002. A new genus and species of rabbitfish (Acanthuroidei: Siganidae) from the Eocene of Monte Bolca, Italy. Studi e Ricerche sui Giacimenti Terziari di Bolca, Museo Civico di Storia Naturale di Verona, Miscellania Paleontologica 9: 37–45.

Bannikov, A.F., J.C. Tyler and C. Sorbini. 2010. Two new taxa of Eocene rabbitfishes (Perciformes, Siganidae) from the North Caucasus (Russia), with redescription of *Acanthopygaeus agassizi* (Eastman) from Monte Bolca (Italy) and a phylogenetic analysis of the family. Bollettino del Museo Civico di Storia Naturale di Verona, Geologia, Paleontologia, Preistoria 34: 3–21.

Bellwood, D.R. 1996. The Eocene fishes of Monte Bolca: the earliest coral reef fish assemblage. Coral Reefs 15: 11–19.

Bellwood, D.R. 2003. Origins and escalation of herbivory in fishes: a functional perspective. Paleobiology 29: 71–83.

Bellwood, D.R., C.H.R. Goatley and O. Bellwood. 2017. The evolution of fishes and corals on reefs: form, function and interdependence. Biological Reviews 92: 878–901.

Bellwood, D.R., C.H.R. Goatley, S.J. Brandl and O. Bellwood. 2014. Fifty million years of herbivory on coral reefs: fossils, fish and functional innovations. Proceedings of the Royal Society B: Biological Sciences 281: 20133046.

Bellwood, D.R., O. Schultz, A.C. Siqueira and P.F. Cowman. 2019. A review of the fossil record of the Labridae. Annalen des Naturhistorischen Museums in Wien, Serie A 121: 125–193.

Bellwood, D.R. and P.C. Wainwright. 2002. The history and biogeography of fishes on coral reefs. pp. 5–32. *In*: Sale, P.F. [ed.]. Coral Reef Fishes: Dynamics and Diversity on a Complex Ecosystem. Academic Press.

Bonaldo, R.M., A.S. Hoey and D.R. Bellwood. 2014. The ecosystem roles of parrotfishes on tropical reefs. Oceanogr. Mar. Biol. Annu. Rev. 52: 81–132.

Borsa, P., S. Lemer and D. Aurelle. 2007. Patterns of lineage diversification in rabbitfishes. Molecular Phylogenetics and Evolution 44: 427–435.

Brandl, S.J. and D.R. Bellwood. 2013. Morphology, sociality, and ecology: can morphology predict pairing behavior in coral reef fishes? Coral Reefs 32: 835–846.

Brandl, S.J. and D.R. Bellwood. 2014. Individual-based analyses reveal limited functional overlap in a coral reef fish community. Journal of Animal Ecology 83: 661–670.

Cantalice, K.M. and J. Alvarado-Ortega. 2016. *Eekaulostomus cuevasae* gen. and sp. nov., an ancient armored trumpetfish (Aulostomoidea) from Danian (Paleocene) marine deposits of Belisario Domínguez, Chiapas, southeastern Mexico. Palaeontologia Electronica 2016: 19.3.53A.

Cantalice, K.M., J. Alvarado-Ortega and A. Alaniz-Galvan. 2018. *Paleoserranus lakamhae* gen. et sp. nov., a Paleocene seabass (Perciformes: Serranidae) from Palenque, Chiapas, southeastern Mexico. Journal of South American Earth Sciences 83: 137–146.

Cantalice, K.M., J. Alvarado-Ortega and D.R. Bellwood. 2020. *Chaychanus gonzalezorum* gen. et sp. nov.: a damselfish fossil (Percomorphaceae: Pomacentridae), from the early Paleocene outcrop of Chiapas, southeastern Mexico. Journal of South American Earth Sciences 98(3): 102322–1726.

Cantalice, K.M., J. Alvarado-Ortega, D.R. Bellwood and A.C. Siqueira. 2022. Rising from the ashes: the biogeographic origins of modern coral reef fishes. BioScience 72: 769–777.

Cheal, A., M. Emslie, I. Miller and S. Sweatman. 2012. The distribution of herbivorous fishes on the Great Barrier Reef. Marine Biology 159: 1143–1154.

Clements, K.D., D.P. German, J. Piché, A. Tribollet and J.H. Choat. 2016. Integrating ecological roles and trophic diversification on coral reefs: multiple lines of evidence identify parrotfishes as microphages. Biological Journal of the Linnean Society 120(4): 729–751.

Cowman, P.F. and D.R. Bellwood. 2011. Coral reefs as drivers of cladogenesis: expanding coral reefs, cryptic extinction events, and the development of biodiversity hotspots. Journal of Evolutionary Biology 24: 2543–2562.

Cowman, P.F. and D.R. Bellwood. 2013. Vicariance across major marine biogeographic barriers: temporal concordance and the relative intensity of hard versus soft barriers. Proceedings of the Royal Society B: Biological Sciences 280: 20131541.

Cowman, P.F., D.R. Bellwood and L. van Herwerden. 2009. Dating the evolutionary origins of wrasse lineages (Labridae) and the rise of trophic novelty on coral reefs. Molecular Phylogenetics and Evolution 52: 621–631.

Cowman, P.F., V. Parravicini, M. Kulbicki and S.R. Floeter. 2017. The biogeography of tropical reef fishes: endemism and provinciality through time. Biological Reviews 92: 2112–2130.

DiBattista, J.D., M.E. Alfaro, L. Sorenson, J.H. Choat, J.P.A Hobbs, T.H. Sinclair-Taylor et al. 2018. Ice ages and butterflyfishes: phylogenomics elucidates the ecological and evolutionary history of reef fishes in an endemism hotspot. Ecology and Evolution 8: 10989–11008.

Ebrahim, A., J.P. Bijoux, P.J. Mumby and I.R. Tibbetts. 2020. The commercially important shoemaker spinefoot, *Siganus sutor*, connects coral reefs to neighbouring seagrass meadows. Journal of Fish Biology 96: 1034–1044.

Floeter, S.R., M.G. Bender, A.C. Siqueira and P.F. Cowman. 2018. Phylogenetic perspectives on reef fish functional traits. Biological Reviews 93: 131–151.

Fox, R.J. and D.R. Bellwood. 2013. Niche partitioning of feeding microhabitats produces a unique function for herbivorous rabbitfishes (Perciformes, Siganidae) on coral reefs. Coral Reefs 32: 13–23.

Fox, R.J., T.L. Sunderland, A.S. Hoey and D.R. Bellwood. 2009. Estimating ecosystem function: contrasting roles of closely related herbivorous rabbitfishes (Siganidae) on coral reefs. Marine Ecology Progress Series 385: 261–269.

Friedman, M. and G. Carnevale. 2018. The Bolca Lagerstätten: Shallow marine life in the Eocene. Journal of the Geological Society 175: 569–579.

Friedman M., J.V. Andrews, H. Saad and S. El-Sayed. 2023. The Cretaceous-Paleogene transition in spiny-rayed fishes: surveying Patterson's gap in the acanthomorph skeletal record. Geologica Belgica 26: 1–23.

Hassan, M., M. Harmelin-Vivien and F. Bonhomme. 2003. Lessepsian invasion without bottleneck: example of two rabbitfish species (*Siganus rivulatus* and *Siganus luridus*). Journal of Experimental Marine Biology and Ecology 291: 219–232.

Heath, T.A., J.P. Huelsenbeck and T. Stadler. 2014. The fossilized birth–death process for coherent calibration of divergence-time estimates. Proceedings of the National Academy of Sciences of the United States of America 111: E2957–E2966.

Henderson, C.J., T. Stevens, S.Y. Lee, B.L. Gilby, T.A. Schlacher, R.M. Connolly et al. 2019. Optimising seagrass conservation for ecological functions. Ecosystems 22: 1368–1380.

Hoey, A.S., S.J. Brandl and D.R. Bellwood. 2013. Diet and cross-shelf distribution of rabbitfishes (f. Siganidae) on the northern Great Barrier Reef: implications for ecosystem function. Coral Reefs 32: 973–984.

Hou, Z. and S. Li. 2018. Tethyan changes shaped aquatic diversification. Biological Reviews 93: 874–896.

Hutchinson, D.K., H.K. Coxall, D.J. Lunt, M. Steinthorsdottir, A.M. de Boer, M. Baatsen et al. 2021. The Eocene-Oligocene transition: A review of marine and terrestrial proxy data, models and model-data comparisons. Climate of the Past 17: 269–315.

Kiessling, W. 2009. Geologic and biologic controls on the evolution of reefs. Annual Review of Ecology Evolution and Systematics 40: 173–192.

Kuriiwa, K., N. Hanzawa, T. Yoshino, S. Kimura and M. Nishida. 2007. Phylogenetic relationships and natural hybridization in rabbitfishes (Teleostei: Siganidae) inferred from mitochondrial and nuclear DNA analyses. Molecular Phylogenetics and Evolution 45: 69–80.

Lacson, J.M. and S.G. Nelson. 1993. Genetic distances among fishes of the from the Western Pacific Ocean. Marine Biology 116: 187–192.

Lemer S., D. Aurelle, L. Vigliola, J.D. Durand and P. Borsa. 2007. Cytochrome b barcoding, molecular systematics and geographic differentiation in rabbitfishes (Siganidae). Comptes Rendus - Biologies 330: 86–94.

Ludt, W.B. and L.A. Rocha. 2015. Shifting seas: the impacts of Pleistocene sea-level fluctuations on the evolution of tropical marine taxa. Journal of Biogeography 42: 25–38.

Marramà, G., C. Garbelli and G. Carnevale. 2016. A clade-level morphospace for the Eocene fishes of Bolca: patterns and relationships with modern tropical shallow marine assemblages. Bollettino della Società Paleontologica Italiana 55: 139–156.

Moleana, T., L.D. Patrona, T. Meziane and Y. Letourneur. 2016. First record of *Siganus randalli* (Teleostei, Siganidae) in New Caledonia, and comments on its diet. Marine Biodiversity Records 9: 81.

Nanami, A. 2018. Spatial distributions, feeding ecologies, and behavioral interactions of four rabbitfish species (*Siganus unimaculatus*, *S. virgatus*, *S. corallinus*, and *S. puellus*). PeerJ 6: e6145.

Nicholson, G.M. and K.D. Clements. 2021. Ecomorphological divergence and trophic resource partitioning in 15 syntopic Indo-Pacific parrotfishes (Labridae: Scarini). Biological Journal of the Linnean Society 132: 590–611.

Ng, I., D.R. Bellwood and A.C. Siqueira. 2022. Do currents shape global patterns of hybrid richness in coral reef fishes? Global Ecology and Biogeography 31: 2524–2540.

Rabosky, D.L. 2014. Automatic detection of key innovations, rate shifts, and diversity-dependence on phylogenetic trees. PLoS ONE 9: e89543.

Randall, J.E. and M. Kulbicki. 2005. *Siganus woodlandi*, new species of rabbitfish (Siganidae) from New Caledonia. Cybium 29: 185–189.

Ravago-Gotanco, R.G. and M.A. Juinio-Meñez. 2010. Phylogeography of the mottled spinefoot *Siganus fuscescens*: Pleistocene divergence and limited genetic connectivity across the Philippine archipelago. Molecular Ecology 19: 4520–4534.

Renema, W., D.R. Bellwood, J.C. Braga, K. Bromfield, R. Hall, K.G. Johnson et al. 2008. Hopping hotspots: global shifts in marine biodiversity. Science 321: 654–657.

Renema, W., J.M. Pandolfi, W. Kiessling, F.R. Bosellini, J.S. Klaus, C. Korpanty et al. 2016. Are coral reefs victims of their own past success? Science Advances 2: e150085.

Robertson, D.R. 1998. Do coral-reef fish faunas have a distinctive taxonomic structure? Coral Reefs 17: 179–186.

Russ, G.R. 1984. Distribution and abundance of herbivorous grazing fishes in the central Great Barrier Reef. I. Levels of variability across the entire continental shelf. Marine Ecology Progress Series 20: 23–34.

Schwarzhans, W.W., G. Carnevale and G.L. Stringer. 2024. The diversity of teleost fishes during the terminal Cretaceous and the consequences of the K/Pg boundary extinction event. Netherlands Journal of Geosciences 103: e5.

Semmler, R.F., S.J. Brandl, S. Keith and D.R. Bellwood. 2021. Fine-scale foraging behavior reveals differences in the functional roles of herbivorous reef fishes. Ecology and Evolution 11(9): 4898–4908.

Siqueira, A.C., D.R. Bellwood and P.F. Cowman. 2019a. The evolution of traits and functions in herbivorous coral reef fishes through space and time. Proceedings of the Royal Society B: Biological Sciences 286: 20182672.

Siqueira, A.C., D.R. Bellwood and P.F. Cowman. 2019b. Historical biogeography of herbivorous coral reef fishes: The formation of an Atlantic fauna. Journal of Biogeography 46: 1611–1624.

Siqueira, A.C., W. Kiessling and D.R. Bellwood. 2022. Fast-growing species shape the evolution of reef corals. Nature Communications 13: 2426.

Siqueira, A.C., R.A. Morais, D.R. Bellwood and P.F. Cowman. 2020. Trophic innovations fuel reef fish diversification. Nature Communications 11: 2669.

Siqueira, A.C., P. Muruga and D.R. Bellwood. 2023. On the evolution of fish-coral interactions. Ecology Letters 26: 1348–1358.

Steininger, F.F. and F. Rögl. 1979. The paratethys history. A contribution towards the Neogene geodynamics of the alpine orogene. Annales Geologiques des Pays Helleniques 3: 1153–1165.

Tea, Y.K., X. Xu, J.D. DiBattista, N. Lo, P.F. Cowman and S.Y.W. Ho. 2018. Phylogenomic analysis of concatenated ultraconserved elements reveals the recent evolutionary radiation of the fairy wrasses (Teleostei: Labridae: *Cirrhilabrus*). Systematic Biology 71: 1–12.

Tebbett, S.B., A.C. Siqueira and D.R. Bellwood. 2022. The functional roles of surgeonfishes on coral reefs: past, present and future. Reviews in Fish Biology and Fisheries 32: 387–439.

Tyler, J.C. and A.F. Bannikov. 1997. Relationships of the fossil and recent genera of rabbitfishes (Acanthuroidei: Siganidae). Smithsonian Contributions to Paleobiology 84: 1–35.

Woodland, D.J. 1983. Zoogeography of the Siganidae (Pisces). an interpretation of distribution and richness. Bulletin of Marine Sciences 33: 713–717.

Woodland, D.J. 1990. Revision of the fish family Siganidae with description of two new species and comments on distribution and biology. Bernice Pauahi Bishop Museum.

Woodland, D.J. and R.C. Anderson. 2014. Description of a new species of rabbitfish (Perciformes: Siganidae) from southern India, Sri Lanka and the Maldives. Zootaxa 3811: 129–136.

Zolkaply, S.Z., T.D. Do, M. Asaduzzaman, Y.G. Seah, D. Hurwood, P. Mather et al. 2021. Evolutionary history and taxonomic reappraisal of coral reef rabbitfishes (Siganidae): Patterns of lineage diversification and speciation. Biology 10: 1109.

4

Nutritional Ecology of Rabbitfish (Siganidae)

Kendall D. Clements

Introduction

The family Siganidae (rabbitfishes) is less diverse and morphologically disparate than the other two ubiquitous taxa of Indo-Pacific herbivorous coral reef fish, the scarine labrids (parrotfishes) and the family Acanthuridae (surgeonfishes). Perhaps as a result, rabbitfishes have received less attention than parrotfishes and surgeonfishes. Nonetheless, rabbitfishes are interesting in terms of nutritional ecology for several reasons: (i) they are one of few taxa that include species capable of digesting and assimilating phaeophyte brown algae; (ii) they appear capable of living in a wide range of reef and non-reef habitats, including seagrass and mangrove areas (Hoey et al. – Chapter 7); (iii) they utilize a wide range of photoautotrophic taxa including cyanobacteria, eukaryotic macroalgae and seagrasses, and some species target sessile invertebrates; and (iv) they may be the only fish taxon containing species that employ fermentative hindgut symbioses to digest vascular plants, i.e., angiosperms. All these factors suggest that rabbitfishes display plasticity in nutritional ecology in both ecological and evolutionary terms.

Here I use the term nutritional ecology to refer to the biochemical composition, detection, acquisition, digestion, and metabolic utilization of food resources (following Clements et al. 2009). Information on the biochemical composition of relevant photoautotrophic food resources is available elsewhere (Clements and Choat 2018 and references therein), and so in this chapter I focus mainly on factors influencing the selection of foods and their subsequent processing by rabbitfishes. Following a brief section on trophic morphology (little information is available on this topic for rabbitfishes), I discuss feeding behaviour, diet, inferences from

Schools of Biological Sciences, University of Auckland, Private Bag 92019, Auckland 1142, New Zealand.
Email: k.clements@auckland.ac.nz

nutritional biomarkers, digestion and assimilation, and finally the composition and role of hindgut microbiota in rabbitfishes. Commentary on the literature is illustrative rather than comprehensive, and designed to focus the reader's attention on what I see are some gaps in knowledge and some conceptual problems with work to date. In doing the research for this chapter I was struck by the lack of information available on many rabbitfish species, and perhaps as a result this has led to several assumptions about nutritional ecology in the group that may not stand the test of time. It is quite ironic that some of the most detailed work available on feeding and diet in rabbitfishes involves two species of Red Sea rabbitfishes outside their native range in the eastern Mediterranean, where they have become established after moving through the Suez Canal (e.g., Lundberg and Lipkin 1979, Stergiou 1988, Azzurro et al. 2007, Sala et al. 2011).

Trophic Morphology

The comparative trophic morphology of rabbitfish has not been extensively studied, but a few comments can be made to provide some context in relation to diet selection. Rabbitfishes exhibit broadly similar trophic morphologies, although there is some variation in the morphology of the head, teeth, gill arches, and alimentary tract. Rabbitfish exhibit some variation in head shape, with several species having an elongated or tubular snout, e.g., *S. unimaculatus*, *S. vulpinus*, *S. uspi*, *S. magnificus*, and *S. corallinus* (Borsa et al. 2007, Zolkaply et al. 2021, Huertas and Hemmingson – Chapter 2). Elongated and narrow snouts clearly evolved independently in separate clades of deep-bodied rabbitfish and are thought to allow feeding in concealed microhabitats that are inaccessible to some other herbivorous fish taxa (Borsa et al. 2007, Fox and Bellwood 2013, Brandl and Bellwood 2016, Nanami 2018, Zolkaply et al. 2021). This difference in microhabitats likely also reflects differences in targeted food resources, with concealed microhabitats typically exhibiting later succession autophototrophic assemblages (Carpenter 1986).

Most rabbitfish appear to have bicuspid and/or tricuspid teeth, with 10–14 teeth on each side of the premaxilla and dentary (Tyler and Bannikov 1997). These combined with a non-protrusible upper jaw (the premaxilla and maxilla are attached to one another; Starks 1907, Woodland 1990) facilitate a 'shearing' bite. *Siganus canaliculatus* and *S. doliatus* have similar bicuspid teeth, with a larger primary and smaller lateral cusp in both upper and lower jaws (Streit et al. 2015). The alignment of these paired cusps is reversed between the upper and lower jaw, i.e., that the large cusps from one jaw align with small cusps in the other, and the tooth arcade forms a serrated but continuous cutting edge. In contrast, *S. rivulatus* have tricuspid teeth on the premaxilla and bicuspid teeth on the dentary (Sayed et al. 2019). In *S. canaliculatus*, *S. doliatus* and *S. spinus*, the bicuspid teeth from the upper jaw pass in front of the teeth in the lower jaw during the bite, creating a 'shearing' bite (Bryan 1975, Streit et al. 2015). The articular in rabbitfish is very small (Starks 1907) and not fused with the dentary, forming an intra-mandibular joint capable of limited movement (Konow et al. 2008).

Rabbitfishes have four gill arches with the fifth modified into upper and lower pharyngeal tooth plates (Bryan 1975, Sayed et al. 2019), similar to the pattern in other

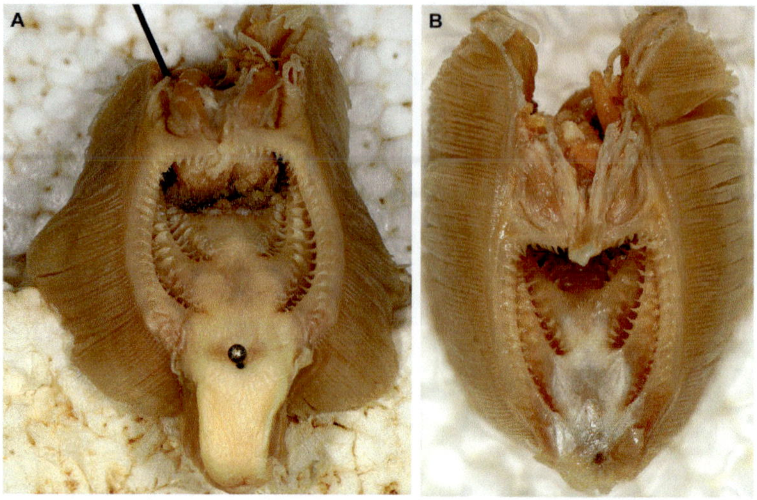

Fig. 1. Focus-stacking photographs of oblique anterodorsal view of branchial basket showing arrangement of gill rakers in (A) *S. guttatus*, and (B) *S. virgatus*.

non-pharyngognath percomorph taxa. The gill rakers in *S. guttatus* and *S. virgatus* are fleshy, short, variable in form and relatively widely spaced, becoming broader on the more posterior gill arches (Fig. 1). A similar pattern is seen in *S. spinus* (Bryan 1975) and *S. luridus* (Alsafy et al. 2023), and gill raker number is relatively invariant among species in the four trophic groups identified by Hoey et al. (2013) (Table 1). In both *S. spinus* and *S. rivulatus,* the lower pharyngeal plates on the fifth gill arch bear papilliform teeth that are curved or oriented posteriorly, unlike the papilliform teeth on the upper pharyngeal plates which are straight and pointed (Bryan 1975, Sayed et al. 2019).

The macroscopic alimentary morphology in *S. fuscescens*, *S. guttatus*, *S. punctatus* and *S. spinus* is very similar (Fig. 2), although relative gut length is slightly shorter in *S. guttatus* than the other three species. These four species represent three of the four main clades of rabbitfishes (Siqueira et al. 2019, Siqueira and Cowman – Chapter 3), so this comparison of gut structure is broadly representative of much of rabbitfish alimentary disparity. The stomach in rabbitfish is thin-walled and elongate (Pillans et al. 2004), with ~ 5 pyloric caecae at the pyloric junction. The intestine is moderately long (relative gut length ~ 3.2–4.2) and arranged in complex coils (Bryan 1975, Fox et al. 2009). For comparison, relative gut length in turf-feeding *Acanthurus* and *Zebrasoma* species is ~ 3.7–5.2, and ~ 3.2 in browsing *Naso* species (Jones 1968). Unlike the gut in most species of kyphosid chubs, the intestine of rabbitfishes is not compartmentalised (Clements 1997). Despite the similarity in alimentary morphology among species, some intra-specific plasticity in relative gut length has been reported. Gut anatomy in captive *S. canaliculatus* juveniles has been shown to change in response to diet, with fish fed *Enteromorpha* and *Gracilaria* having significantly greater relative intestine length and more mucosal folds after eight weeks than fish fed formulated diets or animal material (Xie et al. 2018).

Table 1. Anterior gill raker number in *Siganus* species of the four trophic groups identified in Hoey et al. (2013). Data from Woodland (1990).

Species	Trophic group	Anterior gill raker number
S. canaliculatus	browsers	4–6 + 16–21
S. javus	browsers	4–5 + 17–19
S. argenteus	croppers	4–6 + 17–21
S. corallinus	croppers	4–5 + 18–21
S. doliatus	croppers	4–5 + 17–21
S. spinus	croppers	4–5 + 17–20
S. lineatus	mixed feeders	4–6 + 17–21
S. punctatissimus	mixed feeders	4–5 + 17–20
S. punctatus	mixed feeders	4–5 + 18–22
S. vulpinus	mixed feeders	4–5 + 17–19
S. puellus	spongivores	4–5 + 15–20

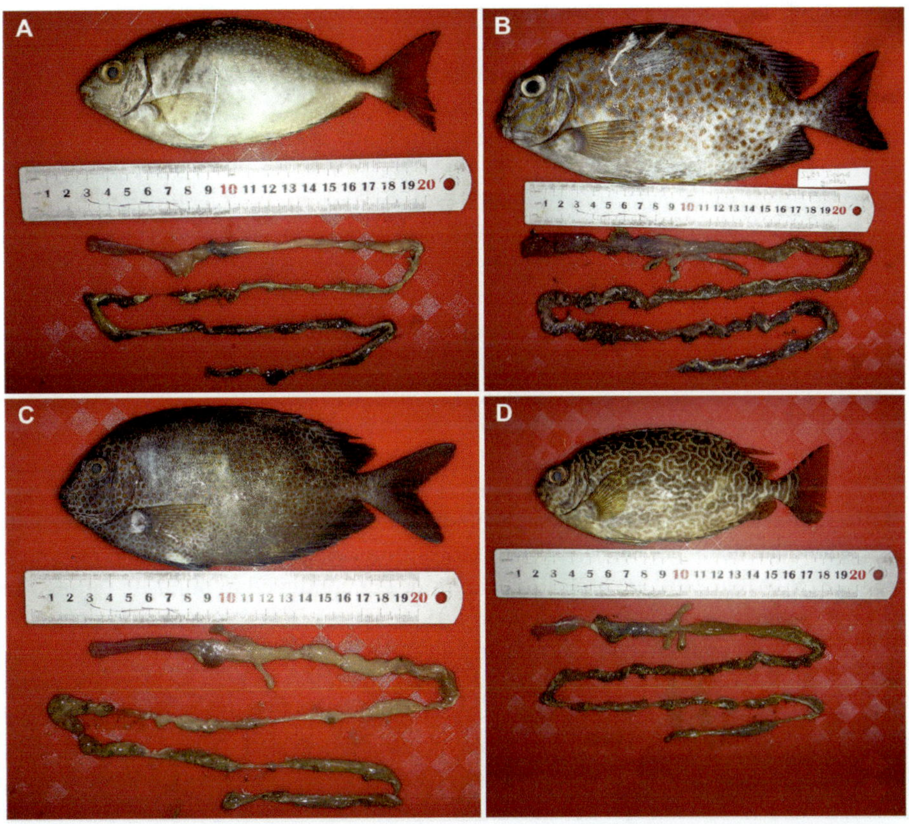

Fig. 2. Alimentary tract morphology in (A) *S. fuscescens*, (B) *S. guttatus*, (C) *S. punctatus* and (D) *S. spinus*. These species represent three of the four main clades of rabbitfishes (Siqueira et al. 2019).

Field Observations of Daily Feeding Periodicity and Feeding Rate

Here I discuss the daily pattern and rate of feeding before work on diet choice, as feeding rate in fish varies depending on several factors including the type of food involved. Nanami (2018) recorded bite rate in four species of rabbitfishes under natural conditions in Okinawa between 0900 and 1600 h. Average feeding rates for *S. unimaculatus*, *S. virgatus*, *S. corallinus*, and *S. puellus* were 8.9, 7.1, 9.5, and 3.0 bites per min, respectively. Average feeding rates for *S. vulpinus*, *S. corallinus* and *S. puellus* at Lizard Island, Great Barrier Reef were slightly lower at ~ 5.0, 6.0 and 2.4 bites per min, respectively (Fox and Bellwood 2013).

Siganus doliatus on reefs at Lizard Island, Great Barrier Reef exhibited the typical diel pattern of feeding between dawn and dusk, with feeding increasing through the morning and peaking in early afternoon before dropping off towards evening (Fox et al. 2009). Interestingly, *S. lineatus* observed over the same time during the day did not feed at all. Stomach contents of these reef-dwelling *S. doliatus* and *S. lineatus* were dominated by thallose rhodophytes and organic detritus, respectively (Fox et al. 2009; see below). The same study also measured the distribution of digesta throughout the alimentary tract of both species at three different time periods over the day: 06:00 to 07:30 h (morning), 11:30 to 13:00 h (midday) and 16:30 to 18:00 h (evening) (Fox et al. 2009). In the morning, digesta was mainly concentrated in the hindgut of *S. doliatus*, whereas at this time, material was also present in the stomach and anterior intestine of *S. lineatus*. The stomach and anterior intestine of *S. doliatus* filled up towards midday, whereas by midday in *S. lineatus* the relative proportions of material had decreased in the stomach and anterior intestine and increased in the hindgut (Fox et al. 2009). The stomach of *S. lineatus* contained the greatest proportion of total ingesta during the morning period. In a subsequent study, individuals of *S. lineatus* in a lagoonal boulder shoreline habitat at Lizard Island were active during the day and rested in holes during the night, whereas individuals at two nearby reef sites remained inactive near coral bommies during the day and foraged out over sand at night, especially just after dark (Fox and Bellwood 2011). No feeding was observed during daylight hours in any *S. lineatus* individuals at the two reef sites, whereas individuals at the shoreline site were observed feeding at relatively uniform rates throughout the day (Fox and Bellwood 2011). The highest volume of digesta in the stomach and anterior intestine of individuals collected from the reef sites was found in individuals collected early in the morning, whereas in individuals collected from the shoreline site, material in the anterior gut continued to build-up through to the early afternoon (Fox and Bellwood 2011). The greatest amount of material in the gut in individuals from both habitats was consistently in the posterior intestine throughout the day. The authors speculated that the *S. lineatus* in the reef sites may have been targeting epifaunal crustaceans in their nocturnal foraging out over the sand, that this increased protein intake may have compensated for the apparent lower rate of foraging in the reef sites (Fox and Bellwood 2011). This may be the case, but the study did not resolve whether the diet of *S. lineatus* differed between the two different habitats, i.e. whether fish were more herbivorous/detritivorous in the shoreline habitat and more omnivorous in the reef habitat.

The average proportions of bites taken on a selection of benthic food resource categories by four *Siganus* species in the Seychelles were reported by Ebrahim et al. (2020). *Siganus argenteus* and *S. sutor* fed predominantly on turf algae, but also targeted seagrass epiphytes, seagrass blades and fleshy macroalgae. *Siganus corallinus* and *S. stellatus* appeared to have a more specialised diet, feeding largely over turf substrata with only a minor proportion of bites on other resources (but see section below on diet). The daily feeding patterns of all four of these *Siganus* species showed the typical diel pattern for grazing fishes, with bite rates increasing from dawn to a peak in the early afternoon followed by a decline in feeding rate to dusk. Overall feeding rates were highest in *S. argenteus* and *S. sutor*, the two species that targeted the broadest variety of trophic resources (Ebrahim et al. 2020). *Siganus vulpinus* exhibited the largest foraging range of a group of comparable herbivorous taxa that included *S. doliatus* and a range of acanthurid and parrotfish species (Chow et al. 2021). This and the large inter-foray distances could indicate a higher degree of patch selectivity or more sparsely distributed food resources, although it is unclear which elements of the diet this involved (see below).

Feeding Assays

The feeding behaviour of rabbitfish in response to macroalgal tether feeding assays has been reported in numerous studies, where it is considered a replicable technique to measure the rate of algal removal (e.g., Vergés et al. 2012, Plass-Johnson et al. 2015, Tebbett et al. 2023). However, macroalgae in this form presents fish with a novel resource that is unlikely to replicate natural conditions in terms of either the spatial distribution or the variety of food resources. One example of this novelty problem is the consumption of *Sargassum* tether assays by the browsing *Naso unicornis* on the Great Barrier Reef, Australia. The fish consume the assays on mid-shelf reefs, where the fish encounter this alga, while on nearby outer-shelf reefs, *N. unicornis* inspected but then ignored similar assays (Hoey and Bellwood 2010). Conversely, some species of coral reef parrotfish eagerly consumed assays of mangrove sponge taxa that they did not normally encounter, while ignoring assays of sponge taxa from their normal reef habitat (Wulff 2021). Responses to feeding assays by herbivorous fish is therefore highly context dependent. The extent to which feeding on macroalgal tether assays represents rates of feeding or qualitative food selection under natural conditions thus remains an open question, but this technique can sometimes provide useful comparative data on diet preferences based on the range of foods provided.

The most widely used macroalga in tether assays is *Sargassum*. Feeding on *Sargassum* assays has been reported in several rabbitfish species including *S. canaliculatus* (Mantyka and Bellwood 2007a, Fox and Bellwood 2008, Bennett and Bellwood 2011, Vergés et al. 2012), *S. fuscescens* (Zarco-Perello et al. 2017, Barrientos et al. 2021), *S. doliatus* (Lefèvre and Bellwood 2011), *S. virgatus* (Müller et al. 2021, Seah et al. 2021), *S. puelloides* (Chong-Seng et al. 2014), *S. sutor* (Chong-Seng et al. 2014), and *S. javus* (Cvitanovic and Bellwood 2009; Vergés et al. 2012), however the extent to which *Sargassum* is a natural component of their diet is unclear. For example, captive *S. canaliculatus* did not prefer *Sargassum* when provided with other algal options (You et al. 2014). Feeding on tethered *Sargassum* has also been reported in *S. doliatus*, but consumption decreased markedly from August to December when thallae were largely epiphyte free, suggesting that the fish were mainly targeting epiphytes (Lefèvre and Bellwood 2011). When feeding on tethered *Sargassum*, *S. canaliculatus* and *S. doliatus* consumed mainly the thallae of *Sargassum* and avoided the stipes, unlike *Kyphosus vaigiensis* (Kyphosidae) and *Naso unicornis* (Acanthuridae) which consumed the whole seaweed (Streit et al. 2015). These differences were suggested to be driven by body size, tooth shape and feeding behaviour (Streit et al. 2015), although it is possible they may also reflect post-ingestive processes involving the breakdown of storage versus cell wall carbohydrates that are differentially distributed in the plants. *Siganus fuscescens* preferentially targeted reproductive receptacles of *Sargassum* (Yamano et al. 2019), similar to the seasonal preference for fucoid receptables in the odacine labrid *Odax pullus* (Clements and Choat 1993), indicating that reproductive status of macroalgae is another factor that can influence feeding. Consumption of tethered *Ecklonia* by *S. fuscescens* was also reported in experiments conducted off Perth in Western Australia (Zarco-Perello et al. 2017) and in the Solitary Islands off the coast of New South Wales, Australia (Verges et al. 2016).

Müller et al. (2021) deployed both single- and multiple-choice feeding assays in the field using four different phaeophyte taxa: *Sargassum*, *Padina*, *Turbinaria* and *Lobophora*. *Siganus virgatus* preferred *Sargassum* and *Padina* with a lower preference for *Turbinaria*, while *Lobophora* was avoided. *Siganus javus* took most of its bites on *Sargassum*, and *S. guttatus* and *S. corallinus* only took bites from *Sargassum*. The study did not remove epiphytes, and indeed the authors concluded that it was likely that *S. virgatus* were targeting epibionts in the feeding assays rather than the macroalgae itself. This suggestion was supported by the lack of a relationship between bites and biomass loss in the *Turbinaria* experiment, indicating that the fish were removing epibionts. Seah et al. (2021) also found that *S. virgatus* was a major consumer of *Sargassum* in Singapore, with greatest feeding activity occurring October–February during the period of peak algal growth. *Sargassum* becomes senescent and heavily epiphytised from January through to May in this location (Seah et al. 2021), suggesting that peak consumption by *S. virgatus* did not involve epiphytes. Plass-Johnson et al. (2015) reported consumption of tethered *Sargassum* by *S. virgatus* and *S. margaritiferus* in Sulawesi, Indonesia. Both rabbitfishes also consumed tethered *Padina*, but at lower rates, especially in *S. virgatus*. *Siganus puelloides* and *S. sutor* were responsible for most bites on *Sargassum* in tether feeding assays conducted on coral-dominated reefs in the Seychelles Islands, although there was a discrepancy between the number of bites taken and the amount of *Sargassum* biomass removed (Chong-Seng et al. 2014). The authors speculated that the cause of the discrepancy was fish tearing off more of the assay than they ingested. Another possibility is that the browsing *S. sutor* removed *Sargassum* biomass while the grazing *S. puelloides* at least partially targeted epiphytes rather than the macroalga itself, but unfortunately epiphytes were not considered in the study.

Balakrishnan et al. (2022) conducted feeding assays on *S. canaliculatus* and *S. javus* housed in enclosures using the following algae: the phaeophytes *Lobophora*, *Padina*, *Sargassum* and *Turbinaria*, and the chlorophytes *Caulerpa* spp., *Chaetomorpha*, *Halimeda* and *Enteromorpha*. The authors conducted consumption experiments during both day and night, and biomass loss was greater at night. There were significant interactions between macroalgal species, fish species and feeding rates. During the day *S. canaliculatus* mainly consumed *Enteromorpha compressa*, *Chaetomorpha* and *Lobophora*, whereas *S. javus* mainly consumed *Padina*, *Caulerpa racemosa*, *Chaetomorpha*, *Lobophora* and *Enteromorpha compressa*. At night both rabbitfish species mainly consumed *Caulerpa racemosa*, with lesser amounts of *Padina*, *Enteromorpha compressa* and *Chaetomorpha*. Interestingly, both rabbitfishes consumed all the *Enteromorpha compressa* provided during both day and night, and during the day *S. javus* also consumed all the *Caulerpa racemosa* and *Chaetomorpha* provided. Some profound differences are apparent between the results of Balakrishnan et al. (2022) and those of Mantyka and Bellwood (2007b). For example, *S. canaliculatus* showed a far higher preference for *Turbinaria* and *Sargassum* in the latter study than in Balakrishnan et al. (2022), highlighting the subjective and contextualised nature of feeding assay studies. Neither study considered the influence of epiphytes, which may have been present to varying degrees on different algal taxa, and the use of captive fish in Balakrishnan et al. (2022) is also likely to have influenced intake rates. Captive *Siganus javus* consumed macroalgae at a significantly higher rate than *S. canaliculatus*, 0.9 ± 0.7 to 10.1 ± 1.8 g.h^{-1} and 0.5 ± 0.14 to 4.7 ± 0.9 g.h^{-1}, respectively (Balakrishnan et al. 2022). Again, it's hard to draw firm conclusions on what this difference may mean in terms of nutritional ecology because the two species were not eating diets of uniform composition. A similar problem applies to a comparison of rates of *Sargassum* removal from assays conducted in the Keppel Islands on the Great Barrier Reef and at Ningaloo Reef in Western Australia (Vergés et al. 2012). *Siganus doliatus* and *S. canaliculatus* were responsible for 10.5% and 8.8%, respectively, of all mass standardised bites (product of the number of bites and the body mass of the fish) on *Sargassum* assays in the Keppel Islands, while *S. argenteus* took the fifth highest number of mass-standardised bites at the Ningaloo site. However, the authors acknowledge that bites by some species (especially parrotfish) appeared to target epibionts rather than the macroalgal thallus itself (Vergés et al. 2012). Finally, feeding assays on the two invasive Lessepsian migrant rabbitfish taxa in the Eastern Mediterranean showed that *S. rivulatus* fed exclusively on algal turf, whereas rates of consumption by *S. luridus* were highest on the phaeophyte taxa *Cystoseira* and *Sargassum*, an order of magnitude lower on *Posidonia* seagrass, and intermediate on turf algae (Vergés et al. 2014). Some of this variation in responses to *Sargassum* assays may reflect variation in nutritional quality, especially when comparing the results of assays using different *Sargassum* taxa (e.g., Steinberg et al. 1991).

Overall, the intraspecific variation in rabbitfish feeding preferences seen in different tether assay studies and especially the failure to control for feeding on epibionts limits the capacity to either draw firm conclusions about patterns of interspecific variation in diet choice or to generalise about rates of ingestion on specific foods. Some differences in preference appear consistent with data from stomach contents, e.g., the trophic differences seen between *S. rivulatus* and *S. luridus* (see below). Future studies using tether assays should: (i) control for the role of macroalgal epibionts in food choice, (ii) consider that both epibionts and the nutritional quality of macroalgae vary with phenological factors such as season and reproductive status, depending on location, (iii) ensure that assays are deployed at times when fish taxa of interest are known to feed actively in the wild, and ideally (iv) compare results with stomach content data on wild, free-feeding fish.

Diet

Available diet information on species of *Siganus* is presented in Table 2. Only quantitative data on stomach contents were used for this purpose, apart from in two species (*S. guttatus* and *S. margaritiferus*) where such data were unavailable but tether assay and captive feeding trial data allowed a reasonably definitive categorisation of preferences for potential foods. Rabbitfish lack a mechanical trituration mechanism such as a pharyngeal mill or gizzard-like stomach, so unlike in parrotfishes for example (Clements et al. 2017), conventional stomach content analysis, especially when conducted at microscopic level, should provide a reliable picture of the foods ingested. A detailed survey of the primary literature shows that the diets of wild rabbitfish are complex, vary with geographic location and habitat, and often include some animal material.

The most comprehensive treatment of diet in various rabbitfish species is Hoey et al. (2013), which presents quantitative stomach content data on 11 species from the northern Great Barrier Reef, Australia. Hoey and colleagues split these species into four main dietary categories: (i) browsers of leathery phaeophyte macroalgae (e.g., *Sargassum* and *Turbinaria*), (ii) croppers of rhodophyte and chlorophyte turf, (iii) mixed feeders of diverse algal material, cyanobacteria, detritus and sediment, and (iv) spongivores (Table 1). These four groups do not correspond to phylogenetic structure (Siqueira et al. 2019), and a broader survey of the literature on rabbitfish diets suggests that some of these category assignments may reflect temporal and spatial dietary variation rather than trophic diversification. Nonetheless, these four groups provide a basis to structure a brief discussion of stomach content data on rabbitfishes. Readers interested in species-level detail should consult Table 2.

Taxa that feed on leathery phaeophytes, i.e. the "browser" group, include *S. canaliculatus*, *S. fuscescens*, *S. javus*, *S. luridus*, *S. margaritiferus*, *S. vermiculatus* and *S. virgatus* (Hoey et al. 2013; Table 2). *Siganus fuscescens* feed on the phaeophytes *Ecklonia* spp., *Eisenia arborea* and *Sargassum fusiforme* in Japan and leave characteristic regular bite marks that can be distinguished from bites of *Calotomus japonicus* and *Kyphosus* spp. (Kiriyama et al. 2001, 2005, Hasegawa 2010). *Siganus fuscescens* appear to browse phaeophytes throughout the year when temperatures are above 20°C (Yamaguchi et al. 2010), significantly preferring *Sa. fusiforme* (Kiriyama et al. 2005), unlike *C. japonicus* which feed on phaeophytes mainly in late summer and autumn when the algae are senescent and have lost thallae (Hasegawa 2010). In an experiment involving transplanted phaeophyte sporophytes in Hainan, Japan, *S. fuscescens* appeared to prefer *Eisenia arborea* over *Ecklonia cava* (Nimura et al. 2007). *Siganus fuscescens* in Japan also ingested small amounts of animal material, mainly sponges and ascidians (Yatsuya et al. 2015). Two gut content studies conducted in New Caledonia showed that *S. fuscescens* can also eat considerable amounts of foliose phaeophytes, e.g., *Dictyota* and *Lobophora* (Debenay et al. 2011) and rhodophytes (Moleana et al. 2016). Depending upon

Table 2. Major dietary taxa of *Siganus* spp. as determined by gut content analysis, showing study location.

Species & study location[1]	Main dietary categories	Main dietary taxa	Source
Siganus argenteus Great Barrier Reef	Corticated & filamentous rhodophytes	*Hypnea*, *Laurencia*, *Gracilaria*, *Chondria*, *Polysiphonia*, *Jania*	Hoey et al. (2013)
S. argenteus Red Sea	Filamentous phaeophytes, rhodophytes and chlorophytes, some corticated rhodophytes	*Sphacelaria*, *Enteromorpha*, *Hypnea*, *Polysiphonia*, *Gracilaria*	Lundberg and Lipkin (1979)
S. canaliculatus Great Barrier Reef	Leathery & foliose phaeophytes, some corticated and filamentous rhodophytes	*Sargassum*, *Turbinaria*, *Padina*, *Lobophora*, *Dictyota*, *Hydroclathrus*, *Gracilaria*	Hoey et al. (2013)
S. canaliculatus Philippines	Filamentous chlorophytes and phaeophytes, some thalloid rhodophytes	*Rhizoclonium*, *Ectocarpus*	Von Westernhagen (1973)
S. corallinus Great Barrier Reef	Corticated & filamentous rhodophytes, foliose phaeophytes, organic detritus	*Hypnea*, *Laurencia*, *Gracilaria*, *Chondria*, *Polysiphonia*, *Dictyota*,	Hoey et al. (2013)
S. corallinus Okinawa, Japan	Foliose phaeophytes, foliose rhodophytes, benthic invertebrates		Nanami (2018)
S. doliatus Great Barrier Reef	Corticated & filamentous rhodophytes, foliose phaeophytes	*Hypnea*, *Laurencia*, *Gracilaria*, *Chondria*, *Dictyota*	Hoey et al. (2013)
S. fuscescens New Caledonia	Calcareous and filamentous rhodophytes, some seagrass	*Amphiroa*, *Gelidiella*, *Thalassia*	Moleana et al. (2016)
S. fuscescens Honshu, Japan	Leathery phaeophytes, some rhodophytes and foliose chlorophytes	*Sargassum*, *Eisenia*, *Ecklonia*, Ulvophyceae	Yamano et al. (2019)
S. guttatus[2] Philippines	Filamentous chlorophytes and rhodophytes[3]	*Enteromorpha*, *Gracilaria*	Von Westernhagen (1974)
S. insomnis[4] Sri Lanka	Diatoms, cyanobacteria, seagrass, filamentous chlorophytes and rhodophytes	*Licmophora*, *Gomphonema*, *Navicula*, Oscillatoriaceae, *Cladophora*, *Chaetomorpha*, *Gracilaria*, *Laurencia*, *Thalassia*	Chitravadivelu and Sivapalan (1984)
S. javus Great Barrier Reef	Leathery phaeophytes, some corticated and filamentous rhodophytes, organic detritus	*Sargassum*, *Turbinaria*, *Gracilaria*	Hoey et al. (2013)
S. javus Philippines	Sponge, filamentous chlorophytes, filamentous and corticated rhodophytes, leathery phaeophytes, seagrass	*Enteromorpha*, *Gelidiella*, *Gracilaria*, *Mastophora*, *Kappaphycus*, *Sargassum*, *Cymodocea*	Perpetua et al. (2013)

Table 2 contd. ...

...Table 2 contd.

Species & study location[1]	Main dietary categories	Main dietary taxa	Source
S. lineatus Great Barrier Reef	Corticated & filamentous rhodophytes, some foliose phaeophytes, cyanobacteria, organic detritus, sediment	*Hypnea*, *Gracilaria*, *Jania*, *Dictyota*	Hoey et al. (2013)
S. lineatus New Caledonia	Foliose phaeophytes, filamentous rhodophytes, filamentous chlorophytes	*Dictyota*, *Bostrychia*, *Polysiphonia*, *Chlorodesmis*	Moleana et al. (2016)
S. luridus Mediterranean	Foliose & filamentous phaeophytes, corticated rhodophytes, some seagrass	*Dictyota*, *Dictyopteris*, *Cystoseira*, *Sphacelaria*, *Kuckuckia*, *Padina*, *Gelidium*	Stergiou (1988)
S. luridus Red Sea	Leathery and foliose phaeophytes, some seagrass	*Lobophora*, *Cystoseira*, *Sargassum*, *Ectocarpus*, *Hydroclathrus*, *Dictyota*	Lundberg and Lipkin (1979)
S. margaritiferus Indonesia	Leathery and foliose phaeophytes[5]		Plass-Johnson et al. (2015)
S. puelloides Maldive Islands	Colonial tunicates, sponges, “benthic algae”		Woodland and Randall (1979)
S. puellus Great Barrier Reef	Sponges, corticated rhodophytes, organic detritus	Porifera, Ascidiacea, *Hypnea*, cyanobacteria	Hoey et al. (2013)
S. puellus Okinawa, Japan	Sponges, foliose rhodophytes, benthic invertebrates		Nanami (2018)
S. punctatissimus Great Barrier Reef	Foliose phaeophytes, corticated & filamentous rhodophytes, cyanobacteria, organic detritus, sediment	*Dictyota*, *Lobophora*, *Jania*	Hoey et al. (2013)
S. punctatus Great Barrier Reef	Sessile invertebrates, articulated calcareous rhodophytes, some foliose phaeophytes, cyanobacteria, organic detritus, sediment	Gastropoda, Porifera, *Galaxaura*, *Jania*, *Dictyota*, *Gracilaria*	Hoey et al. (2013)
S. randalli New Caledonia	Foliose phaeophytes, filamentous and corticated rhodophytes	*Dictyota*, *Bostrychia*, *Lomentaria*	Moleana et al. (2016)
S. rivulatus Red Sea	Filamentous chlorophytes, phaeophytes & rhodophytes, some corticated rhodophytes, leathery and foliose phaeophytes	*Enteromorpha*, *Sphacelaria*, *Laurencia*, *Hypnea*	Lundberg and Lipkin (1979)
S. rivulatus Libya, Mediterranean	Foliose & corticated chlorophytes, foliose, corticated, filamentous and calcareous rhodophytes, foliose, corticated & filamentous phaeophytes, seagrasses[6]	*Codium*, *Cladophora*, *Ulva*, *Griffithsia*, *Corallina*, *Jania*, *Contarinia*, *Rhodophyllis*, *Dictyota*, *Sauvageaugloia*, *Halopteris*, *Stypocaulon*[6]	Shakman et al. (2009)

S. spinus Great Barrier Reef	Filamentous rhodophytes, some foliose phaeophytes and chlorophytes, organic detritus	*Hypnea, Laurencia, Gracilaria, Enteromorpha, Dictyota*	Hoey et al. (2013)
S. spinus Guam	Corticated chlorophytes, filamentous rhodophytes and phaeophytes	*Boodlea, Gelidiopsis, Centroceras, Gelidium, Sphacelaria, Feldmannia*	Bryan (1975)
S. spinus[7] Philippines	Filamentous chlorophytes, rhodophytes and phaeophytes, cyanobacteria	*Schizomeris*, Lyngbya, *Cladophora, Enteromorpha, Rhizoclonium, Ulva, Ceramium, Hypnea, Ectocarpus*	Von Westernhagen (1973)
S. stellatus Red Sea	Sponges, cyanobacteria, some filamentous rhodophytes		Tietbohl (2022)
S. sutor Zanzibar	Thallose and filamentous rhodophytes, filamentous chlorophytes	*Laurencia, Ceramium, Chaetomorpha*	De la Torre-Castro et al. (2008)
S. unimaculatus Okinawa, Japan	Foliose phaeophytes, foliose & stalked rhodophytes, some chlorophytes		Nanami (2018)
S. vermiculatus Fiji	Leathery phaeophytes, calcareous chlorophytes	*Sargassum, Halimeda*	Gunderman et al. (1983)
S. virgatus Okinawa, Japan	Foliose phaeophytes, some stalked rhodophytes		Nanami (2018)
S. vulpinus Great Barrier Reef	Cyanobacteria, corticated rhodophytes, some foliose phaeophytes	*Calothrix*, Oscillatoriales, *Laurencia, Hypnea, Jania, Dictyota*	Hoey et al. (2013) Pers. obs.

[1] Diet information is presented for 24 of the 32 valid species of *Siganus* listed on the California Academy of Sciences website (https://researcharchive.calacademy.org/research/ichthyology/catalog/fishcatmain.asp). The eight species for which I was unable to locate definitive diet data were *S. labyrinthodes*, *S. laqueus*, *S. magnificus*, *S. niger*, *S. studeri*, *S. trispilos*, *S. uspi* and *S. woodlandi*.

[2] Referred to as *S. concatenata* by von Westernhagen (1974); *S. concatenata* is a junior synonym of *S. guttatus* (Woodland 1990).

[3] Based on electivity indices from captive adult S. *guttatus* fed a wide variety of chlorophytes, rhodophytes and phaeophytes (von Westernhagen 1974).

[4] Referred to as *S. lineatus* (non-Valenciennes 1835; Woodland and Anderson 2014) by Chitravadivelu & Sivapalan (1984).

[5] Based on consumption of tethered algal assays in the Spermonde Archipelago, Indonesia involving a wide range of eukaryotic algae (Plass-Johnson et al. 2015).

[6] Shakman et al. (2009) observed strong seasonal differences in taxa consumed, so the taxa listed represent preferences influenced by seasonal variation in availability or nutritional content.

[7] Referred to as *S. striolata* by von Westernhagen (1973); *S. striolata* is a junior synonym of *S. spinus* (Woodland 1990).

habitat this rabbitfish also eats seagrasses including *Zostera*, *Halophila* and *Thalassia* (Arnold et al. 2014, Moleana et al. 2016), although it is possible that some seagrass consumption could be associated with epiphytes. The diet of another "browsing" species, *S. luridus*, can display considerable seasonal variation, with rhodophytes, chlorophytes and seagrass varying in importance through the year (Stergiou 1988, Shakman et al. 2009). Overall, therefore, this dietary category may overlap with croppers depending on the timing and location of sampling.

The "cropper" group, i.e., the group said to target rhodophyte and chlorophyte turf, include *S. argenteus*, *S. corallinus*, *S. doliatus*, *S. randalli*, *S. rivulatus*, *S. spinus*, *S. sutor* and *S. unimaculatus* (Table 2). Again, members of this group overlap with other groups based on which source is used to define diet. For example, *S. doliatus* has been reported to feed on *Sargassum* in tether assays (Streit et al. 2015; see above), although some reports of this involve epiphyte feeding (Lefèvre and Bellwood 2011). *Siganus rivulatus* primarily feeds on filamentous and foliose rhodophytes and chlorophytes, but also on filamentous and foliose phaeophytes (Shakman et al. 2009), perhaps reflecting the differences in protein and lipid between taxa such as *Dictyota* and *Sphacelaria* and leathery phaeophytes such as *Sargassum* and *Turbinaria* (McDermid et al. 2007, Olsson et al. 2020). Like *S. luridus*, invasive *S. rivulatus* in the Mediterranean also display significant seasonal variation in diet (Shakman et al. 2009). *Siganus rivulatus* in the Red Sea have also been reported to feed on gelatinous zooplankton (Bos et al. 2017). The dietary distinction between *S. luridus* and *S. rivulatus* that is apparent in the Red Sea and the northern Mediterranean was less apparent in the southern Mediterranean, where macroalgae were less abundant and both species fed on algal turf (Lundberg and Golani 1995, see also Sala et al. 2011). Smaller individuals of *S. sutor* caught by drag nets (mean TL = 12.7 cm) in seagrass beds off Zanzibar contained substantially more animal material (mainly amphipods, sponges and hydrozoans) than larger individuals caught by basket traps (mean TL = 26.25 cm) (de la Torre-Castro et al. 2008). The smaller fish contained a range of algal species, mainly *Ceramium* spp., *Eucheuma* spp. and *Champia* spp., while the stomach contents of the larger individuals were dominated (~ 66%) by *Laurencia* spp. with much lower proportions of other algal taxa (de la Torre-Castro et al. 2008).

The third dietary group, "mixed feeders", are characterised by ingesting cyanobacteria and sediment in addition to mixed algae (Hoey et al. 2013), and include *S. insomnis*, *S. lineatus*, *S. punctatus*, *S. punctatissimus* and *S. vulpinus* (Table 2). Some species, including *S. punctatus* and *S. lineatus*, also sometimes ingest some animal material, including sponges. The interesting variation in both foraging behaviour and diet of *S. lineatus* at Lizard Island (Fox et al. 2009, Fox and Bellwood 2011) was discussed above. Stomach contents of *S. vulpinus* collected on the northern Barrier Reef were dominated by filamentous cyanobacteria (Hoey et al. 2013), including Nostocales such as *Calothrix* and *Lyngbya*-like Oscillatoriales, some of which are epiphytic on macroalgae (Fig. 3).

Interestingly, *S. vulpinus* has been reported to have a greater foraging range than *S. doliatus* (Chow et al. 2021) and *S. corallinus* (Semmler et al. 2021), with the authors of the latter study suggesting this difference may be due to the spatial organisation of the main food resources on the reef (i.e., *S. vulpinus*: corticated

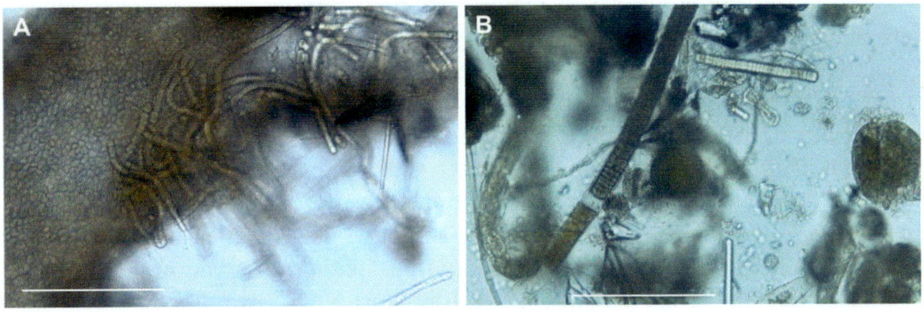

Fig. 3. Microscope photographs (20x) of filamentous cyanobacteria from the stomach contents of *S. vulpinus* collected at Lizard Island, Great Barrier Reef, Australia. (A) *Calothrix* epiphytes on macroalgae. (B) *Lyngbya*-like Oscillatoriales. Scale bars 200 mm.

rhodophytes and filamentous cyanobacteria; *S. corallinus*: foliose phaeophytes and a variety of rhodophytes). These authors go on to speculate that since cyanobacteria are considered unpalatable to many fish species due to the presence of toxins, *S. vulpinus* may need to diet-switch to food sources that contain less toxin, e.g., sparser patches of cyanobacteria that contain less toxin than cyanobacterial mats. However, it is important to note that cyanobacteria vary in the production of toxins, even within taxa (Zubia et al. 2019, Kang et al. 2022). The chemical composition of many coral reef cyanobacteria is yet to be examined (Ford et al. 2021), and several cyanobacterial taxa are used as foods in aquaculture (Nicholson and Clements 2023). Filamentous cyanobacteria are widely distributed over the reef amongst turf, on crustose coralline algae and on bare surfaces, whereas cyanobacterial mats are generally more spatially restricted. For example, a study conducted in the Maldive Islands using 1.3cm diameter sample punches of turf-covered reef rock found that filamentous cyanobacteria comprised 30% of sampled surface and occurred on 93.4% of sample punches, whereas cyanobacterial mats comprised < 1% of surface cover and occurred on only 5.5% of sample punches (Harris et al. 2015). Moreover, at least some of the cyanobacteria ingested by *S. vulpinus* are epiphytic Oscillatoriales (see Fig. 3), and I am unaware of evidence that *S. vulpinus* feeds on cyanobacterial mats. Complementary feeding on a mixture of cyanobacteria and various kinds of algae is more likely to be associated with balancing intake of macronutrients and micronutrients (see Raubenheimer et al. 2005). Cyanobacteria are protein-rich but are generally lacking in long-chain polyunsaturated fatty acids (LC-PUFA) such as 20:5n-3 eicosapentaenoic acid (EPA) and 22:6n-3 docosahexaenoic acid (DHA) (Bergé and Barnathan 2005, Guedes et al. 2011, Liénart et al. 2023). EPA is present in many types of macroscopic algae, and both EPA and DHA are present in Bacillarophyceae (diatoms) and Dinophyceae (dinoflagellates) (Bergé and Barnathan 2005, Guedes et al. 2011, Liénart et al. 2023). *Siganus canaliculatus* contain high levels of 20:4n-6 arachidonic acid in their tissues and in stomach contents, consistent with the high levels of this polyunsaturated fatty acid in dietary algae (Saito et al. 1999, Osako et al. 2006). Feeding on a mixture of cyanobacteria and eukaryotic microalgae or macroalgae would therefore provide a more balanced intake of macronutrients, essential amino acids and essential LC-PUFA.

The fourth trophic group of rabbitfishes feed mainly on sessile invertebrates, particularly sponges, and include *S. puelloides*, *S. puellus* and *S. stellatus* (Table 2). Foraging observations suggested that *S. stellatus* fed on turf algae (Ebrahim et al. 2020), but stomach contents of this species were dominated by sponges and cyanobacteria, with only minor proportions of eukaryotic algae (Tietbohl 2022; Table 2). The sister species of *S. stellatus* is *S. punctatus* (Siqueira et al. 2019), although the diet of the former species more closely resembles that of *S. puellus* and *S. puelloides*. It is noteworthy that quantitative gut content data contradict the apparent food resource of *S. stellatus* based on feeding observations. This pattern of feeding highlights the problem of treating "turf algae" as a homogeneous food resource. Turf assemblages can vary considerably over small spatial scales, enabling fish such as *S. stellatus* and many parrotfishes to target patches of cyanobacteria (and sponges in this case) within what superficially appears as "turf" (Clements et al. 2017, Nicholson and Clements 2020, 2023). This also highlights the problems associated with inferring diet based on behavioural foraging observations of herbivores feeding on heterogenous resource categories such as "turf algae" or "Epilithic Algal Matrix" (Mendes et al. 2018, Leitão et al. 2023; cf. Tebbett et al. 2023).

Several studies show that some rabbitfish modify their feeding behaviour in relation to levels of secondary metabolites in both macrophytes and cyanobacteria. Juvenile *S. spinus* changed feeding preference over time based on exposure to two secondary metabolites from the cyanobacterium *Lyngbya majuscula*, while showing consistent negative preference towards a third compound (Thacker et al. 1997). This suggests that rabbitfish modify their feeding behaviour based on prior experience. This avoidance of toxic cyanobacteria blooms coincident with depletion of dietary algae has been associated with starvation of rabbitfish recruits following mass recruitment events (Nagle and Paul 1998). Notably, *S. fuscescens* consumed *L. majuscula* in equal quantities to the preferred dietary alga *Acanthophora spicifera* when lyngbyatoxin-a was not detectable but avoided *L. majuscula* in simulated bloom conditions, indicating that these rabbitfish avoid this cyanobacteria when toxic secondary metabolites are present (Capper et al. 2006). Consumption of the palatable alga *Acanthophora* by *S. doliatus* was reduced by 53% when presented in close association to the calcified rhodophyte *Galaxaura*, which produces both allelopathic and toxic secondary metabolites (Loffler et al. 2015). In contrast, pairing *Acanthophora* with *Turbinaria* did not reduce feeding by *S. doliatus* on the former species. *Siganus fuscescens* preferentially consumed shoots of the seagrass *Zostera muelleri* collected near a low pH/high CO_2 freshwater spring emerging into Moreton Bay, Queensland, Australia (Arnold et al. 2014). Proximity to the spring did not influence levels of carbon and nitrogen in *Z. muelleri*, but it did reduce soluble phenolics by $\geq$ 87%. In a paired choice feeding trial juvenile *S. fuscescens* consumed 55% of *Z. muelleri* shoots collected 5–10 m from the spring outflow and only 25% of those collected 30–50 m from the outflow, showing that the fish selectively avoided high levels of phenolics (Arnold et al. 2014). Aragonite (calcium carbonate) also proved a feeding deterrent when added to *Enteromorpha* fed to captive *S. spinus*

(Schupp and Paul 1994), and *S. canaliculatus* and *S. doliatus* strongly avoided calcified macroalgae in feeding assays (Mantyka and Bellwood 2007a).

Nutritional Biomarkers

While the previous section on diet provides detail on specific foods ingested, nutritional biomarkers such as tissue stable isotope and fatty acid analyses can provide added insight into the relative importance of different foods in providing nutrition to rabbitfish, and thus inform our understanding of the role of rabbitfish in trophodynamic pathways in reef, seagrass and mangrove habitats. Nutritional biomarkers also allow us to compare the trophic position of rabbitfishes with that of other fish taxa. Unfortunately, the few studies that present biomarker data on rabbitfishes tend to include few species, making these data of limited use in inferring trophic differences among syntopic rabbitfish species.

Siganus doliatus on lagoonal reefs in southwestern New Caledonia displayed a similar $d^{15}N$ signature to that of *Pomacentrus adelus* and *Ctenochaetus striatus* and were slightly depleted in $d^{15}N$ compared to *Acanthurus nigrofuscus* (Carassou et al. 2007). This may reflect the selection of turf assemblages that are slightly richer in diazotrophic cyanobacteria at this location than those selected by the grazing surgeonfish *Acanthurus nigrofuscus*, possibly because of differences in a capacity to access microtopographic refuges for photoautotrophs (Brandl and Bellwood 2016). *Siganus vulpinus* exhibited a broad range of $d^{13}C$ variation and a relatively depleted $d^{15}N$ signature in samples from Kimbe Bay, New Guinea, suggesting a diet containing a mixture of eukaryotic algae and cyanobacteria (Eurich et al. 2019). Juvenile *S. sutor* appear to be herbivorous from settlement, and although they forage widely in seagrass areas their isotopic signal indicates that they gain some nutrition from mangrove-derived material (Marguillier et al. 1997).

Siganus fuscescens has been the subject of stable isotope work in several geographic locations. On reefs within 500m of mangroves in Moreton Bay, Queensland, Australia, *S. fuscescens* obtained a minimum of ~ 44% of their carbon intake from these habitats, but this dropped to ~ 25–30% over distances of 1.5–2 km from mangroves (Davis et al. 2014). This indicates that even though mangroves do not appear to act as nurseries for *S. fuscescens*, mangroves nonetheless provide a significant nutritional subsidy to this species in nearby seagrass and reefs (Davis et al. 2014). *Siganus fuscescens* in a seagrass bed in the northwestern Philippines undergo an ontogenetic dietary shift in which they broaden the diet from POM, macroalgae, seagrass epiphytes, zooplankton and other small invertebrate epiplanktonic animals in juveniles to a wider diet of macroalgae, epiphytes on seagrass, and the seagrass blades in adults (Metillo et al. 2019). Adults at this location also eat some zooplankton, including polychaete larvae, mysids, and ostracods. In a mangrove habitat in New Caledonia *S. fuscescens* had a higher $d^{15}N$ than *S. randalli* and *S. lineatus*, whereas in reef habitat *S. punctatus* was enriched in $d^{15}N$ compared to *S. lineatus* and depleted compared to *S. puellus* (Moleana et al. 2016).

Wu et al. (2022) examined anterior intestinal content of five *S. argenteus*, three *S. punctatissimus* and three *S. puellus* collected from the Xisha Islands, China using 18S rRNA diet metabarcoding to amplify eukaryote sequences. The 18S rRNA sequences from *S. argenteus* were dominated in decreasing order of importance by the seagrass *Halophila ovalis*, Cnidaria (mainly anthozoan corals) and Bacillariophyta (diatoms), and minor proportions of chlorophyte, rhodophyte and ochrophyte sequences. The 18S sequences from *S. punctatissimus* were dominated by opalozoan protozoans, which could be parasitic, with lower proportions of anthozoan and seagrass sequences, while the sequences from *S. puellus* comprised a mixture of anthozoans and Porifera. Fragments of some algal and seagrass species were identified microscopically from the gut samples, but unfortunately no quantitative data were presented based on visual gut content analysis. Because of this it is unclear how well the 18S rRNA data reflect the proportions of what was actually present in the gut, and it should also be noted that the authors did not sequence 16S rRNA which could have identified prokaryotic elements in the diet, i.e., cyanobacteria. While diet metabarcoding is a very useful tool to improve the taxonomic resolution of dietary taxa (e.g., Chelsky Budarf et al. 2011, Nicholson and Clements 2023) it should not be considered strictly quantitative due to biases inherent in the extraction and amplification of target DNA sequences (Nielsen et al. 2018, Alberdi et al. 2019, Deagle et al. 2019). Unlike stomach content data, diet metabarcoding does not provide quantitative parameters relevant to determining macronutrient intake from various food sources.

Digestion and Assimilation

While there is a copious aquaculture literature on digestion and assimilation in cultured rabbitfishes most of this relates to optimising the growth of larvae and juveniles, and hence will not be considered here. Rabbitfish are widely grown in aquaculture as several species including *S. argenteus*, *S. canaliculatus*, *S. guttatus*, *S. javus* and *S. rivalutus* tolerate the environmental variation, handling and crowding associated with culture conditions (Boonyaratpalin 1997). A few studies have examined post-ingestive nutritional ecology in rabbitfishes, although as with the biomarker literature few of these are useful in a comparative context.

In captivity, even herbivorous rabbitfish become omnivorous and will readily eat a range of animal food sources (Boonyaratpalin 1997). As in most fish, growth in several rabbitfishes has been shown to be proportional to dietary protein intake. For example, when fed isocaloric diets, the growth of *S. guttatus* is proportional to dietary protein-to-energy ratio (Parazo 1990). Similarly, in juvenile *S. rivulatus,* growth was proportional to dietary protein level up to the highest level offered (40%), and daily feed intake, apparent protein utilisation and feed conversion ratio decreased as dietary protein increased (El-Dakar et al. 2011). Pillans et al. (2004) showed that preference by *S. fuscescens* for different macrophytes was related to assimilation efficiency, and that preferred macrophytes passed through the gut more rapidly than less preferred species. Mean gut transit times (± SE) following starvation in subadult *S. fuscescens* (i.e., 90–110 mm SL) varied between 146.3 ± 7.5 min with *Acanthophora* and

507.8 ± 29.8 min with *Dictyota* (Pillans et al. 2004), although these likely underestimate gut transit rates in fish that do not have empty guts prior to feeding.

Rabbitfish produce the standard complement of carbohydrases seen in fish (Sabapathy and Teo 1993, 1994). However, evidence for endogenous utilization of the phaeophyte storage compounds laminarin, a β-1,3-glucan, and the sugar alcohol mannitol is equivocal (Choat and Clements 1998). Laminarin was hydrolyzed by extracts of the intestine, pyloric ceca, esophagus, and stomach of *S. canaliculatus* (Sabapathy and Teo 1993), however some (1-3)-β-glucan hydrolysis in fish gut extracts may be the result of exogenous activity from ingested and symbiotic bacteria. Protease and lipase activity in the liver, pyloric caecae and intestine of juvenile *S. canaliculatus* increased significantly in fish fed animal material compared to those fed algae (Xie et al. 2018). Conversely, amylase activity in liver, stomach, pyloric caeca end intestine was significantly higher in the same fish fed algal diet treatments than animal material. Interestingly, the same study demonstrated that amylase activity in the mid-intestine was greater in juvenile *S. canaliculatus* fed the *Gracilaria* diet treatment than in those fed the *Enteromorpha* diet treatment (Xie et al. 2018). This difference could reflect a difference in both the amount or the digestibility of the starch present in these two algal taxa (Zemke-White and Clements 1999).

Multiple-choice feeding trials on captive juvenile *S. canaliculatus* in Guangdong Province, China revealed highest preference for *Enteromorpha prolifera*, *Gracilaria lemaneiformis* and *Chaetomorpha linum*, lower preference for *Enteromorpha pertusa* and *Porphyra haitanensis*, and least preference for *Sargassum fusiforme* and *Corallina sessilis* (You et al. 2014). The authors indicated that levels of crude protein content and soluble sugar were significantly higher in the three most preferred algae than the two least preferred algae, although the highest levels of crude protein and total soluble sugar were in one of the algal species that were intermediate in preference, i.e., *P. haitanensis*. This indicates that factors other than crude protein and soluble sugar must have influenced feeding preferences among the seven algal species used in the experiment. The authors note the low nutritional value of the least preferred alga, the calcareous rhodophyte *C. sessilis* (You et al. 2014), but it is noteworthy that the phaeophyte *Sargassum fusiforme* had the highest carbohydrate levels of the seven algal taxa. Adult *S. canaliculatus* in some areas feed mainly on leathery and foliose phaeophytes including *Sargassum* (Hoey et al. 2013), suggesting that there is an ontogenetic shift in diet from protein-rich chlorophytes and rhodophytes in juveniles to carbohydrate-rich phaeophytes in adults. The importance of protein-rich dietary algae to juvenile rabbitfish has also been demonstrated in a demographic study on wild fish. Recently settled *S. spinus* in a bay in Guam where algal resources were rich in both nitrogen and phosphorus were 10–20% larger than in a second location where algal resources were less nutrient-rich (Priest et al. 2016). The hypothesis that this difference in growth rate was caused by the availability of nutrient-rich algae was confirmed in a factorial experiment where captive *S. spinus* recruits were fed diets corresponding to those present in the two locations (Priest et al. 2016). Clearly, juvenile and adult rabbitfish face different nutritional challenges.

Siganus canaliculatus has been shown to possess the genomic and metabolic capacity to synthesise the long chain polyunsaturated fatty acids (LC-PUFA) 20:5n-3 eicosapentaenoic acid (EPA) and 22:6n-3 docosahexaenoic acid (DHA) from

C_{18}-PUFA 18:3n-3 α-linolenic and 18:2n-6 linoleic acid precursors (Li et al. 2010, 2020, Monroig et al. 2012). This LC-PUFA biosynthesis pathway involves sequential desaturation and elongation steps from C_{18} PUFA substrates catalyzed by the fatty acyl desaturases Δ6/Δ5 and Δ4 Fad and elongation by the very-long-chain fatty acid elongases Elovl5 and Elovl4 (Li et al. 2010, Monroig et al. 2012; Fig. 4). Before this discovery it was thought that vertebrates produced 22:6n-3 from 20:5n-3 only via the alternative "Sprecher" pathway involving Δ6/Δ5 Fad, and that vertebrates did not possess Δ4 Fad activity (Li et al. 2010). This finding is very significant for rabbitfish aquaculture as it means that the fish can produce their own LC-PUFA endogenously from feeds containing C_{18}-rich vegetable oils (Li et al. 2010).

Prior to this work on *S. canaliculatus,* it was thought that marine fish required dietary EPA and DHA, i.e., that they could not produce these from C18-PUFA, a view associated with the greater abundance of EPA and especially DHA in the marine environment (Tocher 2010). Vascular plants almost exclusively synthesise C18-PUFA, and thus fish utilising these resources in freshwater environments require the capacity to synthesise LC-PUFA such as EPA and DHA (Tocher 2009). Marine eukaryotic phytoplankton, especially dinoflagellates, are rich in EPA and DHA (Hanson et al. 2010), and hence these LC-PUFA are generally not a dietary requirement in marine food webs (Tocher 2009), although DHA in seagrass and most macroalgae is either present at low levels or absent altogether (Hanson et al. 2010, Pereira et al. 2012). Prior to the work on LC-PUFA synthesis in *S. canaliculatus*, all of the marine fish studied in this context were omnivorous or carnivorous, i.e., none of the taxa examined were strict herbivores, in contrast to the work on freshwater taxa which included various herbivorous cyprinids. The work by Li, Tocher and

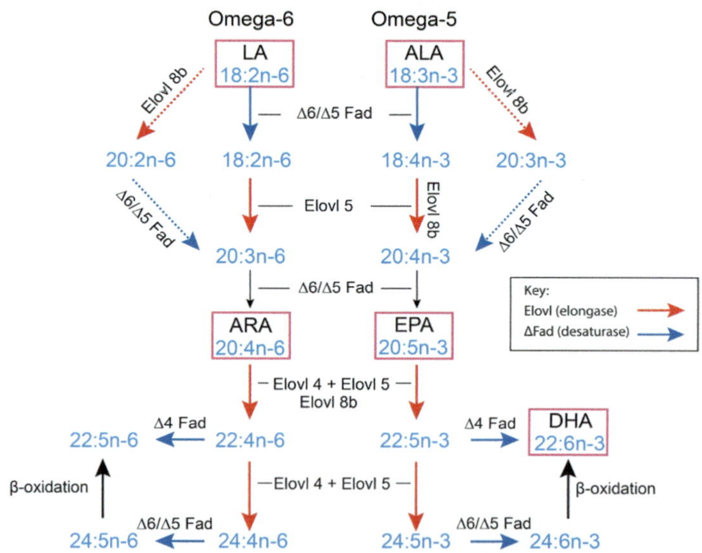

Fig. 4. Biosynthesis pathways for long-chain polyunsaturated fatty acids (LC-PUFA) from 18:2n-6 linoleic and 18:3n-3 α-linolenic acids. Solid lines show pathways confirmed in *S. canaliculatus*, dashed lines indicate pathways confirmed in other organisms, but not yet confirmed in teleosts. Figure redrawn from Li et al. (2010) and Monroig et al. (2012). See text. Abbreviations: LA – linoleic acid, ALA – α-linolenic acid, ARA – arachidonic acid, EPA – eicosapentaenoic acid, DHA – docosahexaenoic acid.

colleagues demonstrated that *S. canaliculatus* has evolved the metabolic capacity to overcome the deficiency of EPA and especially DHA in a diet consisting of marine macrophytes (i.e., macroalgae and seagrasses). The presence of Δ6/Δ5 Fad activity in *S. canaliculatus* shows that this species has the desaturation activities necessary to produce EPA and ARA from 18:3n-3 and 18:2n-6, respectively, and moreover the presence of Δ4 Fad provides this rabbitfish with the endogenous capacity to produce DHA from EPA (Li et al. 2010). These fatty acid desaturases and elongases are the subject of active ongoing biochemical, genomic and transcriptomic research (e.g. Xie et al. 2020, Monroig et al. 2022, Dong et al. 2023).

Gut Microbiota and their Role in Nutrient Supply

Gut microbiota are increasingly seen as critical to health and nutrition in vertebrates and other animals, and literature on the composition and function of gut communities in herbivorous fishes, including rabbitfishes, is growing rapidly. Gut microbiota play an important role in some herbivorous fish in converting refractory algal carbohydrate into short-chain fatty acids (SCFA) that can be assimilated and metabolised by host fish (Clements and Choat 1995, Clements et al. 2014, 2017). The total levels of SCFA in rabbitfish measured to date (Table 3) are at the lower end of the range seen in herbivorous surgeonfish, and much lower than in kyphosid chubs, but higher than levels in parrotfishes and detritivorous surgeonfish (Clements and Choat 1995, Clements et al. 2017). *Siganus lineatus* had the lowest levels of hindgut SCFA of any of the eight rabbitfish species measured by Clements and Choat (1995) and was the only species in which total SCFA concentration was higher in the penultimate intestine section than the distal gut section (Table 3). All the *S. lineatus* sampled for this SCFA work were collected from reefs in the vicinity of Lizard Island and were inactive (i.e., not foraging) when collected during the day, similar to the reef-

Table 3. Levels of short-chain fatty acids (SCFA) and branch-chain fatty acids (BCFA) as % total SCFA in posterior intestine sections of rabbitfish species. Section IV and section V are defined as the penultimate and distal 25% of intestine length, respectively. Source of data for *S. fuscescens* is Clements unpublished, and Clements and Choat (1995) for all other species.

Species (n)	Section IV		Section V	
	Total SCFA mM	BCFA as % total SCFA	Total SCFA mM	BCFA as % total SCFA
S. argenteus (9)	5.53 ± 1.31	3.99 ± 0.79	16.96 ± 0.99	3.20 ± 0.37
S. corallinus (4)	6.48 ± 3.20	7.22 ± 4.42	10.79 ± 1.93	2.57 ± 0.65
S. doliatus (11)	11.76 ± 2.43	7.32 ± 1.83	18.44 ± 1.58	1.72 ± 0.25
S. fuscescens (2)	16.95 ± 6.58	0.3 ± 0.3	20.93 ± 10.10	1.1 ± 0.1
S. lineatus (8)	10.11 ± 1.19	4.53 ± 1.04	9.78 ± 1.50	2.77 ± 0.65
S. puellus (4)	16.75 ± 1.83	8.69 ± 2.02	23.01 ± 4.24	5.25 ± 1.37
S. punctatissimus (3)	12.18 ± 1.36	3.33 ± 0.92	21.86 ± 1.36	4.44 ± 0.74
S. punctatus (8)	9.01 ± 4.76	6.96 ± 2.26	14.29 ± 3.01	4.57 ± 0.90
S. vulpinus (5)	5.76 ± 0.97	7.63 ± 0.73	20.70 ± 4.54	3.48 ± 0.26

dwelling *S. lineatus* examined by Fox and Bellwood (2011). It is therefore likely that these fish were also feeding on sediment-associated microphytobenthos and detritus rather than macroscopic algae. The proportion of total SCFA made up by the branched-chain fatty acids isobutyrate and isovalerate was comparable in *S. lineatus* to that of cropping rabbitfish species (Table 3), i.e., *S. argenteus*, *S. corallinus* and *S. doliatus*, indicating that the proportion of protein in the diet was not that dissimilar among these rabbitfish species (Clements et al. 2017). Interestingly, levels of SCFA in the posterior intestine were higher in the mixed feeders and spongivore groups than in croppers and browsers. This may reflect lower gut passage rates in omnivorous than herbivorous taxa, and thus greater accumulation of SCFA in the posterior gut, consistent with feeding rates in these species (Nanami 2018).

In addition to the concentration and production rate of SCFA in the gut, the daily yield of SCFA from hindgut fermentation is heavily influenced in different host fish species by the mass of digesta in the posterior intestine relative to total body mass (Mountfort et al. 2002). The maximum mass of hindgut contents was ~ 6% of total body mass in *S. doliatus* (in early morning) and ~ 3.8% of total body mass in *S. lineatus* (at midday) (Fox et al. 2009). By comparison, mass of hindgut contents in adult *Kyphosus sydneyanus* is ~ 9.5% of total body mass (Mountfort et al. 2002). Therefore, the relative mass of fermentative digesta in the hindgut of *S. doliatus* and *S. lineatus* is ~ 60% and ~ 40%, respectively, of that in *K. sydneyanus*. Overall, therefore, the importance of hindgut fermentation is likely to be lower in rabbitfish than in taxa such as kyphosid chubs and browsing *Naso* unicornfishes.

The hindgut communities of rabbitfish include a similar range of bacterial taxa to those found in the hindgut communities of other algivorous taxa such as surgeonfish and kyphosid chubs (Miyake et al. 2015, Pisaniello et al. 2022). Accordingly, hindgut communities from rabbitfish are generally rich in bacterial taxa associated with fermentative digestion (Cheutin et al. 2021). The 16S rRNA sequences amplified from the hindgut microbiota of three *S. stellatus* collected from the central Red Sea were proportionally dominated by the bacterial phyla Firmicutes (44%), Proteobacteria (35%) and Bacteroidetes (~ 12%) (Miyake et al. 2015); the valid names for these three phyla are now Bacillota, Pseudomonadota and Bacteroidota, respectively (Oren and Garrity 2021). Of the other nominally herbivorous Red Sea fish taxa included in the study, this pattern of phylum-level relative sequence abundance most closely matched the parrotfish *Scarus niger* (see Miyake et al. 2015), although this species was lower in the proportional abundance of Bacteroidota. This resemblance might appear superficially surprising given the differences between parrotfish and rabbitfish in levels of hindgut fermentation (Clements and Choat 1995, Clements et al. 2017), however the diet of *S. stellatus* consists mainly of sponges and cyanobacteria, with only very minor proportions of eukaryotic algae (Tietbohl 2022; Table 1). In this respect the protein-rich diet of *S. stellatus* is convergent with that of some parrotfishes (Clements et al. 2017), although at present we lack the data on SCFA to compare levels hindgut fermentation in these taxa.

The hindgut microbiota of *S. fuscescens* from the southern Great Barrier Reef was dominated by the bacterial phyla Pseudomonadota and Bacillota with smaller proportions of Bacteroidota (Nielsen et al. 2017), as reported for *S. stellatus* (Miyake et al. 2015). The diet of the *S. fuscescens* sampled by Nielsen and colleagues was observed to be dominated by algae (i.e., presumably not seagrass), although this was not quantified. The same three bacterial phyla dominated the hindgut microbiota of *S. fuscescens* collected across a 2000 km latitudinal gradient in Western Australia (Jones et al. 2018), although the proportion of Bacteroidota was considerably higher than in the study by Nielsen and colleagues. Thépot et al. (2022) reanalysed the data from Nielsen (2017), Jones et al. (2018), and a third set of samples from captive samples collected in southern Queensland (Thépot et al. 2021). The ratio of Pseudomonadota, Bacillota and Bacteroida varied among the locations analysed and the communities differed significantly between studies and locations, but interpreting this variation in the absence of dietary information on all the sampled individuals is problematic. Thépot et al. (2022) reported that 55 of 174 bacterial taxa detected in their overall dataset were present at all four sample locations, suggesting some commonality in community composition. Thépot et al. (2022) also examined variation in hindgut community structure in response to 15 diet treatments involving 3% additions of different algae and supplements. Treatments only had subtle effects on the diversity and composition of hindgut communities, but the diet supplementation may have been too limited to elicit community responses and there may be priority effects (i.e., historical contingency; Costello et al. 2012) that limit changes in stable gut populations. The hindgut microbiota of migrating postlarval *S. guttatus* (average length 17.1–17.9 mm) collected from river mouths in Vietnam were also dominated by Bacillota, with lower proportions of Pseudomonadota, Actinomycetota and Verrucomicrobiota (formerly Proteobacteria, Actinobacteria and Verrucomicrobia, respectively; Oren and Garrity 2021) (Le et al. 2020). At this stage the postlarvae have commenced feeding on the benthos as herbivores (Le et al. 2020), presumably on filamentous algae.

Interestingly, the gut microbiota of the Lessepsian migrants *S. luridus* and *S. rivulatus* in the Mediterranean Sea was greater in taxonomic and phylogenetic diversity than in conspecifics from their native range in the Red Sea, indicating that shifts in diet and/or environment can have a strong effect on community assembly in rabbitfishes (Escalas et al. 2022). Micronutrients such as selenium can also influence the diversity and stability of hindgut communities in rabbitfish. Captive *S. oramin* (a junior synonym of *S. canaliculatus*, sensu Woodland 1990) were fed four dietary treatments of *Gracilaria lemaneiformis* that had been cleaned of epiphytes (Su et al. 2023). These were fed to a control group of rabbitfish with natural levels of selenium and to three treatment groups with *G. lemaneiformis* enriched with selenium at low, medium, and high concentrations. The proportions of Desulfovibrionaceae in intestinal communities decreased in rabbitfish in the low selenium treatment compared to the control, medium and high treatments, while the low selenium treatment elicited an increase in the proportion of Rikenellaceae compared to the other three treatments (Su et al. 2023). Moreover, the rabbitfish in the low selenium treatment exhibited significantly enhanced growth and maintained ecological greater stability of their hindgut communities. In particular, the low selenium treatment regulated Bacillota in the families Lachnospiraceae and Ruminococcaceae which appear to be most important in maintaining microbial stability and providing SCFA to the host fish (Su et al. 2023).

Simora et al. (2015) examined enzymatic activity in cultures of aerobic and facultatively anaerobic bacteria isolated from the gut of *S. guttatus*, and report both cellulolytic and carrageenolytic enzyme activity. Zhang et al. (2016) conducted a similar study on the capacity of hindgut cultures from *S. oramin* (a junior synonym of *S. canaliculatus*, sensu Woodland 1990) to degrade polysaccharides from *Enteromorpha*. However, these cultures almost certainly do not reflect activity *in vivo*, as the hindgut of these fish is highly anaerobic. More recent metagenomic work on the microbiome of captive *S. oramin* suggested that various bacterial taxa degraded polysaccharides from *Enteromorpha*, including cellulose (Xu et al. 2021). This suggests that hindgut microbiota may salvage energy from the cellulolytic cell walls of seagrass that is eaten by rabbitfish. In addition to some *Siganus* species several other fish taxa inhabit seagrass areas and feed on seagrass and associated epiphytes, e.g., the sparisomatinine labrids *Calotomus spinidens*, *Leptoscarus vaigiensis* and *Sparisoma radians* (Lobel and Ogden 1981, Nakamura et al. 2003, Gullström et al. 2011) and the sparid *Sarpa salpa* (Goldenberg and Erzini 2014). There are also many freshwater fish taxa that feed on aquatic angiosperms, most notably cyprinids such as the grass carp *Ctenopharyngodon idella*. However, of these rabbitfish appear to be the only taxa in which hindgut fermentation is quantitatively important (see Clements et al. 2014, Hao et al. 2017). If microbial cellulolysis is a viable strategy in rabbitfish feeding on seagrasses then why is this nutritional strategy not exploited by any freshwater fish species, let alone other taxa of herbivorous marine fish?

Conclusions and Questions for Future Research

The most striking aspect of the available literature on nutritional ecology of rabbitfishes is the relative lack of comparative empirical studies compared to work on comparable herbivorous taxa such as surgeonfish, parrotfish and even kyphosid chubs. There is a great need for future studies such as Hoey et al. (2013) and Nanami (2018) that present stomach content data on syntopic rabbitfish species and that accordingly allow an evaluation of interspecific trophic variation and diversification. Similarly, very little information is available that allow multiple rabbitfish taxa to be compared in terms of nutritional biomarkers such as stable isotopes and fatty acids. Most of the information on nutritional ecology in rabbitfishes available is fragmentary and relates to single species in a variety of geographic locations and habitat settings. The exception is in the Mediterranean Sea, where the invasive *S. luridus* and *S. rivulatus* have been studied extensively.

Much research effort in recent years has gone into field tether assays and behavioural observations (either direct or by fixed video) that yield contextual information of an inductive nature but do not generate the transferable parameters required to (a) understand evolutionary aspects of trophic diversification, (b) the trophodynamic role of different rabbitfish species in their respective habitats, or even (c) reliably quantify rates of algal removal that can be compared across systems. Several examples described above demonstrate the problematic nature of making generalised inferences on trophodynamics based on these techniques, especially in cases that involve the use of heterogeneous resource categories such as "turf algae". Clearly, rabbitfish species partition different elements within these categories, as do surgeonfish (Tebbett et al. 2017) and parrotfishes (Nicholson and Clements 2020, 2023).

Much information of value could be obtained by estimating both interspecific variation in diet and relative gut content mass in multiple rabbitfish species, e.g., as in Fox et al. (2009). Studies examining intraspecific variation would also be very interesting in taxa such as *S. fuscescens* that occupy a range of habitats utilising a variety of distinct trophic resources, e.g., as in Pillans et al. (2004). The finding that the browsing rabbitfish species *S. canaliculatus* has the metabolic capacity to synthesise essential LC-PUFA (Li et al. 2010, Monroig et al. 2012) raises the question of how this capacity is distributed among other rabbitfish species and trophic groups. Is it only present in rabbitfish species that feed largely on phaeophytes whose diets are poor in some LC-PUFA? Comparative work on the composition and function and gut microbiota in different rabbitfish taxa is also required, and the use of metagenomics and transcriptomics offers much potential here. This work is being undertaken on some cultured species in captivity, but to understand the role of gut symbionts in wild fish, it will be essential to sample communities in wild rabbitfish.

Based on the information available extant rabbitfish appear to have evolved a flexible nutritional ecology that allows them to utilise a wide range of diets and habitats. Apart from some differences in body shape and snout shape that influence the capacity to feed in different reef microhabitats (Fox and Bellwood 2013, Zolkaply et al. 2021), they appear to be relatively conservative in trophic morphology. Most species crop turf (of varying composition in terms of algae versus cyanobacteria), browse macrophytes, or a mixture of the two, with a few species targeting sessile invertebrates, especially sponges. Rabbitfish appear to be intermediate in terms of hindgut fermentation, displaying intermediate gut passage times and levels of SCFA in the hindgut compared to taxa such as kyphosid chubs, parrotfishes, and surgeonfishes. In this sense, they appear to be neither strict rate maximisers nor yield maximisers (Clements and Raubenheimer 2006, German et al. 2015), but can shift either way along this spectrum depending on the macronutrient composition of their diet. In other words, they are mostly generalists in terms of nutritional ecology, which is consistent with the comparatively high levels of intraspecific variation in diet. This flexibility also explains their widespread use in aquaculture and their success as invasive species, as seen in the two Lessepsian species that have colonised the eastern Mediterranean Sea. Testing this hypothesis of rabbitfish nutritional ecology requires more detailed work on their post-ingestive physiology, especially in response to variations in diet composition.

Acknowledgments

I am grateful to Howard Choat and Georgina Nicholson for helpful comments, Georgina Nicholson for the cyanobacteria photos from *S. vulpinus* stomach contents,

Matt Tietbohl for allowing me to use diet data on *Siganus stellatus* from his PhD thesis, Andy Hoey for some unpublished detail on stomach contents from Great Barrier Reef rabbitfish species, Aimee van der Reis for the focus-stacking images and Viv Ward for graphics. I also thank Andy Hoey and Brett Taylor for the opportunity to contribute this chapter.

References

Alberdi, A., O. Aizpurua, K. Bohmann, S. Gopalakrishnan, C. Lynggaard, M. Nielsen et al. 2019. Promises and pitfalls of using high-throughput sequencing for diet analysis. Mol. Ecol. Res. 19: 327–348.

Alsafy, M.A., S.A. El-Gendy, N.E. El-Bakary, B.M. Kamal, A. Derbalah and K. Roshdy. 2023. Morphological comparison of the detailed structure of gill rakers from three different feeding habits of marine fish species. Zoomorphology 142: 87–97.

Arnold, T., G. Freundlich, T. Weilnau, A. Verdi and I.R. Tibbetts. 2014. Impacts of groundwater discharge at Myora Springs (North Stradbroke Island, Australia) on the phenolic metabolism of eelgrass, *Zostera muelleri*, and grazing by the juvenile rabbitfish, *Siganus fuscescens*. PloS One 9: e104738.

Azzurro, E., E. Fanelli, E. Mostarda, M. Catra and F. Andaloro. 2007. Resource partitioning among early colonizing *Siganus luridus* and native herbivorous fish in the Mediterranean: an integrated study based on gut-content analysis and stable isotope signatures. J. Mar. Biol. Assoc. U.K. 87: 991–998.

Balakrishnan, M., Y. Selvam, V. Anto and M. Chellandi. 2022. Macroalgal susceptibility to grazing by herbivore fish *Siganus* spp. at Gulf of Mannar reef, Indian Ocean. Envir. Biol. Fish. 105: 167–177.

Barrientos, S., S. Zarco-Perello, C. Piñeiro-Corbeira, R. Barreiro and T. Wernberg. 2021. Feeding preferences of range-shifting and native herbivorous fishes in temperate ecosystems. Mar. Envir. Res. 172: 105508.

Bennett, S. and D.R. Bellwood. 2011. Latitudinal variation in macroalgal consumption by fishes on the Great Barrier Reef. Mar. Ecol. Prog. Ser. 426: 241–252.

Bergé, J.-P. and G. Barnathan. 2005. Fatty acids from lipids of marine organisms: molecular biodiversity, roles as biomarkers, biologically active compounds, and economical aspects. Adv. Biochem. Engin. Biotechnol. 96: 49–125.

Boonyaratpalin, M. 1997. Nutrient requirements of marine food fish cultured in Southeast Asia. Aquaculture 151: 283–313.

Borsa, P., S. Lemer and D. Aurelle. 2007. Patterns of lineage diversification in rabbitfishes. Mol. Phyl. Evol. 44: 427–435.

Bos, A.R., E. Cruz-Rivera and A.M. Sanad. 2017. Herbivorous fishes *Siganus rivulatus* (Siganidae) and *Zebrasoma desjardinii* (Acanthuridae) feed on Ctenophora and Scyphozoa in the Red Sea. Mar. Biodiv. 47: 243–246.

Brandl, S.J. and D.R. Bellwood. 2016. Microtopographic refuges shape consumer-producer dynamics by mediating consumer functional diversity. Oecologia 182: 203–217.

Bryan, P.G. 1975. Food habits, functional digestive morphology, and assimilation efficiency of the rabbitfish *Siganus spinus* (Pisces, Siganidae) on Guam. Pacific Sci. 29: 269–277.

Capper, A., I.R. Tibbetts, J.M. O'Neil and G.R. Shaw. 2006. Feeding preference and deterrence in rabbitfish *Siganus fuscescens* for the cyanobacterium *Lyngbya majuscula* in Moreton Bay, south-east Queensland, Australia. J. Fish Biol. 68: 1589–1609.

Carassou, L., M. Kulbicki, T.J. Nicola and N.V. Polunin. 2008. Assessment of fish trophic status and relationships by stable isotope data in the coral reef lagoon of New Caledonia, southwest Pacific. Aquat. Living Resour. 21: 1–12.

Carpenter, R.C. 1986. Partitioning herbivory and its effects on coral reef algal communities. Ecol. Monogr. 56: 345–364.

Chelsky Budarf, A., D.D. Burfeind, W.K.W. Loh and I.R. Tibbetts. 2011. Identification of seagrasses in the gut of a marine herbivorous fish using DNA barcoding and visual inspection techniques. J. Fish Biol. 79: 112–121.

Cheutin, M.C., S. Villéger, C.C. Hicks, J.P. Robinson, N.A. Graham, C. Marconnet et al. 2021. Microbial shift in the enteric bacteriome of coral reef fish following climate-driven regime shifts. Microorganisms 9: 1711.

Chitravadivelu, K. and A. Sivapalan. 1984. Food and feeding *of Siganus lineatus* from waters around northern Sri Lanka. J. Nat. Sci. Found. Sri Lanka 12: 129–139.

Choat, J.H. and K.D. Clements. 1998. Vertebrate herbivores in marine and terrestrial environments: a nutritional ecology perspective. Annu. Rev. Ecol. Syst. 29: 375–403.

Chong-Seng, K.M., K.L. Nash, D.R. Bellwood and N.A.J. Graham. 2014. Macroalgal herbivory on recovering versus degrading coral reefs. Coral Reefs 33: 409–419.

Chow, C.F., E. Wassénius, M. Dornelas and A.S. Hoey. 2021. Species differences drive spatial scaling of foraging patterns in herbivorous reef fishes. Oikos 130. 2217–2230.

Clements, K.D. 1997. Fermentation and gastrointestinal microorganisms in fishes. pp. 156–198. *In*: Mackie, R.I. and B.A. White [eds.]. Gastrointestinal microbiology. Vol. 1. Gastrointestinal ecosystems and fermentations. Chapman and Hall, New York, NY, USA.

Clements, K.D. and J.H. Choat. 1993. Influence of season, ontogeny and tide on the diet of the temperate marine herbivorous fish *Odax pullus* (Odacidae). Mar. Biol. 117: 213–220.

Clements, K.D. and J.H. Choat. 1995. Fermentation in tropical marine herbivorous fishes. Physiological Zoology 68: 355–378.

Clements, K.D. and D. Raubenheimer. 2006. Feeding and nutrition. pp. 47–82. *In*: Evans, D.H. and J.B. Claiborne [eds.]. The Physiology of Fishes. Taylor and Francis Group, Boca Raton, FL, USA.

Clements, K.D., D. Raubenheimer and J.H. Choat. 2009. Nutritional ecology of marine herbivorous fishes: ten years on. Func. Ecol. 23: 79–92.

Clements, K.D., E.R. Angert, W.L. Montgomery and J.H. Choat. 2014. Intestinal microbiota in fishes: what's known and what's not. Mol. Ecol. 23: 1891–1898.

Clements, K.D., D.P. German, J. Piché, A. Tribollet and J.H. Choat. 2017. Integrating ecological roles and trophic diversification on coral reefs: multiple lines of evidence identify parrotfishes as microphages. Biol. J. Linn. Soc. 120: 729–751.

Clements, K.D. and J.H. Choat. 2018. Nutritional ecology of parrotfishes (Scarinae, Labridae). pp. 42–68. *In*: Hoey, A.S. and R.M. Bonaldo [eds.]. Biology of parrotfishes. CRC Press, Taylor and Francis Group, Boca Raton, FL, USA.

Costello, E.K., K. Stagaman, L. Dethlefsen, B.J.M. Bohannan and D.A. Relman. 2012. The application of ecological theory toward an understanding of the human microbiome. Science 336: 1255–1262.

Davis, J.P., K.A. Pitt, B. Fry, A.D. Olds and R.M. Connolly. 2014. Seascape-scale trophic links for fish on inshore coral reefs. Coral Reefs 33: 897–907.

Deagle, B.E., A.C. Thomas, J.C. McInnes, L.J. Clarke, E.J. Vesterinen, E.L. Clare et al. 2019. Counting with DNA in metabarcoding studies: How should we convert sequence reads to dietary data? Mol. Ecol. 28: 391–406.

Debenay, J.P., A. Sigura and J.L. Justine. 2011. Foraminifera in the diet of coral reef fish from the lagoon of New Caledonia: Predation, digestion, dispersion. Rev. Micropaleontol. 54: 87–103.

De la Torre-Castro, M., J.S. Eklöf, P. Rönnbäck and M. Björk. 2008. Seagrass importance in food provisioning services: fish stomach content as a link between seagrass meadows and local fisheries. West. Indian Ocean J. Mar. Sci. 7: 95–110.

Dong, Y., Z. Xie, C. You, M. Li, Y. Li, J. Zhao et al. 2023. GPR120–ERK1–Srebp1c signalling pathway regulates long-chain polyunsaturated fatty acids biosynthesis in marine teleost *Siganus canaliculatus*. Comp. Biochem. Physiol. B, Biochem. Mol. Biol. 264: 110815.

Ebrahim, A., T.S. Martin, P.J. Mumby, A.D. Olds and I.R. Tibbetts. 2020. Differences in diet and foraging behaviour of commercially important rabbitfish species on coral reefs in the Indian Ocean. Coral Reefs 39: 977–988.

El-Dakar, A.Y., S.M. Shalaby and I.P. Saoud. 2011. Dietary protein requirement of juvenile marbled spinefoot rabbitfish *Siganus rivulatus*. Aquacult. Res. 42: 1050–1055.

Escalas, A., J.-C. Auguet, A. Avouac, J. Belmaker, T. Dailianis, M. Kiflawi et al. 2022. Shift and homogenization of gut microbiome during invasion in marine fishes. Anim. Microbiome 4: 37.

Eurich, J.G., J.K. Matley, R. Baker, M.I. McCormick and G.P. Jones. 2019. Stable isotope analysis reveals trophic diversity and partitioning in territorial damselfishes on a low-latitude coral reef. Mar. Biol. 166: 17.

Ford, A.K., P.M. Visser, M.J. van Herk, E. Jongepier and V. Bonito. 2021. First insights into the impacts of benthic cyanobacterial mats on fish herbivory functions on a nearshore coral reef. Scientific Reports 11: 7147.

Fox, R.J. and D.R. Bellwood. 2008. Remote video bioassays reveal the potential feeding impact of the rabbitfish *Siganus canaliculatus* (f: Siganidae) on an inner-shelf reef of the Great Barrier Reef. Coral Reefs 27: 605–615.

Fox, R.J., T.L. Sunderland, A.S. Hoey and D.R. Bellwood. 2009. Estimating ecosystem function: contrasting roles of closely related herbivorous rabbitfishes (Siganidae) on coral reefs. Mar. Ecol. Prog. Ser. 385: 261–269.

Fox, R.J. and D.R. Bellwood. 2011. Unconstrained by the clock? Plasticity of diel activity rhythm in a tropical reef fish, Siganus lineatus. Funct. Ecol. 25: 1096–1105.

Fox, R.J. and D.R. Bellwood. 2013. Niche partitioning of feeding microhabitats produces a unique function for herbivorous rabbitfishes (Perciformes, Siganidae) on coral reefs. Coral Reefs 32: 13–23.

German, D.P., A. Sung, P. Jhaveri and R. Agnihotri. 2015. More than one way to be an herbivore: convergent evolution of herbivory using different digestive strategies in prickleback fishes (Stichaeidae). Zoology 118: 161–170.

Goldenberg, S.U. and K. Erzini. 2014. Seagrass feeding choices and digestive strategies of the herbivorous fish *Sarpa salpa*. J. Fish Biol. 84: 1474–1489.

Guedes, A.C., H.M. Amaro, C.R. Barbosa, R.D. Pereira and F.X. Malcata. 2011. Fatty acid composition of several wild microalgae and cyanobacteria, with a focus on eicosapentaenoic, docosahexaenoic and α-linolenic acids for eventual dietary uses. Food Res. Int. 44: 2721–2729.

Gullström, M., C. Berkström, M.C. Öhman, M. Bodin and M. Dahlberg. 2011. Scale-dependent patterns of variability of a grazing parrotfish (*Leptoscarus vaigiensis*) in a tropical seagrass-dominated seascape. Mar. Biol. 158: 1483–1495.

Gundermann, N., D.M. Popper and T. Lichatowich. 1983. Biology and life cycle of *Siganus vermiculatus* (Siganidae, Pisces). Pac. Sci. 37: 165–180.

Hanson, C.E., G.A. Hyndes and S.F. Wang. 2010. Differentiation of benthic marine primary producers using stable isotopes and fatty acids: Implications to food web studies. Aquat. Bot. 93: 114–122.

Hao, Y.T., S.G. Wu, I. Jakovlić, H. Zou, W.X. Li and G.T. Wang. 2017. Impacts of diet on hindgut microbiota and short-chain fatty acids in grass carp (*Ctenopharyngodon idellus*). Aquacult. Res. 48: 5595–5605.

Harris, J.L., L.S. Lewis and J.E. Smith. 2015. Quantifying scales of spatial variability in algal turf assemblages on coral reefs. Mar. Ecol. Prog. Ser. 532: 41–57.

Hasegawa, M. 2010. Isoyake studies in Shizuoka Prefecture, Japan. Bull. Fish. Res. Agen. 32: 109–114.

Hoey, A.S. and D.R. Bellwood. 2010. Cross-shelf variation in browsing intensity on the Great Barrier Reef. Coral Reefs 29: 499–508.

Hoey, A.S., S.L. Brandl and D.R. Bellwood. 2013. Diet and cross-shelf distribution of rabbitfishes (f. Siganidae) on the northern Great Barrier Reef: implications for ecosystem function. Coral Reefs 32: 973–984.

Jones, J., J.D. DiBattista, M. Stat, M. Bunce, M.C. Boyce, D.V. Fairclough et al. 2018. The microbiome of the gastrointestinal tract of a range-shifting marine herbivorous fish. Front. Microbiol. 9: 2000.

Jones, R.S. 1968. Ecological relationships in Hawaiian and Johnston Island Acanthuridae (surgeonfishes). Micronesica 4: 309–361.

Kang, J., H.F. Mohamed, X. Liu, L. Pei, S. Huang, X. Lin et al. 2022. Combined culture and DNA metabarcoding analysis of cyanobacterial community structure in response to coral reef health status in the South China Sea. J. Mar. Sci. Eng. 10: 1984.

Kiriyama, T., M. Noda and A. Fujii. 2001. Grazing and bite marks on *Ecklonia kurome*, caused by several herbivorous fishes. Aquacult. Sci. 49: 431–438. (in Japanese with English abstract)

Kiriyama, T., A. Fujii and Y. Fujita. 2005. Feeding and characteristic bite m arks on *Sargassum fusiforme* by several herbivorous fishes. Aquacult. Sci. 53: 355–365. (in Japanese with English abstract)

Konow, N., D.R. Bellwood, P.C. Wainwright and A.M. Kerr. 2008. Evolution of novel jaw joints promote trophic diversity in coral reef fishes. Biol. J. Linn. Soc. 93: 545–555.

Le, D., P. Nguyen, D. Nguyen, K. Dierckens, N. Boon, T. Lacoere et al. 2020. Gut microbiota of migrating wild rabbit fish (*Siganus guttatus*) larvae have low spatial and temporal variability. Microb. Ecol. 79: 539–551.

Lefèvre, C.D. and D.R. Bellwood. 2011. Temporal variation in coral reef ecosystem processes: herbivory of macroalgae by fishes. Mar. Ecol. Prog. Ser. 422: 239–251.

Leitão, M., C.W. Hackradt, Í. Lima e Silva, J.E. Ciancio and F.C. Félix-Hackradt. 2023. Effect of human impact on coral reef herbivorous fish niche. Mar. Biol. 170: 59.

Li, Y., O. Monroig, L. Zhang, S. Wang, X. Zheng, J.R. Dick et al. 2010. Vertebrate fatty acyl desaturase with Δ4 activity. Proc. Nat. Acad. Sci. USA 107: 16840–16845.

Li, Y., Z. Wen, C. You, Z. Xie, D.R. Tocher, Y. Zhang et al. 2020. Genome wide identification and functional characterization of two LC-PUFA biosynthesis elongase (elovl8) genes in rabbitfish (*Siganus canaliculatus*). Aquaculture 522: 735127.

Liénart, C., M. Tedengren, A. Garbaras, H. Lim, M. Chynel, C. Robinson et al. 2023. Diet quality determines blue mussel physiological status: A long-term experimental multi-biomarker approach. J. Exp. Mar. Biol. Ecol. 563: 151894.

Lobel, P.S. and J.C. Ogden. 1981. Foraging by the herbivorous parrotfish *Sparisoma radians*. Mar. Biol. 64: 173–183.

Loffler, Z., D.R. Bellwood and A.S. Hoey. 2015. Associations among coral reef macroalgae influence feeding by herbivorous fishes. Coral Reefs 34: 51–55.

Lundberg, B. and Y. Lipkin. 1979. Natural food of the herbivorous rabbitfish (*Siganus* spp.) in northern Red Sea. Bot. Mar. 22: 173–181.

Lundberg, B. and D. Golani. 1995. Diet adaptations of Lessepsian migrant rabbitfishes, *Siganus luridus* and *S. rivulatus*, to the algal resources of the Mediterranean coast of Israel. Marine Ecology 16: 73–89.

Mantyka, C.S. and D.R. Bellwood. 2007a. Macroalgal grazing selectivity among herbivorous coral reef fishes. Mar. Ecol. Prog. Ser. 352: 177–185.

Mantyka, C.S. and D.R. Bellwood. 2007b. Direct evaluation of macroalgal removal by herbivorous coral reef fishes. Coral Reefs 26: 435–442.

Marguillier, S., G. Van der Velde, F. Dehairs, M.A. Hemminga and S. Rajagopal. 1997. Trophic relationships in an interlinked mangrove-seagrass ecosystem as traced by $d^{13}C$ and $d^{15}N$. Marine Ecology Progress Series 151: 115–121.

McDermid, K.J., B. Stuercke and G.H. Balazas. 2007. Nutritional composition of marine plants in the diet of the green sea turtle (*Chelonia mydas*) in the Hawaiian Islands. Bull. Mar. Sci. 81: 55–71.

Mendes, T.C., C.E.L. Ferreira and K.D. Clements. 2018. Discordance between diet analysis and dietary macronutrient content in four nominally herbivorous fishes from the Southwestern Atlantic. Mar. Biol. 165: 180.

Metillo, E.B., R. Villanueva, K.I. Hayashizaki, S. Tamada, M. Sano and S. Nishida. 2019. Stable C and N isotope analysis elucidated the importance of zooplankton in a tropical seagrass bed of Santiago Island, Northwestern Philippines. Chem. Ecol. 35: 143–163.

Miyake, S., D.K. Ngugi and U. Stingl. 2015. Diet strongly influences the gut microbiota of surgeonfishes. Mol. Ecol. 24: 656–672.

Moleana, T., L. Della Patrona, T. Meziane and Y. Letourneur. 2016. First record of *Siganus randalli* (Teleost, Siganidae) in New Caledonia, and comments on its diet. Mar. Biodiv. Rec. 9: 81.

Monroig, Ó., S. Wang, L. Zhang, C. You, D.R. Tocher and Y. Li. 2012. Elongation of long-chain fatty acids in rabbitfish *Siganus canaliculatus*: Cloning, functional characterisation and tissue distribution of Elovl5-and Elovl4-like elongases. Aquaculture 350: 63–70.

Monroig, Ó., A.C. Shu-Chien, N. Kabeya, D.R. Tocher and L.F.C. Castro. 2022. Desaturases and elongases involved in long-chain polyunsaturated fatty acid biosynthesis in aquatic animals: From genes to functions. Prog. Lipid Res. 101157.

Mountfort, D.O., J. Campbell and K.D. Clements. 2002. Hindgut fermentation in three species of marine herbivorous fish. Appl. Envir. Microbiol. 68: 1374–1380.

Müller, M., C.F. Staab, L.D. Puk, E.M. Schoenig, S.C. Ferse and C. Wild. 2021. The rabbitfish *Siganus virgatus* as key macroalgae browser in coral reefs of the Gulf of Thailand. Diversity 13: 123.

Nagle, D.G. and V.J. Paul. 1998. Chemical defense of a marine cyanobacterial bloom. J. Exp. Mar. Biol. Ecol. 225: 29–38.

Nakamura, Y., M. Horinouchi, T. Nakai and M. Sano. 2003. Food habits of fishes in a seagrass bed on a fringing coral reef at Iriomote Island, southern Japan. Ichthyol. Res. 50: 15–22.

Nanami, A. 2018. Spatial distributions, feeding ecologies, and behavioral interactions of four rabbitfish species (*Siganus unimaculatus*, *S. virgatus*, *S. corallinus*, and *S. puellus*). PeerJ 6: e6145.

Nicholson, G.M. and K.D. Clements. 2020. Resolving resource partitioning in parrotfishes (Scarini) using microhistology of feeding substrata. Coral Reefs 39: 1313–1327.

Nicholson, G.M. and K.D. Clements. 2023. Micro-photoautotroph predation as a driver for trophic niche specialization in 12 syntopic Indo-Pacific parrotfish species. Biol. J. Linn. Soc. https://doi.org/10.1093/biolinnean/blad005.

Nielsen, J.M., E.L. Clare, B. Hayden, M.T. Brett and P. Kratina. 2018. Diet tracing in ecology: method comparison and selection. Methods Ecol. Evol. 9: 278–291.

Nielsen, S., J.W. Walburn, A. Verges, T. Thomas and S. Egan. 2017. Microbiome patterns across the gastrointestinal tract of the rabbitfish *Siganus fuscescens*. PeerJ 5: e3317.

Nimura, K., H. Takatsuji, S. Masuda and J. Shimamoto. 2007. Growth and maturation of *Ecklonia cava* and *Eisenia arborea* seedlings transplanted along the coast of Hainan, Shizuoka Prefecture and the grazing caused by herbivorous fish *Siganus fuscescens*. Aquacult. Sci. 55: 541–546. (in Japanese with English abstract).

Olsson, J., G.B. Toth and E. Albers. 2020. Biochemical composition of red, green and brown seaweeds on the Swedish west coast. J. Appl. Phycol. 32: 3305–3317.

Oren, A. and G.M. Garrity. 2021. Valid publication of the names of forty-two phyla of prokaryotes. Int. J. Syst. Evol. Microbiol. 71: 005056.

Osako, K., H. Saito, K. Kuwahara and A. Okamoto. 2006. Year-round high arachidonic acid levels in herbivorous rabbit fish *Siganus fuscescens* tissues. Lipids 41: 473–489.

Parazo, M.M. 1990. Effect of dietary protein and energy level on growth, protein utilization and carcass composition of rabbitfish, *Siganus guttatus*. Aquaculture 86: 41–49.

Pereira, H., L. Barreira, F. Figueiredo, L. Custódio, C. Vizetto-Duarte, C. Polo et al. 2012. Polyunsaturated fatty acids of marine macroalgae: potential for nutritional and pharmaceutical applications. Mar. Drugs 10: 1920–1935.

Perpetua, M.D., J.G. Gorospe, M.A. Torres and C.G. Demayo. 2013. Diet composition based on stomach content of the Streaked spinefoot (*Siganus javus*) from three coastal bays in Mindanao, Philippines. Adv. Envir. Sci. 5: 49–61.

Pillans, R.D., C.E. Franklin and I.R. Tibbetts. 2004. Food choice in *Siganus fuscescens*: influence of macrophyte nutrient content and availability. J. Fish Biol. 64: 297–309.

Pisaniello, A., L.D. Bojarski, K.M. Handley, W.L. White, E.R. Angert and K.D. Clements. 2022. Sources of variation in community composition of the hindgut microbiota in two tropical *Kyphosus* species. Coral Reefs 41: 1523–1535.

Plass-Johnson, J.G., S.C. Ferse, J. Jompa, C. Wild and M. Teichberg. 2015. Fish herbivory as key ecological function in a heavily degraded coral reef system. Limnol. Oceanogr. 60: 1382–1391.

Priest, M.A., A.R. Halford, K.D. Clements, E. Douglas, S.L. Abellana and J.L. McIlwain. 2016. What a difference a bay makes: natural variation in dietary resources mediates growth in a recently settled herbivorous fish. Coral Reefs 35: 1187–1199.

Raubenheimer, D., W.L. Zemke-White, R.J. Phillips and K.D. Clements. 2005. Algal macronutrients and food selection by the omnivorous marine fish *Girella tricuspidata*. Ecology 86: 2601–2610.

Sabapathy, U. and L.H. Teo. 1993. A quantitative study of some digestive enzymes in the rabbitfish, *Siganus canaliculatus* and the sea bass, *Lates calcarifer*. J. Fish Biol. 42: 595–602.

Sabapathy, U. and L.H. Teo. 1994. Some kinetic properties of amylase from the intestine of the rabbitfish, *Siganus canaliculatus* (Park). Comp. Biochem. Physiol. B 109: 139–144.

Saito, H., R. Yamashiro, C. Alasalvar and T. Konno. 1999. Influence of diet on fatty acids of three subtropical fish, subfamily Caesioninae (*Caesio diagramma* and *C. tile*) and family Siganidae (*Siganus canaliculatus*). Lipids 34: 1073–1082.

Sala, E., Z. Kizilkaya, D. Yildirim and E. Ballesteros. 2011. Alien marine fishes deplete algal biomass in the eastern Mediterranean. PloS One 6: e17356.

Sayed, A.E.D.H., U.M. Mahmoud and F. Essa. 2019. The microstructure of buccal cavity and alimentary canal of *Siganus rivulatus*: Scanning electron microscope study. Microsc. Res. Tech. 82: 443–451.

Schupp, P.J. and V.J. Paul. 1994. Calcium carbonate and secondary metabolites in tropical seaweeds: variable effects on herbivorous fishes. Ecology 75: 1172–1185.

Seah, J.C., A.G. Bauman and P.A. Todd. 2021. Temporal variation in macroalgal removal: insights from an impacted equatorial coral reef system. Mar. Biol. 168: 2.

Semmler, R.F., S.J. Brandl, S.A. Keith and D.R. Bellwood. 2021. Fine-scale foraging behavior reveals differences in the functional roles of herbivorous reef fishes. Ecol. Evol. 11: 4898–4908.

Shakman, E., C. Boedeker, M. Bariche and R. Kinzelbach. 2009. Food and feeding habits of the Lessepsian migrants *Siganus luridus* Ruppell, 1828 and *Siganus rivulatus* Forsskal, 1775 (Teleostei: Siganidae) in the southern Mediterranean (Libyan coast). J. Biol. Res.-Thessalon. 12: 115–124.

Simora, R.M.C., R.F.M. Traifalgar and F.S. Legario. 2015. Characterization of extracellular enzymes from culturable autochthonous gut bacteria in rabbitfish (*Siganus guttatus*). ELBA Bioflux 7: 67–76.

Siqueira, A.C., D.R. Bellwood and P.F. Cowman. 2019. The evolution of traits and functions in herbivorous coral reef fishes through space and time. Proc. Roy. Soc. B 286: 20182672.

Starks, E.C. 1907. On the relationship of the fishes of the family Siganidae. Biol. Bull. 13: 211–218.

Steinberg, P.D., K. Edyvane, R. De Nys, R. Birdsey and I.A. Van Altena. 1991. Lack of avoidance of phenolic-rich brown algae by tropical herbivorous fishes. Mar. Biol. 109: 335–343.

Stergiou, K.I. 1988. Feeding habits of the Lessepsian migrant *Siganus luridus* in the eastern Mediterranean, its new environment. J. Fish Biol. 33: 531–543.

Streit, R.P., A.S. Hoey and D.R. Bellwood. 2015. Feeding characteristics reveal functional distinctions among browsing herbivorous fishes on coral reefs. Coral Reefs 34: 1037–1047.

Su, E., Y. Wu, P. Chen, H. Yu, S. Liu, H. Luo et al. 2023. Dietary selenium regulates the diversity and stability of microbial communities in stomach and intestine of rabbitfish (*Siganus oramin*). Aquaculture 563: 738979.

Tebbett, S.B., C.H.R. Goatley and D.R. Bellwood. 2017. Clarifying functional roles: algal removal by the surgeonfishes *Ctenochaetus striatus* and *Acanthurus nigrofuscus*. Coral Reefs 36: 803–813.

Tebbett, S.B., S. Bennett and D.R. Bellwood. 2023. A functional perspective on the meaning of the term 'herbivore': patterns versus processes in coral reef fishes. Coral Reefs https://doi.org/10.1007/s00338-023-02378-4.

Thacker, R.W., D.G. Nagle and V.J. Paul. 1997. Effects of repeated exposures to marine cyanobacterial secondary metabolites on feeding by juvenile rabbitfish and parrotfish. Mar. Ecol. Prog. Ser. 147: 21–29.

Thépot, V., A.H. Campbell, N.A. Paul and M.A. Rimmer. 2021. Seaweed dietary supplements enhance the innate immune response of the mottled rabbitfish, *Siganus fuscescens*. Fish Shellfish Immunol. 113: 176–184.

Thépot, V., J. Slinger, M.A. Rimmer, N.A. Paul and A.H. Campbell. 2022. Is the intestinal bacterial community in the Australian rabbitfish *Siganus fuscescens* influenced by seaweed supplementation or geography?. Microorganisms 10: 497.

Tietbohl, M. 2022. The ecology of herbivorous fishes in the Red Sea. Unpublished PhD thesis, King Abdullah University of Science and Technology Thuwal, Kingdom of Saudi Arabia, https://doi.org/10.25781/KAUST-74527.

Tocher, D.R. 2009. Issues surrounding fish as a source of omega-3 long-chain polyunsaturated fatty acids. Lipid Technol. 21: 13–16.

Tocher, D.R. 2010. Fatty acid requirements in ontogeny of marine and freshwater fish. Aquacult. Res. 41: 717–732.

Tyler, J.C. and A.F. Bannikov. 1997. Relationships of the fossil and Recent genera of rabbitfishes (Acanthuroidei: Siganidae). Smithson. Contrib. Paleobiol. No. 14., Smithsonian Institution Press, Washington DC.

Vergés, A., S. Bennett and D.R. Bellwood. 2012. Diversity among macroalgae-consuming fishes on coral reefs: a transcontinental comparison. PLoS ONE 7: e45543.

Vergés, A., F. Tomas, E. Cebrian, E. Ballesteros, Z. Kizilkaya, P. Dendrinos et al. 2014. Tropical rabbitfish and the deforestation of a warming temperate sea. J. Ecol. 102: 1518–1527.

Vergés, A., C. Doropoulos, H.A. Malcolm, M. Skye, M. Garcia-Pizá, E.M. Marzinelli et al. 2016. Long-term empirical evidence of ocean warming leading to tropicalization of fish communities, increased herbivory, and loss of kelp. Proc. Nat. Acad. Sci. USA 113: 13791–13796.

von Westernhagen, H. 1973. The natural food of the rabbitfish *Siganus oramin* and *S. striolata*. Mar. Biol. 22: 367–370.

von Westernhagen, H. 1974. Food preferences in cultured rabbitfishes (Siganidae). Aquaculture 3: 109–117.

Woodland, D.J. 1990. Revision of the fish family Siganidae with descriptions of two new species and comments on distribution and biology. Indo-Pac. Fishes 19: 1–136.

Woodland, D.J. and J.E. Randall. 1979. *Siganus puelloides*, a new species of rabbitfish from the Indian Ocean. Copeia 1979: 390–393.

Woodland, D.J. and R.C. Anderson. 2014. Description of a new species of rabbitfish (Perciformes: Siganidae) from southern India, Sri Lanka and the Maldives. Zootaxa 3811: 129–136.

Wu, P., T. Wang, Y. Liu, C. Li, Y. Xiao, S. Xu et al. 2022. Differences of macroalgal consumption by eight herbivorous coral reef fishes from the Xisha Islands, China. Front. Mar. Sci. 9: 882196.

Wulff, J.L. 2021. Targeted predator defenses of sponges shape community organization and tropical marine ecosystem function. Ecol. Monogr. 91: e01438.

Xie, D., S. Xu, Q. Wu, F. Chen, S. Wang, C. You and Y. Li. 2018. Changes of visceral properties and digestive enzymes in the herbivorous marine teleost *Siganus canaliculatus* fed on different diets. Acta Oceanol. Sin. 37: 85–93.

Xie, D., J. Ye, M. Lu, S. Wang, C. You and Y. Li. 2020. Comparison of activities of fatty acyl desaturases and elongases among six teleosts with different feeding and ecological habits. Front. Mar. Sci. 7: 117.

Xu, Y., J. Li, X. Han, Z. Zhang, M. Zhong and Z. Hu. 2021. *Enteromorpha prolifera* diet drives intestinal microbiome composition in *Siganus oramin*. Curr. Microbiol. 78: 229–237.

Yamaguchi, A., K. Furumitsu, N. Yagishita and G. Kume. 2010. Biology of herbivorous fish in the coastal areas of Western Japan. pp. 181–190. *In*: Ishimatsu, A. and H.-J. Lie [eds.]. Coastal environmental and ecosystem issues of the East China Sea. TERRAPUB and Nagasaki University, Japan.

Yamano, S., S. Akita, Y. Hayakawa and D. Fujita. 2020. Finding of regeneration in a large number of wounded receptacles in *Sargassum ringgoldianum* (Phaeophyceae, Fucales). Phycol. Res. 68: 183–186.

Yatsuya, K., S. Kiyomoto and T. Yoshimura. 2015. Seasonal changes in dietary composition of the herbivorous fish *Kyphosus bigibbus* in southwestern Japan. Fish. Sci. 81: 1025–1033.

You, C., F. Zeng, S. Wang and Y. Li. 2014. Preference of the herbivorous marine teleost *Siganus canaliculatus* for different macroalgae. J. Ocean Univ. China 13: 516–522.

Zarco-Perello, S., T. Wernberg, T.J. Langlois and M.A. Vanderklift. 2017. Tropicalization strengthens consumer pressure on habitat-forming seaweeds. Sci. Rep. 7: 820.

Zemke-White, W.L. and K.D. Clements. 1999. Chlorophyte and rhodophyte starches as factors in diet choice by marine herbivorous fish. J. Exp. Mar. Biol. Ecol. 240: 137–149.

Zhang, Z., X. Han, Y. Xu, J. Lin, Y. Li and Z. Hu. 2016. Biodegradation of *Enteromorpha* polysaccharides by intestinal micro-community from *Siganus oramin*. J. Ocean Univ. China 15: 1034–1038.

Zolkaply, S.Z., T.D. Do, M. Asaduzzaman, Y.G. Seah, D. Hurwood, P. Mather et al. 2021. Evolutionary history and taxonomic reappraisal of coral reef rabbitfishes (Siganidae): Patterns of lineage diversification and speciation. Biology 10: 1109.

Zubia, M., C. Vieira, K.A. Palinska, M. Roué, J.C. Gaertner, I. Zloch et al. 2019. Benthic cyanobacteria on coral reefs of Moorea Island (French Polynesia): diversity response to habitat quality. Hydrobiologia 843: 61–78.

5

Rabbitfish Life Histories

Unique Features that Support Resilient Fisheries

Brett M. Taylor,[1,*] *Leilani Sablan,*[2] *Mark A. Priest*[3] and *J. Howard Choat*[4]

Introduction

A life-history trait is any feature that affects the reproductive potential of individuals within populations (Cole 1954). As such, these traits are the cornerstone to population dynamics because they dictate the way a population responds to its surrounding environment, and to potential disturbances, by influencing the probability of survival and successful reproduction at different life stages. Life histories across the animal kingdom are incredibly diverse. This diversity reflects the breadth of evolutionary outcomes to the seemingly endless array of selective pressures acting on a multitude of genotypes across the earth's biosphere (Hutchings 2021). Individual traits—such as the life span, growth rate, size and age at maturation, fecundity, offspring size, and mortality rate—vary widely across taxa, but are often limited by physiological and environmental constraints (Partridge and Silby 1991). For example, the metabolic cost of cellular maintenance constrains the reproductive output in fishes (Charnov et al. 2001). Operating within the bounds of these constraints, species ultimately evolve specific life-history strategies to optimize their fitness within their environment.

[1] Sea Grant and Marine Laboratory, University of Guam, Mangilao, Guam.
[2] Sea Grant, University of Guam, Mangilao, Guam.
Email: sablanl9009@triton.uog.edu
[3] Institute for Marine and Antarctic Studies, University of Tasmania, Hobart, Tasmania, Australia; Email: markandrewpriest@hotmail.com
[4] College of Science and Engineering, James Cook University, Townsville, Queensland, Australia; Email: john.choat@jcu.edu.au
* Corresponding author: brettmtaylor@gmail.com

Commonly, life-history variation among species or taxonomic groups has been classified on a fast-slow continuum (Stearns 1983). Here, species will occupy a position along a spectrum of natural mortality and rates of population increase. Many correlated traits, such as body size and life span, are associated with this spectrum. However, it is important to define, and be cognizant of, the taxonomic scale at which species are compared. Tropical coastal teleost fishes show an incredible diversity of life histories, even within a narrow range of temperature. Body sizes of species existing in the same habitats range from tens of millimeters to over two meters, while maximum life spans vary from less than ten weeks to over 80 years (Depczynski and Bellwood 2005, Taylor et al. 2021).

Life-history strategies generally dictate how vulnerable a species is to overexploitation, as they determine the intrinsic rate of population recovery—a key feature that defines the speed at which a population can recover at low densities, and therefore withstand high exploitation pressure. However, life histories can also define the nature in which a species or population is harvested. For example, a species may follow predictable patterns of reproductive behavior and subsequent recruitment that lend themselves to being harvested in large quantities. In the coastal tropical regions of the Indo-Pacific, this is true for many species that form large aggregations for the sole purpose of spawning. Many tropical snappers (Lutjanidae) and groupers (Epinephelidae) form spawning aggregations, but particular concomitant aspects of their evolved life-histories—such as slow growth, late maturity, and long life span—make many of these species susceptible to rapid population depletion, and subsequent collapse, when overharvested. However, imagine a fish population that has predictable timing of reproduction and recruitment, is easy to harvest in large numbers across multiple life phases, and yet has a rapid enough rate of population turnover to facilitate resilience to fishery exploitation. Some small, seasonal coastal pelagic species such as many clupeids (e.g., sardines and herrings), scombrids (e.g., mackerel), or scads come to mind, but there are very few demersal coral reef-associated fishes that fit this description. Rabbitfishes (Siganidae), however, represent one of the most dynamic fishery resources in coral reef fisheries. They are common components to otherwise highly diverse fisheries (some regions target up to 300 different fish and invertebrate species commercially or for subsistence), but they are rarely a dominant feature. However, for many targeted rabbitfish species, harvest patterns are intricately linked to phenological patterns and processes in their life-history strategies.

Life Histories

General Biology

Rabbitfishes are a unique and widespread component of Indo-Pacific coral reef assemblages and are now increasingly common in some temperate regions through the expansion of their ranges. Rabbitfishes are typically considered to be diurnal herbivores, but both their activity patterns (Fox and Bellwood 2011) and their diets (Hoey et al. 2013) show surprising diversity, with the latter extending beyond herbivory. They are unique with respect to other groups of piscine herbivores in that the greatest species diversity and numerical abundance occurs in shallow inshore reef habitats and have limited representation on exposed reef crest habitats (Hoey et al. 2013). Rabbitfishes are broadly characterized by rapid turnover rates (compared to other commercially-harvested reef fishes) facilitated through short life spans (less than 10 years), rapid sexual maturation (less than two years), high fecundity (generating lots of larvae), high natural mortality, and fast growth early in the life span (Woodland 1990). Many species spawn in aggregations at predictable times of the year, synchronized to lunar patterns (Takemura et al. 2004), and some species exhibit episodes of mass recruitment of newly settled juveniles (Kami and Ikehara 1976). A unique feature of rabbitfishes among most commercially-harvested coral reef species is that they lay adhesive demersal eggs. Of coral reef fishes known to form spawning aggregations, only three families (Balistidae, Pomacentridae, and Siganidae) have demersal eggs (Choat 2012). These eggs hatch into pelagic larvae, thereby facilitating the potential for wide dispersal to other regions. Of the extant species, only one (*Siganus argenteus*) spawns directly pelagic eggs. Rabbitfishes are gonochores, i.e., they are born either male or female and do not undergo any form of sex reversal throughout their ontogeny, either before or after sexual maturation. They broadly fall under two social modes, pairing and schooling. Schooling species will form schools of various sizes depending on the species and habitat, whereas pairing species are often found in individual pairs, exhibiting coordinated vigilance rather than monogamous mating whereby individuals will watch for predators while their partner feeds. Social modes are largely reflected in species' evolutionary histories and morphologies.

Age-based Biology

The long history of harvesting rabbitfishes and the more recent history of rabbitfish aquaculture has made some aspects of their biology obvious, even prior to formal scientific research (Lam 1974). For instance, fishers in the Pacific Islands and elsewhere are highly attuned to the annual and lunar patterns across which species reproduce and subsequently recruit (e.g., de Freycinet 1828, Johannes 1981). For instance, Johannes (1981) documented fisher accounts of detailed spawning rhythms and behaviours of *Siganus canaliculatus*, *S. lineatus*, *S. argenteus*, and *S. punctatus* and how these rhythms and behaviors facilitate harvests. In Guam, accounts of the harvest of newly recruited *S. spinus* and *S. argenteus* stretch back 200 years and demonstrate an intimate knowledge of spawning and recruitment cycles amongst the indigenous Chamorros (de Freycinet 1828). The Fijian traditional calendar goes back centuries and recognizes the rabbitfish annual runs by naming two months after them: *Vula Nuqa Lailai* (small rabbitfish run month) and the subsequent *Vula Nuqa Levu* (large rabbitfish run month). Decades of aquaculture trials have demonstrated that, compared with other commercially-harvested coastal fish species in the tropics, most rabbitfish species are fast growing, short-lived, and highly fecund. However, our quantitative understanding of the age-based biology of rabbitfishes is extremely nascent.

An initial reason for the dearth of age-based research on rabbitfishes is that scientific research on the population biology of tropical coral reef species in general has lagged far behind that of species in temperate, high-value fisheries (Choat et al. 2009, Taylor et al. 2018). This time lag in effort and understanding stems from both logistical (historical shortage of funding and manpower for scientific research in the developing tropical nations) and theoretical constraints (the belief that tropical fish otoliths do not deposit annual structures; Fowler 2009) (Taylor et al. 2018). Beyond these initial challenges, deriving the ages of individual rabbitfishes is generally more difficult and labor intensive compared to other tropical fish species. There are two reasons for this. First, rabbitfish otoliths are small and fragile often with a long rostrum that can easily break, thus snapping the otolith in half. Moreover, transverse sections of rabbitfish otoliths, especially at the lowest (most tropical) latitudes are not easy to interpret. Figure 1 provides some clear examples of annual increment patterns in rabbitfish otolith sections (see also Grandcourt et al. [2007] from the highly

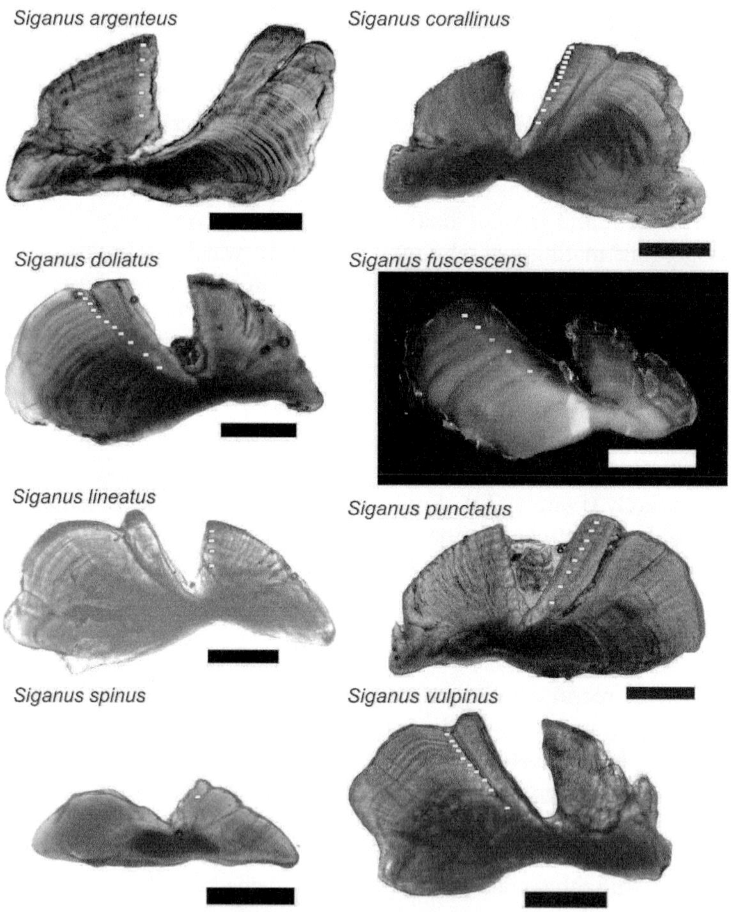

Fig. 1. Examples of transverse sections of sagittal rabbitfish otoliths displaying annual increment patterns (denoted by white marks). Specimens are viewed using a dissecting microscope with transmitted light, except for *S. fuscescens* (reflected light). Geographic locations: *S. argenteus* (Saipan, Mariana Islands); *Siganus corallinus*, *S. doliatus*, *S. lineatus*, *S. vulpinus* (Lizard Island, Great Barrier Reef), *Siganus fuscescens* (Perth, Western Australia), *Siganus punctatus* (Pohnpei, Caroline Islands), and *Siganus spinus* (Guam, Mariana Islands). Scale bars are 500 μm.

climatically seasonal Arabian Gulf). Nearly all these examples are at least four years old and are used to articulate the general microstructure and deposition patterns of annual bands in rabbitfish otoliths. However, this leads us to the second reason why deriving ages in rabbitfish can be difficult. Rabbitfishes four years or older are rare in the natural environment, even more so in the schooling species of general interest to fisheries. Because of high natural mortality rates, most rabbitfishes in scientific collections are very young (e.g., Grandcourt 2002, Bariche 2005, Grandcourt et al. 2007, Taylor et al. 2017, Rhodes et al. 2017). Beyond this fact, fast growth in the early life history makes the deposition of the first annulus somewhat amorphous, and challenging to identify in younger fish. We have found the most effective technique is to invest heavily in the interpretation of daily microband structures and coordinate the timing of deposition with otolith growth (e.g., Taylor et al. 2017). In fact, some of the earliest studies of age-based demography of rabbitfishes used daily microbands to quantify the full life-history of a very short-lived species, *S. sutor*, important to fisheries in the western Indian Ocean (Ntiba and Jaccarini 1988, Grandcourt 2002). Ultimately, overcoming technical challenges like these are nothing new to researchers, but the process of deriving ages for tropical fishes in general has lagged far behind compared with temperate research. Thus, to date there are only a handful of studies that have used reliable methods to empirically derive population level age-based demographic information for rabbitfishes.

As noted above, rabbitfishes are broadly characterized as either pairing or schooling species. The distribution of 'schooling' and 'pairing' species has a strong phylogenetic underpinning (Borsa et al. 2007) and is reflected in species' morphological and ecological traits (Brandl and Bellwood 2013, Brandl and Tebbetts—Chapter 6). Within coral reef fish species, social behaviors can mediate their ecological functions through changes in diet and nutritional ecology (Guerra et al. 2023). Across species, such behavioral adaptations may also have a significant effect on species' life histories, because social group behaviors have a strong influence on many life functions through cooperative advantage (Kasumyan and Pavlov 2018). Species that form schools employ a strength-in-numbers predator avoidance strategy, which has been demonstrated to reduce mortality risk at the individual level. This strategy is often employed by high-mortality, rapid-turnover species that traverse comparatively large home ranges to exploit resources opportunistically, with high numbers also conferring the ability to access defended resources (Foster 1985). Conversely, pairing species maintain small, stable territories that they defend, thereby investing proactively into their resource pool through coordinated vigilance with their long-term partners (Fox and Donelson 2014, Brandl and Bellwood 2015). Pairing in coral reef fishes has been broadly identified as having a stronger basis in food procurement and predation risk than in reproduction (Brandl and Bellwood 2014), and for rabbitfishes, socially monogomous pairing species are known to form spawning aggregations, suggesting that monogomous pairs are not formed for mating purposes (Fox et al. 2015). Interestingly, pairing behavior in rabbitfishes is not associated with specific food resources: pairing species are known to feed on filamentous and turfing red algae (e.g., *S. doliatus*), cyanobacteria (e.g., *S. vulpinus*), or to be largely carnivorous (*S. puellas*) (Hoey et al. 2013). Ultimately, the behavioral tactics employed by pair-forming species to enhance resource acquisition and survival, and the observed longevity of these pairs, demonstrates that individuals risk losing a lot if their partner dies.

We hypothesize that pairing rabbitfish species will exhibit greater longevity (i.e., individual life spans observed at the population level), on average, compared with schooling species. A single individual of the pairing species *Siganus uspi* was reported to survive in aquarium conditions for at least 26 years (Delbeek and Randall 2015), demonstrating that the capacity for high longevity exists among the pairing species. However, the majority of rabbitfish species considered important food fishes are schooling species and, for this reason, almost all the published work on rabbitfish age-based biology to date is focused on schooling species. Hence, this hypothesis has not been tested as the requisite data has not been generated. The reciprocal altruism hypothesis posits that animal longevity will increase local dominance by favoring cooperation among individuals (although this hypothesis is more focused on cooperative breeding; Ridley et al. 2005). Further, Réale et al. (2010) highlighted the importance of integrating behavior into the pace-of-life syndrome concept, which describes a spectrum of fast life-history strategies (prioritization of reproduction over survival) to slow life-history strategies (prioritization of survival over reproduction; Stearns 1983, Montiglio et al. 2018). However, it is difficult to transcribe the 'proactive' and 'reactive' behavioral categories of Koolhaus et al. (1999) to the differences in rabbitfish pairing versus schooling, but it could be argued that pairing species prioritize survival over reproduction compared with their schooling congeners, thus favoring a slower-paced life-history strategy.

To illustrate potential differences in life histories between the pairing and schooling species of rabbitfishes, we assembled published information on life spans of wild populations for all rabbitfish species available based on studies using sectioned otolith analysis. These included schooling species *S. sutor* (Seychelles and Kenya), *S. argenteus* (Saipan, Micronesia), *S. canaliculatus* (Arabian Gulf), *S. rivulatus* (Eastern Mediterranean), *S. fuscescens* (Western Australia), and *S. luridus* (Eastern Mediterranean), and for the pairing species *S. punctatus* (Pohnpei, Micronesia) (Table 1). We then incorporated unpublished collections of mostly pairing rabbitfish species (*S. corallinus*, *S. doliatus*, *S. vulpinus*, *S. puellus*, *S. punctatissimus*, and *S. lineatus* [schooling]) from Lizard Island, Great Barrier Reef (Table 1). As crude as this analysis is (e.g., no correction for life span variability across latitudinal changes in ocean temperature), it demonstrates that pairing species (longevity range: 8–13 yrs) exhibit population-level longevity exceeding that of schooling species (range: 1–7 yrs). However, pairing versus schooling is not a binary trait, and species' proclivity to form pairs or schools exists on a spectrum (Brandl and Bellwood 2013). When we account for this, the percentage of individuals identified in pairs across species is strongly correlated with the species' life spans (Fig. 2). This rudimentary analysis demonstrates two key points. First, there is a coupled phylogenetic-behavioral signature that is reflected in the general life-history strategies of rabbitfishes. Species existing at the extremes of the pairing-schooling spectrum are certainly employing behaviors that facilitate different strategies to avoid predation while procuring resources and therefore manifest different pace-of-life, which is observable through

Table 1. Estimates of maximum age for rabbitfish species derived from transverse sections of sagittal otoliths.

Species	Location	Maximum age	Sociality	References
Siganus sutor	Seychelles	1	Schooling	Grandcourt 2002
Siganus sutor	Kenya	2	Schooling	Ntiba and Jaccarini 1988
Siganus canaliculatus	Arabian Gulf	7	Schooling	Grandcourt et al. 2007
Siganus argenteus	Saipan	7	Schooling	Taylor et al. 2017
Siganus punctatus	Pohnpei	8	Pairing	Rhodes et al. 2017
Siganus rivulatus	E. Mediterranean	6	Schooling	Bariche 2005
Siganus luridus	E. Mediterranean	6	Schooling	Bariche 2005
Siganus fuscescens	Western Australia	6	Schooling	Zarco-Perello et al. 2022
Siganus corallinus	Lizard Island, GBR	13	Pairing	This study
Siganus doliatus	Lizard Island, GBR	13	Pairing	This study
Siganus vulpinus	Lizard Island, GBR	12	Pairing	This study
Siganus lineatus	Lizard Island, GBR	7	Schooling	This study
Siganus puellus	Lizard Island, GBR	11	Pairing	This study
Siganus punctatissimus	Lizard Island, GBR	10	Pairing	This study

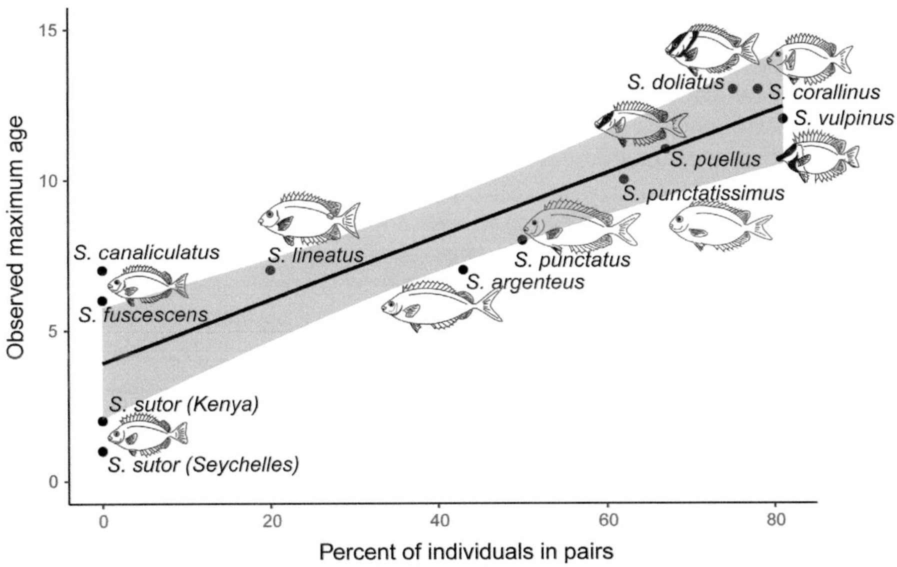

Fig. 2. Relationship between maximum age and the percent of individuals observed in pairs. Maximum age was derived from published estimates based on sectioned sagittal otoliths (Table 1). Percentage of individuals in pairs from observations by Brandl and Bellwood (2013) (percent pairing information was not available for all species in Table 1). R^2 for the linear relationship is 0.81.

empirical estimations of natural mortality and life span. Secondly, these differences highlight how little we know about the diversity of *Siganus* life-history traits, given that coarse and rapid collections of specimens of unstudied species from one location can provide the maximum ages observed anywhere to date. Overall, of 29 species, we have age-based life-history information for about a third of them.

Phenotypic Plasticity

An important feature of many marine ectotherms (i.e., cold-blooded animals) is their ability to adapt to different environmental conditions, such as changes in temperature or food availability. We can better understand the inherent capacity for a species to adapt to different conditions by observing the plasticity of their phenotypic traits (Petersen and Warner 2002). Successful invasive and range expanding species often demonstrate considerable flexibility in traits such as their life span, growth rate, body size, and reproductive timing, thereby tailoring trait values to optimize fitness in prevailing and novel environments (Davidson et al. 2011). This pattern is commonly seen in tropical marine fishes along latitudinal gradients (e.g., Robertson et al. 2005, Munch and Salinas 2009, Taylor et al. 2019, Moore 2022). Some species of rabbitfish have been successful Lessepsian migrants, with *S. rivulatus* and *S. luridus* invading and establishing themselves in the eastern Mediterranean Sea from the Red Sea through the Suez Canal. Another species complex, *S. fuscescens/canaliculatus* has expanded its range into temperate latitudes across its distribution. In both cases, the expansions have facilitated new fisheries for local human populations. These expansions often require the species to adapt to new resources and new temperature regimes. This is achieved by modifying the size at maturity, size-specific fecundity, rates of growth, body size, and life span (Zarco-Perello et al. 2022). Populations in higher latitudes also modify their phenological patterns by shortening the length of the annual reproductive period. Such plasticity in life-history traits allows these species to adapt to a range of conditions over a wide biogeographic scope.

Phenotypic plasticity can also manifest at small spatial scales (e.g., Gust et al. 2002, Paddack et al. 2009, Benkwitt et al. 2021). The early stages of life history can have a large influence on subsequent adult demography and survivability (McCormick and Hoey 2004). Post-settlement growth of newly recruited *S. spinus* from different reef flats in Guam has been shown to reflect the nutritional yield of algal assemblages across the various sites (Priest et al. 2016). This led to vastly different growth rates and, presumably, early-stage survival, and demonstrates that food quality and quantity is a limiting resource for the species, which recruit to the reef in extremely large sporadic settlement pulses (Priest et al. 2016). Hence, at this scale, these differences represent the magnitude of phenotypic plasticity but likely do not reflect an adaptive response. Nevertheless, in this scenario, the variability of resource distribution across reef flats heavily influences the growth and survivability of recruits; after mass recruitment of juveniles, these post-settlement processes dictate the demographic composition of adult populations across different reef flats island-wide.

Fisheries

Rabbitfishes are common components of commercial, recreational, and subsistence catches throughout their range (Lam 1974). Most species that are important to regional fisheries are those that are abundant and form schools, whereas many pairing species though still commonly harvested, are comparatively less common in the fishery. The reasons for this disparity are clear: schooling species are generally more productive and, through their behavior, provide fishers with an opportunity to harvest large numbers of individuals in discrete fishing events. Further, there is a tendency for pairing species to inhabit more coral rich habitats on outer reef slopes whereas schooling species will commonly inhabit nearshore habitats such as mangroves, reef flats, and sea grass beds, making them easier targets for harvest. Across their range, adults are the primary targets of fisheries although in some locations newly settled juveniles are harvested *en masse* (Soliman and Yamaoka 2010) and dried, pickled, fried, or mashed into paste. Adults are generally marketed fresh. Rabbitfishes are harvested in a variety of coastal habitats using many methods including surround nets, gill nets, spears, cast nets, traps, or bottom trawls, depending on the location and the ecology of targeted species.

Overall, rabbitfishes are a common but generally only a modest contributor to total coastal fisheries catches throughout their range. An exception to this is areas of the Western Indian Ocean (primarily Mozambique, Tanzania, Kenya, and the Seychelles), where rabbitfish harvests comprise up to 50% of the total catch (Nzioka 1984). One species endemic to this region dominates as the primary fishery target: *Siganus sutor* (Ntiba and Jaccarini 1988, McClanahan and Mangi 2004, Hornell 1927, Kamukuru 2009). Trap fisheries are commonly used to target seasonally predictable spawning aggregations of *S. sutor* resulting in very high fishery yields (Maina et al. 2013, Robinson et al. 2011). Elsewhere in the Indo-Pacific region, rabbitfishes are common contributors to coastal fisheries catches. In the Red Sea, the important species to fisheries are *S. rivulatus*, *S. luridus*, *S. stellatus*, and *S. argenteus* (Mehanna et al. 2018), of which the first two have made Lessepsian migrations through the Suez Canal to invade the Mediterranean Sea, where they are now common fishery targets (Bariche 2005). From the Arabian Gulf to Indonesia and Western Australia, through the Coral Triangle to subtropical Japan, the most dominant rabbitfish within fisheries harvests is the complex of *S. fuscescens* and *S. canaliculatus* (Grandcourt et al. 2007, Al-Marzouqi et al. 2011, Unsworth et al. 2014, Katayama et al. 2009, Dalzell et al. 1996), which have been identified as synonyms with interbreeding color morphs (Hsu et al. 2011). However, the dominance of a single species or species complex seems to diminish towards the Pacific Islands, where rabbitfish fisheries generally comprise less than seven percent of coastal fisheries (Dalzell et al. 1996). Here, rabbitfish target species are more diverse, likely reflecting the ecology of species, the geomorphology of habitats, and perhaps sociocultural differences among island communities. For instance, *S. spinus* and *S. argenteus* are heavily targeted in the Mariana Islands of Micronesia, whereby a unique fishery targets both adult and newly recruited individuals (Kami and Ikehara 1976). However, in nearby

Yap, while *S. spinus* has the highest contribution among rabbitfishes, people in some regions without modern sanitation infrastructure will avoid this species as they are commonly seen feeding on human feces in intercoastal waterways (B. Taylor, personal observation). In Pohnpei, rabbitfishes have been increasing in the total reef-associated catch (Rhodes et al. 2018) and while *S. spinus* is highly uncommon in catches, the top three species (*S. doliatus*, *S. punctatus*, and *S. puellas*) are pairing species (Rhodes et al. 2008). In the South Pacific, the importance of rabbitfishes to fisheries diminishes from west to east (Dalzell et al. 1996), likely reflecting species diversity and abundance approaching the edge of their biogeographic extent.

Here, we provide case studies from island nations of the Western Pacific Ocean (Guam) and the Western Indian Ocean highlighting the divergent nature of rabbitfish fisheries. In Guam, rabbitfishes represent a moderate but important portion of the overall fishery harvest, but the fishery also has long-standing cultural and economic importance to many island residents. In the Western Indian Ocean, a single species has a large impact in coastal fisheries across the region. However, both locations provide excellent examples of dynamic fisheries that are focused on the biological and phenological patterns of fast-growing, high-turnover species, both of which show signs of being sustainable.

Guam

Guam has an active coral reef-associated fishery which spans commercial, recreational, and subsistence use. Hundreds of species of reef fishes are targeted using a variety of methods. However, the harvest of rabbitfishes on Guam represents arguably the most dynamic part of the fishery whereby multiple life phases are actively targeted using different fishing methods. In addition to the regular harvest of adults of principally four species (*S. argenteus*, *S. punctatus*, *S. randalli*, and *S. spinus*), the fishery is internationally known for the periodic harvest of newly recruited juveniles of *S. argenteus* and *S. spinus en masse* immediately after they metamorphose from their pelagic larval phase and settle onto the reef (Fig. 3A). The nature and economic importance of this fishery is reflected in how rabbitfishes have multiple names in the native Chamorro language to represent different size classes (Kerr 1990): *mañahak* (recently recruited juveniles [< 5 cm], a principal fishery target at particular times of year), *dagge'* (5–10 cm), *seyun* (for *S. spinus* 10–20 cm), and *hiteng* (for *S. argenteus* and *S. punctatus* > 20 cm). Written accounts of *mañahak* harvests on Guam by native Chamorros stretch back over 200 years (de Freycinet 1828, Amesbury and Hunter-Anderson 2008), but the harvest of rabbitfishes on Guam has very likely been sustained for several thousand years. The harvest of *dagge'* is traditionally unpracticed by most Chamorro people (in fact the Chamorro name itself is not widely known), hinting at how indigenous knowledge may have left unharvested the soon-to-be reproductive individuals that would repopulate fishing grounds. Meanwhile, fishing pressure on juvenile (*mañahak*) and adult (*seyun*, *hiteng*) life phases can be intense, and annual yields of juveniles can be highly variable—sometimes biblical in magnitude—yet this fishery persists because of a unique combination of resilient life-history traits and strong seasonal patterns in biological processes. This fishery as a whole spans the commercial, subsistence, and recreational sectors on Guam, as

Fig. 3. (A) A mixed school of newly recruited *mañahak* (scribbled rabbitfish, *Siganus spinus* and forktail rabbitfish, *Siganus argenteus*) in a backreef in Guam (Photo: B. Tibbatts). These schools are seasonally targeted by fishers in Guam. (B) A dissected *Siganus spinus* specimen with large ovaries (Photo: M. Priest).

the motivation for fisher groups is intrinsically diverse. Fishers will often sell part of their catch, give part of their catch to extended family or friends, and keep a portion to feed their immediate families. Regardless of the intent, many fishers also consider the practice to be an activity that facilitates the persistence of cultural practices and an important ritual observed by family units.

The patterns and persistence of this fishery are both inherently linked to life-history strategies and phenological patterns (i.e., periodic or seasonal events in environmental conditions and biological life cycles). The two species (*S. spinus* and *S. argenteus*) are fast-growing and generally short-lived species with fusiform bodies that commonly comprise small to large roving schools on the reef. Both are considered 'croppers' of red and green macroalgae (Hoey et al. 2013) and therefore play a considerable role in consuming macroalgae from reef substrates. Both species form aggregations to spawn twice a year: the largest spawning event occurs earlier in the year between March and May, with a smaller spawning episode from August to September (Kami and Ikehara 1976, Taylor et al. 2017). Spawning is followed by recruitment of juveniles after an approximately one-month duration as larval fishes in the pelagic realm (Chirichetti 1996). This linkage is important as detailed genetics and oceanographic modelling work on *S. spinus* spanning the Western Pacific region suggests that the Mariana Islands are genetically and oceanographically isolated and therefore demographically reliant on self-seeding for population persistence (Priest et al. 2012, Kendall et al. 2019, Wolanski et al. 2021). The spawning and recruitment cycles are synchronized with lunar periods. *Siganus argenteus* from nearby Saipan spawns on or just after full moon (Taylor et al. 2017); the recruitment of the juvenile siganids is documented to occur a few days before or after last quarter moon (Kami and Ikehara 1976).

A common feature among these *S. spinus* and *S. argenteus* is that they both have life histories that yield rapid turnover times. A detailed biological study of wild populations has only been published from Saipan for *S. argenteus* (Taylor et al. 2017), but there have been many observations to gain insights on Guam populations for both species. Both species are short-lived, with maximum life spans up to eight years for *S. argenteus* and probably no more than 4–6 years for *S. spinus*. Despite these potential life spans, both species experience high natural mortality which means that a vast majority of the population at any time is very young. For instance, in the fishery for adult *S. argenteus* on Saipan (not counting any juvenile harvest), 85% of the harvested individuals were under three years old. Further, *S. argenteus* reaches maturity at just 1.3 years. We have observed *S. spinus* individuals on Guam reaching maturity just six months after recruitment, meaning that some individuals can be ready to reproduce in the next mass spawning event directly following recruitment to the reef. Finally, both species are highly fecund and can release a large amount of eggs or sperm during spawning (Fig. 3B). We have observed *S. argenteus* as having a maximum gonadosomatic index (GSI) during spawning times of up to 10%, meaning that the weight of an individual's ovaries represented 10% of the gonad-free bodyweight. For *S. spinus*, we have observed individual GSI values of up to 30%! *Siganus spinus* lays demersal-adhesive eggs that remain on the reef for several days and only disperse later after hatching (Lam 1974, Woodland 1990). Interestingly, *S. argenteus* is the only siganid species that spawns fully pelagic eggs from the outset (Popper et al. 1979). It is hypothesized that demersal-adhesive eggs may increase initial survivability and decrease dispersal potential (Taylor et al. 2017), although the outcomes of this subtly different life-history strategy for these species is uncertain.

A final characteristic of interest in this fishery is that catches of *mañahak* recruits vary substantially from year to year (Kami and Ikehara 1976, Hensley and Sherwood 1993). Given that Guam's fishing community is uniquely in tune with these biological patterns (both the annual and lunar timing of spawning, and offshore pelagic pre-settlement recruits first turn up in the stomachs of harvested tunas), this interannual variability strongly reflects the number of recruits reaching the reef. Recent oceanographic modelling has demonstrated that mesoscale turbulence and ENSO-driven changes in surface water circulation drives the recruitment of rabbitfishes in Guam (Wolanski et al. 2021). As the Mariana Islands are largely self-seeding (Priest et al. 2012, Kendall et al. 2019), it is possible that interannual variability in individual fecundity also contributes to different recruitment magnitudes, given rabbitfishes have a high capacity for phenotypic plasticity. On nearby Saipan, *S. argenteus* individuals were found to be twice as fecund (i.e., by measuring their average GSI values during the spawning season) in one year versus the previous year (Taylor et al. 2017). These rabbitfishes have strong preference for filamentous algal species, some of which can be highly seasonal in their abundance (Tsuda and Bryan

1973, Bryan 1975). Further, experimental aquaculture research has demonstrated that higher quality diets can induce earlier maturation, higher fecundity, and higher spawning frequency (e.g., Soletchnik 1984). Hence, it seems reasonable to presume that per capita fecundity can vary as a function of the quality and abundance of nutritional resources in advance of spawning seasons.

On Guam, most juvenile harvest likely stems from individual fishers using throw nets (*talåya* in local Chamorro language). Adult rabbitfishes are harvested using spears (day and night), nets, and even hook and line baited with filamentous algae, but the most efficient means of harvest is performed by teams of individuals using surround nets or drag nets (*chenchulu*). Adult rabbitfishes will often be more heavily targeted during March-April spawning times as they coincide with lent, when Guam's predominantly Catholic population will refrain from consuming most meats other than fish. Using *chenchulu*, whereby a surround net will be deployed by two individuals in a large semicircle on a shallow reef flat, all other fishers will slap the water and drive fish towards the net opening as the two net guiders will complete the circle, enclosing the fish within. The resulting higher density of fishes within the netted area is then harvested by the collective group of fishers using hand spears.

As noted by Kami and Ikehara (1976) nearly 50 years ago, the unmitigated harvest of juvenile recruits may appear an "unwise conservation practice", yet this fishery persists to this day at consistent levels, albeit with large interannual variability. Despite evidence of regional genetic isolation through larval retention, this persistence is largely attributable to the fast-paced biology of the two targeted species. During exceptionally large recruitment events, extremely high natural mortality is commonly observed through starvation once the schools of recruits have exhausted algal resources on reef flats, indicating that the effects of fishing pressure are weaker than existing density-dependent patterns. During small recruitment events, diminishing returns to fishers reduce the demand for harvest. In both cases, many surviving rabbitfishes are fecund and ready to join the reproductive pool within one to three spawning cycles. As the biological rhythm of these species is highly dependent on seasonal algal resources, a burgeoning question is how climate change might affect these patterns in the future.

Western Indian Ocean

While rabbitfishes represent an important component of subsistence and commercial fisheries on coral reefs throughout the Indo-Pacific region, they generally comprise only a modest portion of what are otherwise highly diverse fisheries. In the Western Indian Ocean (WIO), however, rabbitfishes are among the most important fishery resources. This dominance stems primarily from a single species, the shoemaker spinefoot rabbitfish *Siganus sutor*. This species is endemic to the WIO region (Woodland 1990), is commonly found in moderate to large schools, and spawns in large aggregations at predictable times and locations. This single species has constituted up to 45% of the coastal fishery biomass in Kenya (Hicks and McClanahan 2012), and the species dominates the coastal trap fisheries (up to 60–80%) in the Seychelles and Tanzania (Kamukuru 2009, Robinson et al. 2017). This level of contribution by a single species in tropical fisheries is rare from any family.

Rabbitfishes in the WIO region are harvested using a variety of fishing gears, including traps, beach seines, gill nets, hook and line, and spears. The proportional contribution of gear types varies within and between countries, but the predominant method of harvest across the region is by baited basket traps. Some of the most detailed fishery statistics at the species level comes from Kenya (McClanahan and Mangi 2004). There, *S. sutor* is the most abundant species in the gill net and trap fisheries, makes substantial contributions to the beach seine (~ 12%) and spear fisheries (~ 9%), and is infrequently caught by hook-and-line (~ 1%; McClanahan and Mangi 2004). As an archipelagic island system, the Seychelles fishery dynamics differ from those of continental coastal systems in the WIO region. However, rabbitfishes (principally *S. sutor*) are still a major component of the artisanal fishery of the Seychelles, where they are predominantly captured by basket traps deployed by boats and very infrequently taken by line or gillnets (Seychelles Fishing Authority 2016). Overall, available data demonstrate that fish trap techniques dominate the harvest of *S. sutor* in this region. However, there is another interesting feature common to these fisheries: the periodic and focused harvest of temporally predictable spawning aggregations.

The harvest of *S. sutor* within spawning aggregations is confirmed from the Seychelles, Kenya, and Tanzania (Robinson et al. 2008), and likely occurs elsewhere in the WIO. This species is reported to form the largest (i.e., most abundant) spawning aggregations of any reef fish species in the region (Robinson et al. 2008). Most research on the dynamics of rabbitfish aggregation fisheries stems from the granitic islands of the Seychelles, where aggregations of *S. sutor* have been harvested for at least a century (Hornell 1927). The Seychelles has among the highest per capita seafood consumption rates in the world and artisanal fisheries comprise a large portion of the domestic protein intake (FAO 2022), hence this one species is of prime importance to the region. In the region, peak aggregation times for *S. sutor* are identified as October to December, with spawning activity focused around full moons, but aggregations tends to continue through to May (Robinson et al. 2004). Higher catch rates relative to effort expenditure attract fishers to spawning aggregations (Robinson et al. 2017); however, the targeting of spawning aggregations in the Seychelles accounts for only 15% of the annual catch of *S. sutor* (Robinson et al. 2011). Further, a time series analysis of the fishery found it to be within sustainable limits (Robinson et al. 2011). Despite the importance of *S. sutor* to artisanal fisheries in the Seychelles, modest fishing effort targeting spawning aggregations (only 8% of annual effort resulting in 15% of annual catch) is likely one contributing factor to its sustainability. Another factor relates to individual reproductive dynamics. Bijoux et al. (2013) found high turnover rates of individual fish within aggregations, a behavior that likely enhances individual survivability during spawning periods by limiting individual exposure to targeted fishing effort. A final factor influencing the sustainability of aggregation harvests is the life history of *S. sutor*. While this species is fully recruited to the trap fishery before reaching sexual maturity (Grandcourt and Cesar 2003), it has a short life span of approximately two years (Grandcourt 2002, Ntiba and Jaccarini 1988) and thus has a rapid turnover time fueled by high natural mortality.

The case study of rabbitfish aggregation fishing from the Seychelles demonstrates that aggregation fishing can be sustainable when fishing effort is modest and biological traits are resilient. However, even the most resilient species have their limits. Seychelles experienced a multi-year decline in *S. sutor* trap catches following the 1998 coral bleaching that resulted in mass mortality of corals, suggesting that substantial changes to reef habitat had a deleterious effect on siganid populations (Grandcourt and Cesar 2003, Robinson et al. 2011). Other WIO countries have experienced long-term declines in siganid populations, which are perceived to be influenced by excessive aggregation fishing. Across the coast of Kenya, Kaunda-Arara et al. (2003) reported a ~ 40% decline in siganid populations during 1982–2000 and Maina et al. (2013) reported declines in the highest catches of *S. sutor* from 1989 to 2009. In the southern region of Kenya, *S. sutor* spawning aggregations are targeted using ring nets and are reported to produce landings up to one ton in a single event (Robinson et al. 2008).

The targeting of *S. sutor* spawning aggregations in the WIO represents another example of how rabbitfish fisheries can be intrinsically driven by, or at least linked to, phenological biological patterns. There has been much research on the harvest of spawning aggregations of marine species and the vast majority of this has elicited a precautionary tale of overharvesting through hyperstability (Sadovy and Domeier 2005, Erisman et al. 2011). Evidence from the WIO region potentially shows both the danger of aggregation fishing, but also how this activity might be sustainable when effort is modest and the target species has a resilient biology. We emphasize, however, that the "resilient biology" component of this example is highly unique. Among commercially-important coral reef fish species that form spawning aggregations, rabbitfishes have exceptional life-history traits that facilitate rapid turnover times and confer greater resilience to overexploitation. Nevertheless, a precautionary approach to management is warranted across their range as reported declines in harvested rabbitfish populations appear to be increasing.

Conclusion

Despite their modest diversity and their restriction to more inshore habitats, rabbitfishes manifest a surprising complexity of social behaviors, life histories, nutritional ecology, diet, and temporal foraging patterns, perhaps reflective of the fact that siganids are an ancient group well represented on Eocene reefs (Tyler and Bannikov 1997). Rabbitfishes represent an important component of reef fish faunas across the Indo-Pacific region. Their contribution to coastal fisheries is biogeographically diverse but appears to be increasing in many regions. This increase is likely a reflection of the overharvest of more intrinsically vulnerable species from other reef fish families. Because of their fast-paced life history, rabbitfishes are generally resilient to overexploitation compared to other harvested families. Further, in many tropical coastal fish families, the most desired species are often the most biologically vulnerable because of human preference for large-bodied fishes and their economic return in commercial markets. For rabbitfishes, the opposite pattern appears to emerge, as fast-growing and short-lived schooling species dominate fisheries catches due to the predictable nature of their harvestability. However, there is truly a dearth of studies that document the age-based biology of wild populations of rabbitfishes anywhere across their range; thus, the biological underpinning of this chapter was based on limited empirical information. An enhanced understanding of the diversity inherent across the age-based biology of rabbitfishes is warranted for both fisheries management and ecological research. Trends from several regions demonstrate that while rabbitfish fisheries can be highly resilient, they are not inexhaustible. Hence, research and proactive management is paramount, especially in areas where rabbitfishes represent important sources of protein and micronutrients for coastal populations.

Acknowledgements

We thank Barnaby Acfalle, Felix Reyes, Judy Amesbury, and Brent Tibbatts for insightful discussions regarding traditional harvest of *mañahak* in Guam; Will Robbins and John Ackerman contributed to rabbitfish otolith collections of pairing species from Lizard Island, Great Barrier Reef; Simon Brandl and Mike Gawel for general discussions; and anonymous reviewers for comments on previous drafts.

References

Al-Marzouqi, A., N. Jayabalan, A. Al-Nahdi and I. Al-Anbory. 2011. Reproductive biology of the white-spotted rabbitfish, *Siganus canaliculatus* (Park, 1797) in the Arabian Sea coast of Oman. Western Indian Ocean Journal of Marine Science 10: 73–82.

Amesbury, J.R. and R.L. Hunter-Anderson. 2008. An analysis of archaeological and historical data on fisheries for pelagic species in Guam and the Northern Mariana Islands. Report to Pelagic Fisheries Research Program, Joint Institute for Marine and Atmospheric Research, University of Hawai'i at Manoa. 156 p.

Bariche, M. 2005. Age and growth of Lessepsian rabbitfish from the eastern Mediterranean. J. Appl. Ichthyol. 21: 141–145.

Benkwitt, C.E., B.M. Taylor, M.G. Meekan and N.A.J. Graham. 2021. Natural nutrient subsidies alter demographic rates in a functionally important coral-reef fish. Sci. Rep. 11: 12575.

Bijoux, J.P., L. Dagorn, G. Berke, P.D. Cowley, M. Soria, J.-C. Gaertner et al. 2013. Temporal dynamics, residency and site fidelity of spawning aggregations of a herbivorous tropical reef fish *Siganus sutor*. Mar. Ecol. Prog. Ser. 475: 233–247.

Borsa, P., S.L. Lemer and D. Aurelle. 2007. Patterns of lineage diversification in rabbitfishes. Mol. Phylogenet. Evol. 44: 427–435.

Brandl, S.J. and D.R. Bellwood. 2013. Morphology, sociality, and ecology: can morphology predict pairing behavior in coral reef fishes? Coral Reefs 32: 835–846.

Brandl, S.J. and D.R. Bellwood. 2014. Pair-formation in coral reef fishes: an ecological perspective. Oceanogr. Mar. Biol. 52: 1–80.

Brandl, S.J. and D.R. Bellwood. 2015. Coordinated vigilance provides evidence for direct reciprocity in coral reef fishes. Sci. Rep. 5: 14556.

Bryan, P.G. 1975. Food habits, functional digestive morphology, and assimilation efficiency of the rabbitfish *Siganus spinus* (Pisces, Siganidae) on Guam. Pac. Sci. 29: 269–277.

Charnov, E.L., T.F. Turner and K.O. Winemiller. 2001. Reproductive constraints and the evolution of life histories with indeterminate growth. P. Natl. Acad. Sci. USA 98: 9460–9464.

Chirichetti, P.R. 1996. Otolith Increment Analysis in the Rabbitfish *Siganus spinus*: Validation of Daily Increment Deposition and Patterns of Larval and Otolith Growth. Masters Thesis, University of Guam. 26 pp.

Choat, J.H., J.P. Kritzer and J.L. Ackerman. 2009. Ageing in coral reef fishes: do we need to validate the periodicity of increment formation for every species of fish for which we collect age-based demographic data. pp. 23Y54. *In:* Green, B.S., B.D. Mapstone, G. Carlos and G.A. Begg [eds.]. Tropical Fish Otoliths: Information for Assessment, Management and Ecology. Springer, Netherlands.

Choat, J.H. 2012. Spawning aggregations in reef fishes; ecological and evolutionary processes. pp. 85–116. *In:* Sadovy de Mitcheson, Y. and P.L. Colin [eds.]. Reef Fish Spawning Aggregations: Biology, Research and Management. Springer, Heidelberg.

Cole, L.C. 1954. The population consequences of life history phenomena. Q. Rev. Biol. 29: 103–137.

Dalzell, P., T.J.H. Adams and N.V.C. Polunin. 1996. Coastal fisheries in the Pacific Islands. Oceanogr. Mar. Biol. 34: 395–531.

Davidson, A.M., M. Jennions and A.B. Nicotra. 2011. Do invasive species show higher phenotypic plasticity than native species and, if so, is it adaptive? A meta-analysis. Ecol. Lett. 14: 419–431.

de Freycinet, M.L. 1828. Voyage autour du monde, enterpris par ordre du Roi, sous le ministère et conformément aux instructions de S. Exc. M. le Vicomte du Bouchage, Secrétaire d'État u Département de la Marine, exécuté sur les corvettes de S.M. l'Oranie er la Physicienne, pendant les années 1817, 1818, 1819, et 1829. Paris, France: Pillet Aîné.

Delbeek, J.C. and J.E. Randall. 2015. Aquarium Fish Longevity. Reefs Magazine. https://reefs.com/magazine/aquarium-fish-longevity/.

Depczynski, M. and D.R. Bellwood. 2005. Shortest recorded vertebrate lifespan found in a coral reef fish. Curr. Biol. 15: R288–R289.

Erisman, B.E., Allen, L.G., Claisse, J.T., Pondella, D.J., Miller, E.F. and Murray, J.H. 2011. The illusion of plenty: hyperstability masks collapses in two recreational fisheries that target fish spawning aggregations. Can. J. Fish. Aquat. Sci. 68: 1705–1716.

FAO. 2022. Fishery and Aquaculture Country Profiles. Seychelles, 2019. Country Profile Fact Sheets. Fisheries and Aquaculture Division [online]. Rome. https://www.fao.org/fishery/en/facp/syc?lang=en

Foster, S.A. 1985. Group foraging by a coral reef fish: a mechanism for gaining access to defended resources. Anim. Behav. 33: 782–792.

Fowler, A.J. 2009. Age in years from otoliths of adult tropical fish. pp. 55–92. *In:* Green, B.S., B.D. Mapstone, G. Carlos and G.A. Begg [eds.]. Tropical Fish Otoliths: Information for Assessment, Management and Ecology. Springer, Netherlands.

Fox, R.J. and D.R. Bellwood. 2011. Unconstrained by the clock? Plasticity of diel activity rhythm in a tropical reef fish *Siganus lineatus*. Funct. Ecol. 25: 1096–1105.

Fox, R.J. and J.M. Donelson. 2014. Rabbitfish sentinels: first report of coordinated vigilance in conspecific marine fishes. Coral Reefs 33: 253.

Fox, R.J., D.R. Bellwood and M.D. Jennions. 2015. Why pair? Evidence of aggregative mating in a socially monogamous marine fish (*Siganus doliatus*, Siganidae). Roy. Soc. Open Sci. 2: 150252.

Grandcourt, E.M. 2002. Demographic characteristics of a selection of exploited reef fish from the Seychelles: preliminary study. Mar. Freshwater. Res. 53: 123–130.

Grandcourt, E.M. and H.S.J. Cesar. 2003. The bio-economic impact of mass coral mortality on the coastal reef fisheries of the Seychelles. Fish. Res. 60: 539–550.

Grandcourt, E., T. Al Abdessalaam, F. Francis and A. Al Shamsi. 2007. Population biology and assessment of the white-spotted spinefoot, *Siganus canaliculatus* (Park, 1797), in the southern Arabian Gulf. J. Appl. Ichthyol. 23: 53–59.

Guerra, A.S., J.C. Van Wert, A.J. Haupt, D.J. McCauley, E.J. Eliason, H.S. Young et al. 2023. Differences in the behavior and diet between shoaling and solitary surgeonfish (*Acanthurus triostegus*). Ecol. Evol. 13: e9686.

Gust, N., J.H. Choat and J.L. Ackerman. 2002. Demographic plasticity in tropical reef fishes. Mar. Biol. 140: 1039–1051.

Hensley, R.A. and T.S. Sherwood. 1993. An overview of Guam's inshore fisheries. Mar. Fish. Rev. 55: 129–138.

Hicks, C.C. and T.R. McClanahan. 2012. Assessing gear modifications needed to optimize yields in a heavily exploited, multi-species, seagrass and coral reef fishery. PLoS One 7: e36022.

Hoey, A.S., S.J. Brandl and D.R. Bellwood. 2013. Diet and cross-shelf distribution of rabbitfishes (f. Siganidae) on the northern Great Barrier Reef: implications for ecosystem function. Coral Reefs 32: 973–984.

Hornell, J. 1927. Report on the fisheries and fish resources of the Seychelles Islands. H.M.S.O., London.

Hsu, T.-H., Y.T. Adiputra, C.P. Burridge and J.-C. Gwo. 2011. Two spinefoot colour morphs: mottled spinefoot *Siganus fuscescens* and white-spotted spinefoot *Siganus canaliculatus* are synonyms. J. Fish Biol. 79: 1350–1355.

Hutchings, J.A. 2021. A Primer of Life Histories: Ecology, Evolution, and Application. Oxford University Press, Oxford.

Johannes, R.E. 1981. Words of the Lagoon: Fishing and marine lore in the Palau district of Micronesia. Los Angeles, CA: University of California Press.

Kami, H.T. and I.I. Ikehara. 1976. Notes on the annual juvenile siganid harvest in Guam. Micronesica 12: 323–325.

Kamukuru, A.T. 2009. Trap fishery and reproductive biology of the whitespotted rabbitfish *Siganus sutor* (Siganidae), within the Dar es Salaam Marine Reserves, Tanzania. Western Indian Ocean Journal of Marine Science 8: 75–86.

Kasumyan, A.O. and D.S. Pavlov. 2018. Evolution of schooling behavior in fish. J. Ichthyology 58: 670–678.

Katayama, S., S. Akiyama, M. Naganuma and R. Shibata. 2009. Age and growth of rabbitfish *Siganus fuscescens* in Tateyama Bay, Japan. Aquaculture Science 57: 417–422.

Kaunda-Arara, B, G.A. Rose, M.S. Muchiri and R. Kaka. 2003. Long-term trends in coral reef fish yields and exploitation rates of commercial species from coastal Kenya. Western Indian Ocean Journal of Marine Science 2: 105–116.

Kendall, M.S., M. Poti and A. Winship. 2019. Is Guam a regional source, destination, or stepping-stone for larvae of three fisheries species? Fish. Oceanogr. 28: 159–170.

Kerr, A.M. 1990. Chamorro Fish Names. Micronesica 23: 93–118.

Koolhaus, J.M., S.M. Korte, S.F. De Boer, B.J. Van Der Vegt, C.G. Van Reenen, H. Hopster et al. 1999. Coping styles in animals: current status in behavior and stress-physiology. Neurosci. Biobehav. R. 23: 925–935.

Lam, T.J. 1974. Siganids: Their biology and mariculture potential. Aquaculture 3: 325–354.

Maina, G.W., M. Samoilys, H. Alidina and K. Osuka. 2013. Targeted fishing of the shoemaker spinefoot rabbitfish, *Siganus sutor*, on potential spawning aggregations in southern Kenya. pp. 13–26. *In*: Robinson, J. and M. Samoilys [eds.]. Reef Fish Spawning Aggregations in the Western Indian Ocean: Research for Management. WIOMSA/SIDA/SFA/CORDIO. WIOMSA Book Series.

McClanahan, T.R. and S.C. Mangi. 2004. Gear-based management of a tropical artisanal fishery based on species selectivity and capture size. Fisheries Manag. Ecol. 11: 51–60.

McCormick, M.I. and A.S. Hoey. 2004. Larval growth history determines juvenile growth and survival in a tropical marine fish. Oikos 106: 225–242.

Mehanna, S.F., A.S. Mohammad, S.M. El-Mahdy and Y.A.A. Osman. 2018. Stock assessment and management of the rabbitfish *Siganus rivulatus* from the Southern Red Sea, Egypt. Egyptian Journal of Aquatic Biology & Fisheries 22: 323–329.

Montiglio, P.-O., M. Dammhahn, G.D. Messier and D. Réale. 2018. The pace-of-life syndrome revisited: the role of ecological conditions and natural history on the slow-fast continuum. Behav. Ecol. Sociobiol. 72: 116.

Moore, B.R. 2022. Age-based life history of Pacific longnose parrotfish *Hipposcarus longiceps* from New Caledonia. J. Fish Biol. 100: 997–1008.

Munch, S.B. and S. Salinas. 2009. Latitudinal variation in lifespan within species is explained by the metabolic theory of ecology. P. Natl. Acad. Sci. USA 106: 13860–13864.

Ntiba, M.J. and V. Jaccarini. 1988. Age and growth parameters of *Siganus sutor* in Kenyan marine inshore water, derived from numbers of otolith microbands and fish lengths. J. Fish Biol. 33: 465–470.

Nzioka, R.M. 1984. The evaluation of marine fisheries in Kenya. pp. 27–32. *In*: Iversen, S.A. and S. Myklevoll [eds.]. Proc. Norad-Kenya Seminar to Review the Marine Fish Stocks and Fisheries in Kenya, Mombasa, Kenya, 13–15 March 1984. Bergen: Institute of Marine Research.

Paddack, M.J., S. Sponaugle and R.K. Cowen. 2009. Small-scale demographic variation in the stoplight parrotfish *Sparisoma viride*. Journal of Fish Biology 75: 2509–2526.

Partridge, L. and R. Silby. 1991. Constraints in the evolution of life histories. Philos. T. Roy. Soc. B 332: 3–13.

Petersen, C.W. and R.R. Warner. 2002. The ecological context of reproductive behavior. pp. 103–118. *In*: Sale, P.F. [ed.]. Coral Reef Fishes: Dynamics and diversity in a complex ecosystem. San Diego: Academic.

Popper, D., R. Pitt and Y. Zohar. 1979. Experiments on the propagation of Red Sea siganids and some notes on their reproduction in nature. Aquaculture 16: 177–181.

Priest, M.A., A.R. Halford and J.L. McIlwain. 2012. Evidence of stable genetic structure across a remote island archipelago through self-recruitment in a widely dispersed coral reef fish. Ecol. Evol. 2. 3195–3213.

Priest, M.A., A.R. Halford, K.D. Clements, E. Douglas, S.L. Abellana and J.L McIlwain. 2016. What a difference a bay makes: natural variation in dietary resources mediates growth in a recently settled herbivorous fish. Coral Reefs 35: 1187–1199.

Réale, D., D. Garant, M.M. Humphries, P. Bergeron, V. Careau and P.-O. Montiglio. 2010. Personality and the emergence of the pace-of-life syndrome concept at the population level. Philos. T. Roy. Soc. B 365: 4051–4063.

Ridley, J., D.W. Yu and W.J. Sutherland. 2005. Why long-lived species are more likely to be social: the role of local dominance. Behav. Ecol. 16: 358–363.

Rhodes, K.L., M.H. Tupper and C.B. Wilchilmel. 2008. Characterization and management of the commercial sector of the Pohnpei coral reef fishery, Micronesia. Coral Reefs 27: 443–454.

Rhodes, K., D. Hernandez-Ortiz, M. Ioanis, W. Washington, S. Maxin, K. Olpet et al. 2017. Goldspotted spinefoot *Siganus punctatus* (Siganidae) age-based reproductive life history and fisheries vulnerability. J. Fish Biol. 91: 1392–1406.

Rhodes, K.L., D.X. Hernandez-Ortiz, J. Cuetos-Bueno, M. Ioanis, W. Washington and R. Ladore. 2018. A 10-year comparison of the Pohnpei, Micronesia, commercial inshore fishery reveals an increasingly unsustainable fishery. Fish. Res. 204: 156–164.

Robertson, D.R., J.L. Ackerman, J.H. Choat, J.M. Posada and J. Pitt. 2005. Ocean surgeonfish *Acanthurus bahianus*. I. The geography of demography. Mar. Ecol. Prog. Ser. 295: 229–244.

Robinson, J., M. Isidore, M.A. Marguerite, M.C. Ohman and R.J. Payet. 2004. Spatial and temporal distribution of reef fish spawning aggregations in the Seychelles—an interview-based survey of artisanal fishers. Western Indian Ocean Journal of Marine Science 3: 63–69.

Robinson, J., M. Samoilys and P. Kimani. 2008. Reef fish spawning aggregations in the Western Indian Ocean: current knowledge and implications for management. pp. 263–276. *In*: Obura, D., J. Tamelander and O. Linden [eds.]. Coastal Oceans Research and Development in the Indian Ocean: Status Report 2008. CORDIO.

Robinson, J., M.A. Samoilys, E. Grandcourt, D. Julie, M. Cedras and C. Gerry. 2011. The importance of targeted spawning aggregation fishing to the management of Seychelles' trap fishery. Fish. Res. 112: 96–103.

Robinson, J., N.A.J. Graham, A. Grüss, C. Gerry and J. Bijoux. 2017. Fishery benefits from exploiting spawning aggregations not solely dependent on enhanced fish density. Afr. J. Mar. Sci. 39: 269–278.

Sadovy, Y. and M. Domeier. 2005. Are aggregation-fisheries sustainable? Reef fish fisheries as a case study. Coral Reefs 24: 254–262.

Seychelles Fishing Authority. 2016. Seychelles artisanal fisheries statistics for 2015. Seychelles Fishing Authority Technical Report. Government of Seychelles, Seychelles.

Soletchnik, P. 1984. Aspects of nutrition and reproduction in *Siganus guttatus* with emphasis on applications to aquaculture. Report to the Aquaculture Department, Southeast Asian Fisheries Development Center, Tigbauan, Philippines. 75 p.

Soliman, V.S. and Yamaoka, K. 2010. Assessment of the fishery of siganid juveniles caught by bagnet in Lagonoy Gulf, Southeastern Luzon, Philippines. J. Appl. Ichthyol. 26: 561–567.

Stearns, S.C. 1983. The influence of size and phylogeny on patterns of covariation among life-history traits in the mammals. Oikos 41: 173–187.

Takemura, A., Rahman, M.S., Nakamura, S., Park, Y.J. and Takano, K. 2004. Lunar cycles and reproductive activity in reef fishes with particular attention to rabbitfishes. Fish Fish. 5: 317–328.

Taylor, B.M., J. Gourley and M.S. Trianni. 2017. Age, growth, reproductive biology and spawning periodicity of the forktail rabbitfish (*Siganus argenteus*) from the Mariana Islands. Mar. Freshw. Res. 68: 1088–1097.

Taylor, B.M., E.D.L. Trip and J.H. Choat. 2018. Dynamic demography: investigations of life-history variation in the parrotfishes. pp. 69–98. *In:* Hoey, A.S. and R.M. Bonaldo [eds.]. Biology of Parrotfishes. Taylor and Francis, Boca Raton.

Taylor, B.M., J.H. Choat, E.E. DeMartini, A.S. Hoey, A. Marshell, M.A. Priest et al. 2019. Demographic plasticity facilitates ecological and economic resilience in a commercially important reef fish. J. Anim. Ecol. 88: 1888–1900.

Taylor, B.M., C.B. Wakefield, S.J. Newman, M. Chinkin and M.G. Meekan. 2021. Unprecedented longevity of unharvested shallow-water snappers in the Indian Ocean. Coral Reefs 40: 15–19.

Tsuda, R.T. and P.G. Bryan. 1975. Food preference of juvenile *Siganus rostratus* and *S. spinus* in Guam. Copeia 1973: 604–606.

Tyler, J.C. and A.F. Bannikov. 1997. Relationships of the fossil and recent genera of rabbitfishes (Acanthuroidei: Siganidae). Sm. C. Paleob. 84: 35.

Unsworth, R.K.F., S.L. Hinder, O.G. Bodger and L.C. Cullen-Unsworth. 2014. Food supply depends on seagrass meadows in the coral triangle. Environ. Res. Lett. 9: 094005.

Wolanski, E., R.H. Richmond and Y. Golbuu. 2021. Oceanographic chaos and its role in larval self-recruitment and connectivity among fish populations in Micronesia. Estuar. Coast. Shelf S. 259: 107461.

Woodland, D.J. 1990. Revision of the fish family Siganidae with descriptions of two new species and comments on distribution and biology. Indo-Pacific Fishes 19.

Zarco-Perello, S., D. Fairclough, C. Dowling, J. DiBattista, R. Austin, T. Wernberg et al. 2022. Maximization of fitness by phenological and phenotypic plasticity in range expanding rabbitfish (Siganidae). J. Anim. Ecol. 91: 1666–1678.

6

The Functional Role of Rabbitfishes in Coastal Marine Ecosystems

Fundamental and Realized Ecological Niches

Simon J. Brandl[1,*] and *Sterling B. Tebbett*[2]

Introduction

In the eternal struggle for existence, all species must grow, survive, and reproduce. To do so, each species holds a suite of species-specific traits that permit them to exploit a given set of resources. The resulting biological construct of traits, tolerances, and behaviors is usually termed the *ecological niche* (Hutchinson 1957). In recent years, ecologists have traditionally leaned on the concepts of the Grinnellian (environmental) and Eltonian (functional) niches to identify drivers of species' distribution patterns and the roles that species play within ecosystems. However, more recently, trait-based analyses have paved the way for quantitatively examining the functional structure of even highly diverse species assemblages (McGill et al. 2006), yielding important insights into the assembly and functioning of communities across biomes (Kraft et al. 2008, Stuart-Smith et al. 2013, Gagic et al. 2015). Although functional traits are inherently limited by their resolution and relevance to actual ecological processes (Bellwood et al. 2019, Brandl et al. 2019), they can offer useful means for high-level comparisons among species in diverse communities (Mouillot et al. 2013).

[1] Department of Marine Science, The University of Texas at Austin, Marine Science Institute, Port Aransas, TX, USA.

[2] Research Hub for Coral Reef Ecosystem Functions and College of Science and Engineering, James Cook University, Queensland, Australia; Institute for Marine and Antarctic Studies, University of Tasmania, Hobart, Tasmania, Australia.
Email: sterling.tebbett@utas.edu.au

* Corresponding author: simon.brandl@austin.utexas.edu

On coral reefs, the ecological niches of herbivorous fishes have been studied extensively due to their consumptive effects on algae, which can shape benthic interactions (Bellwood et al. 2004). In short, in the intense competition for benthic space, algae can outcompete, overgrow, or smother corals unless controlled by herbivorous fishes (Hughes et al. 2007, Burkepile and Hay 2010, Rasher and Hay 2010). If this equilibrium is disturbed by, for instance, overexploitation of herbivorous fishes, shifts from coral to algal domination may occur. As a consequence, various components of the ecological niches of herbivorous coral reef fishes have been studied with great vigor over the past decade, ranging from evaluations of their Grinnellian niches (i.e., a species' abilities to persist across geographic, environmental, or habitat gradients) (Russ 1984a, Fox and Bellwood 2007, Cheal et al. 2012, Heenan et al. 2016) to a broad range of work investigating fundamental and realized functional niches of a vast number of herbivorous fish species (Choat et al. 2002, Burkepile and Hay 2008, Price et al. 2011, Brandl and Bellwood 2013a, Rasher et al. 2013, Brandl et al. 2015, Bejarano et al. 2017, Clements et al. 2017). While the literature on herbivorous coral reef fishes is broadly dominated by parrotfishes (scarine Labridae) and surgeonfishes (Acanthuridae) due to their high abundances and global distributions (Cheal et al. 2012, Siqueira et al. 2019), rabbitfishes (Siganidae) appear to have carved out a unique niche on coral reefs and beyond, which merits deeper examination. Below, we discuss the ecological niches of rabbitfishes in coastal reef ecosystems, with specific focus on the Eltonian niche, i.e., the functional role of rabbitfishes. To do this we consider information from both the *fundamental* and *realized* niches, where the former describes the theoretical niche space a species may occupy considering its phenotypic traits, while the latter refers to the empirically observed niche occupied *in situ* when abiotic and biotic constraints are considered (McGill et al. 2006, Devictor et al. 2010).

Fundamental Functional Niches of Rabbitfishes

The deep split dividing rabbitfishes into two main clades (in addition to a small clade of *S. argenteus* (Quoy and Gaimard 1825) and *S. woodlandi* (Randall and Kulbicki 2005)) has been linked to an array of ecological and morphological traits that shape the role of rabbitfishes on reefs (Woodland 1990, Borsa et al. 2007, Lemer et al. 2007). Specifically, rabbitfishes are typically divided into a mostly reef-associated, brightly colored, deep-bodied clade that predominantly occurs in pairs or small groups, and a drab colored, fusiform, schooling clade (Woodland 1990). However, there are additional ecological and morphological differences that may delineate different roles on reefs, both within and among the three clades.

The Categorical Rabbitfish Trait-space

Discrete classification schemes across a series of ecologically-meaningful axes (such as body size, trophic group, or mobility) are often used to gauge species' fundamental Grinnellian or Eltonian niches (McGill et al. 2006, Hadj-Hammou et al. 2021), as well as community-scale characteristics of niche space occupation (D'agata et al. 2014, Brandl et al. 2016, Richardson et al. 2017). Although the majority of traits used in trait-based analyses suffer from taxon-specific idiosyncrasies, limited resolution, and, at best, indirectly inferable linkages to actual function (Hadj-Hammou et al. 2021, Brandl et al. 2022, Streit and Bellwood 2022), trait-based classifications do offer glimpses into the hypothetical niche volume that different species may occupy. Since rabbitfishes appear to diverge across a series of potentially highly correlated niche axes, the multidimensional nature of trait-based analyses offers a useful window into their roles on reefs. Indeed, our trait-based analyses revealed clear delineations of several functional groupings within and among the phylogenetic clades (Fig. 1). Specifically, the categorical trait space: (1) mirrors the deep phylogenetic split in rabbitfishes, (2) supports the unique positioning of *S. argenteus* and *S. woodlandi*, (3) shows complete functional overlap in most (but not all) sister species, (4) suggests greater functional diversification in the deep-bodied clade, and (5) lays bare several fine-scale ecological separations within the deep-bodied clade, with extensive functional overlap among small, pairing, reef-associated species (*S. doliatus* (Guérin-Méneville 1829–38), *S. virgatus* (Valenciennes 1835), *S. corallinus* (Valenciennes 1835), *S. trispilos* (Woodland and Allen 1977), and the cluster of species formerly in the genus *Lo*), and a diversity of functional roles in the rest of the species in the reef-based clade. Interpreting these patterns is aided by the examination of variability within each trait category.

Body Size

Body size is one of the most crucial organismal characteristics, as it correlates broadly with an almost innumerable range of other traits (including, for instance, mortality, metabolic rate, or mobility) (Healy et al. 2019, Brandl et al. 2022). For fishes, body size is a particularly important predictor of species' functional niches, as the lack of maceration ability introduces significant size-based constraints on available prey and potential predators (Gill 2003, Mihalitsis and Bellwood 2017). Rabbitfishes exhibit relatively limited variation in body size, ranging from approximately 20 cm maximum total length (TL) in *Siganus unimaculatus* (Evermann and Seale 1907) to just over 50 cm maximum TL in *S. javus* (Linnaeus 1766), the largest species in the genus. Since most rabbitfishes obtain food by biting off small fragments of benthic prey, size-based gape limitation is unlikely to shape their role as consumers beyond per capita effects of consumption (Kelly et al. 2017). However, the comparatively larger size of species such as *S. javus*, *S. lineatus* (Valenciennes 1835) and its sister species, and *S. fuscescens* (Houttuyn 1782) and *S. sutor* (Valenciennes 1835) may provide increased access to resources by mediating predation risk, thus allowing these fishes to traverse through open water that connects reefs or habitats (Fox and Bellwood 2011, Pickholtz et al. 2018). Notably, body size is the only trait category in which *S. argenteus* and *S. woodlandi* diverge (based on the extremely limited ecological knowledge about *S. woodlandi*), separating the otherwise functionally indistinguishable species pair in the ancestral clade.

Social System

Differences in social groupings clearly manifest in rabbitfishes, delineating the two main clades of mostly schooling (fusiform) and pairing (deep-bodied) species

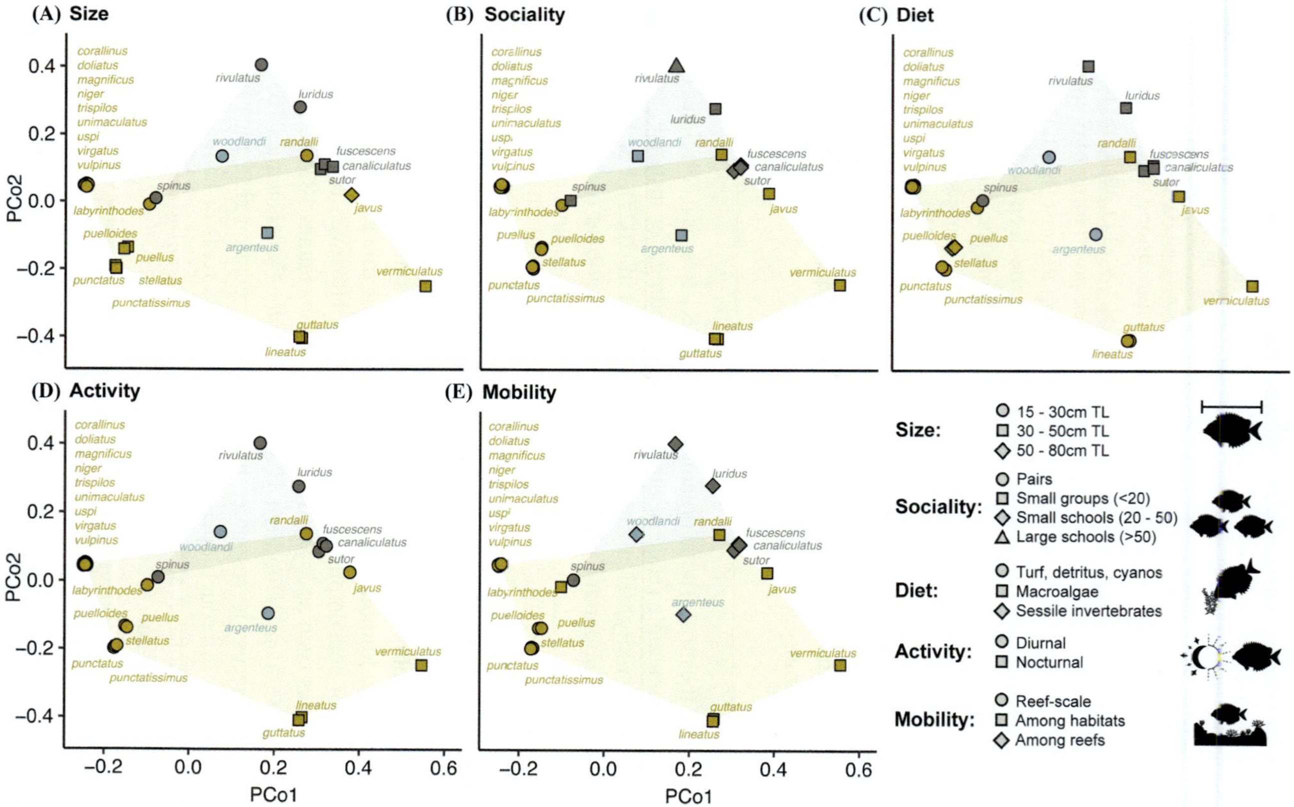

Fig. 1. Categorical trait space of rabbitfishes, displayed across five trait categories (**A–E**). The displayed data represent the two-dimensional output from a Principal Co-ordinates Analysis (PCoA) based on the pairwise Gower's dissimilarity of species across trait categories. Colors correspond to the three clades, while symbols indicate the respective trait values in each category. Species names are indicated for each datapoint and datapoints are jittered due to multiple overlying points.

(Woodland 1990, Borsa et al. 2007). However, there is substantial nuance within both pairing and schooling species. The species most likely to be found in pairs are *S. vulpinus* (Schlegel and Müller 1845) (81 percent of individuals in pairs at Lizard Island on the Great Barrier Reef (GBR)), *S. corallinus* (78 percent), and *S. doliatus* (75 percent) (Brandl and Bellwood 2013), suggesting that their respective sister species (which are all part of the same trait entity [i.e., clusters of points in trait space]) also pair strongly (although there is no empirical data). The pair bond in these species is clearly very strong, as highlighted by their fidelity, early onset of pair bonding, and occurrence of same-sex and multi-species pair bonds at small sizes (Brandl and Bellwood 2013b, Mirbach and Brandl 2016). The strength of pair formation is somewhat weaker in *S. puellus* (Schlegel 1852) (67 percent), *S. punctatissimus* (Fowler and Bean 1929) (62 percent), and *S. punctatus* (Schneider and Forster 1801) (50 percent), which form separate trait entities with their respective sister species, suggesting subtle differences among these primarily pair-forming species (Brandl and Bellwood 2013). These differences in pairing strength and frequency likely relate to the access and defense of resources in coral dominated habitats. Specifically, while territorial defense may promote pair bonding in some reef fishes (Roberts and Ormond 1992, Pratchett et al. 2006), a coordinated, cooperative vigilance system, which allows paired individuals to access resources in reef crevices while minimizing the predation risk associated with obscured field of vision (Fox and Bellwood 2013, Brandl and Bellwood 2014, 2015, Nanami 2018) (Fig. 2), may promote strong pairing in rabbitfish species. Notably, while *S. puellus*, *S. punctatus*, and *S. punctatissimus* also access prey in crevices (Brandl and Bellwood 2014), their slightly larger body size may weaken the importance of pair-bonding by reducing predation risk.

Beyond the strongly pairing species, there are several species in the clade (*S. lineatus*, *S. guttatus* (Bloch 1787), *S. vermiculatus* (Valenciennes 1835), *S. randalli* (Woodland 1990), as well as probably *S. labyrinthus* (Bleeker 1853) and *S. insomnis* (Woodland and Anderson 2014)) that rarely occur in pairs, but tend to form loose groups of a few individuals (Woodland 1990, Brandl and Bellwood 2013a, Woodland and Anderson 2014). These species do not commonly feed on coral reefs, but venture into adjacent habitats such as mangroves or sand flats to feed, often at night (Fox and Bellwood 2011, Woodland and Anderson 2014, Moleana et al. 2016). Thus, in these species, pair formation seems of limited ecological value, as the low complexity foraging grounds favor larger group sizes to dilute predation (Orpwood et al. 2008), and there is limited availability or incentive to exploit concealed microhabitats for prey. This is even more pronounced in the schooling clade, which mostly form small schools (20 to 50 individuals) (Woodland 1990). In these species, schooling is likely crucial to counteract the increased predation pressure associated with more fusiform bodies and occurrence in habitats with low complexity and shelter availability. Notably, while this reduces per capita vulnerability to predation, it increases population-level mortality, which is reflected in the lower longevity in schooling species compared to strongly pairing species (Taylor et al. – Chapter 5). Thus, while pairing provides access to concealed prey on coral reefs through a sophisticated behavioral adaptation, schooling expands access to resources across different habitats by diluting per capita mortality. The schooling species appear

Fig. 2. Coordinated vigilance in four species of pairing rabbitfishes, which permits the exploitation of concealed coral reef habitats (modified from Brandl and Bellwood (2015)). **(A)** Time series of body orientation (described as the angle relative to a vertical line perpendicular to the substratum, where 180 degrees is vertically facing up, and 0 degrees is vertically facing down, while 90 degrees is perfectly horizontal) for one pair of *S. corallinus* on Big Vicki's reef at Lizard Island, northern GBR. While swimming, angles of pair members are largely synchronized. In contrast, during foraging, individuals reciprocally forage (head down) or watch for predators (head up). Axis ticks mark 5 second intervals. **(B)** Body orientation angles of pair members (x- and y-axis denote the two individuals) during foraging, aggregated for 59 pairs across four species (*S. corallinus*, *S. doliatus*, *S. puellus*, and *S. vulpinus*) on Lizard Island. The solid and dashed lines mark the model fit from a Generalized Additive Mixed Model and its uncertainty estimates, revealing the coordination of body orientation between pair members. **(C–F)** Photographs of the coordinated vigilance behavior in *S. vulpinus*, *S. corallinus*, *S. puellus*, and *S. doliatus*.

unable to exploit cracks and crevices in the same manner as pairing species (Zarco-Perello et al. 2021), which may be due to their social structure, morphology, or both.

Trophic Group

As most rabbitfishes feed on sessile, benthic organisms, there is limited variability in the assignment of trophic groups. Broadly, most species in the pairing clade—with the exception of the sibling species *S. puellus* and *S. puelloides* (Woodland and Randall 1979), which feed largely on benthic invertebrates and sometimes macroalgae (Woodland and Randall 1979, Woodland 1997, Hoey et al. 2013, Chong-Seng et al. 2014, Topor et al. 2019)—are classified to feed on small, filamentous turf algae and associated detritus/cyanobacteria, while the schooling clade feeds predominantly on thallous macroalgae (Hoey et al. 2013). Nevertheless, in the schooling clade, *S. spinus* (Linnaeus 1758) is separated from other species by also feeding on finer, filamentous algae (Hoey et al. 2013), while *S. argenteus* and *S. woodlandi* also feed predominantly on filamentous turf algae (Randall and Kulbicki 2005), although the former shows very similar dietary preferences to *S. sutor* (Ebrahim et al. 2020), while the latter has been observed to feed on zooplankton (Randall and Kulbicki 2005).

Indeed, the broad categorization of trophic groups belies substantial variability within the pairing and small group clade (which is discussed in detail in the realized niche section below). For example, *S. lineatus* appears to consume substantial amounts of particulate detrital material (Fox et al. 2009), a pattern not consistently displayed by its closest sibling species (*S. guttatus* and *S. insomnis*) (Woodland and Anderson 2014), the two other group-forming species in the pairing clade (*S. randalli* and *S. vermiculatus*) (Gundermann et al. 1983, Moleana et al. 2016), or the strongly pairing species. In turn, some species in the pairing clade (e.g., *S. virgatus*, *S. doliatus*, and *S. javus*) appear to readily consume thallous macroalgae alongside their diet of smaller, filamentous turf algae (Cvitanovic and Bellwood 2009, Rizzari et al. 2014, Müller et al. 2021), although their functional impact on larger, fleshy algae can be limited (Fox and Bellwood 2008, Streit et al. 2015). In any case, rabbitfishes clearly differ in their functional impact beyond the simple classifications implemented in a trait-based assessment. A detailed account of their realized functional roles as benthic consumers is provided in the second part of this chapter.

Activity

Differences in activity period basically separate *S. lineatus* and its sister species from the remaining species of rabbitfishes. Indeed, there is evidence that all three species (*S. lineatus*, *S. guttatus*, and *S. insomnis*) perform most of their foraging activity at night, consuming benthic algae and detrital matter in off-reef habitats such as mangroves, seagrasses or sand aprons (Fox and Bellwood 2011, Woodland and Anderson 2014). In contrast, all other species of rabbitfishes have been reported to forage during the day (Fox and Bellwood 2011, Brandl and Bellwood 2014, Ebrahim et al. 2020), thus fitting the profile of classic, diurnally-active herbivores. Notably, *S. randalli* showed substantial dietary niche overlap with *S. lineatus* in New Caledonia (Moleana et al. 2016), suggesting that this species may have similar behavioral patterns and thus also forage at night. In addition, the frequent absence of species such as *S. javus* from diurnal visual censuses (Cvitanovic and Bellwood 2009) might suggest that these species may exhibit more plasticity in their activity patterns than currently assumed, although their morphology suggests otherwise (see below). Yet, at this stage, the major functional distinction in the genus is limited to nocturnal vs. diurnal in the few species for which nocturnal foraging has been reported.

Mobility

Categorical classifications of mobility in rabbitfishes essentially reveal three functional groups that largely correspond to the main clades and their sub-divisions, with pairing, reef-associated species limited to small-scale movements at the reef scale (Brandl and Bellwood 2013b, Nanami 2015, 2018, Semmler et al. 2021), group-forming species in the pairing clade moving between reefs and nearby habitats (Fox and Bellwood 2011), and schooling species potentially moving between reef systems to act as mobile links (Fox and Bellwood 2008, Pickholtz et al. 2018). Notably, *S. spinus* may exhibit fairly limited movement due to its association with shallow reef flats, while considerable variability exists in the home range sizes of pairing species (Brandl and Bellwood 2013b, Nanami 2015, 2018, Semmler et al. 2021).

In the latter species, site fidelity appears to be established early in their ontogeny (Bellwood et al. 2016). The extremely limited movements and site fidelity in these species helps to explain the ontogenetic shift in pairing behavior, from mostly solitary individuals in the earliest stages (likely unable to find a partner due to limited densities), to paired individuals in mixed species and same-sex pairs (where mobility is high enough to find a suitable partner, but sometimes limited to sub-optimal choices), to almost exclusively paired individuals that defend fixed home ranges against conspecifics (Brandl and Bellwood 2013b, Mirbach and Brandl 2016, Semmler et al. 2021). Thus, spatial differentiation in functional roles largely follows the broad divisions in the family, while, again, exhibiting nuanced differentiation within each trait entity that is not captured by broad functional categories.

The Morphological Functional Niche Space of Rabbitfishes

Morphological differentiation among closely related species can offer vital clues concerning species' ecosystem roles. In rabbitfishes, there is a clear, diagnostic ecomorphological distinction between the two major clades as well as the *S. argenteus*/*S. woodlandi* clade, with the reef-associated clade generally exhibiting deeper bodies than the schooling clade, and *S. argenteus* and *S. woodlandi* having a strongly forked caudal fin and fusiform body (Fig. 3). Body shape variation along the dorso-ventral axis is common in reef fishes and often associated with a range of other ecological traits such as habitat association or mobility (Friedman et al. 2020). This appears to be the case for rabbitfishes as well, which are separated in terms of their body depth along PC2. Specifically, the distinct clades of rabbitfishes separated completely along PC2, but overlapped broadly on PC1. *S. vulpinus* was located at the extreme positive end of PC2 and *S. argenteus* at the extreme negative end, with all schooling species in our dataset (*S. fuscescens*, *S. canaliculatus* (Park 1797), and *S. spinus*) located towards the negative end as well. Notably, morphological traits that clearly correlated with deeper bodies were long heads and large head angles (a measure of the concavity of the anterior cranial structure), both of which are associated with precise benthic foraging and the exploitation of concealed spaces on coral reefs (Brandl and Bellwood 2013a, Fox and Bellwood 2013, Brandl et al. 2015, Nanami 2018). As such, PC2 appears to primarily separate species based on their ability to feed on discrete prey items in inaccessible habitats, with additional influence of traits such as snout length or the snout angle, both of which again favor access to cracks and crevices and load partially on PC2 and partially on PC1.

PC1, in turn, was strongly affected by a suite of morphological traits including eye diameter, peduncle depth, mouth-eye distance and snout-eye angle, as well as the snout angle. Thus, species on the positive end of PC1, such as *S. guttatus*, *S. randalli*, *S. lineatus* and *S. punctatissimus* are characterized by relatively large eyes that sit far away from the mouth towards the dorso-posterior part of the head, blunt snouts, and relatively thick caudal peduncles. This supports the notion that *S. lineatus* and its sister species, *S. randalli* and *S. vermiculatus*, as well as possibly *S. punctatissimus* are well adapted to nocturnal activity that requires large eyes (Goatley and Bellwood 2009, Schmitz and Wainwright 2011), but with little need to access interstitial spaces (as indicated by the blunt snouts) (Brandl and Bellwood 2013a) or an ability to display

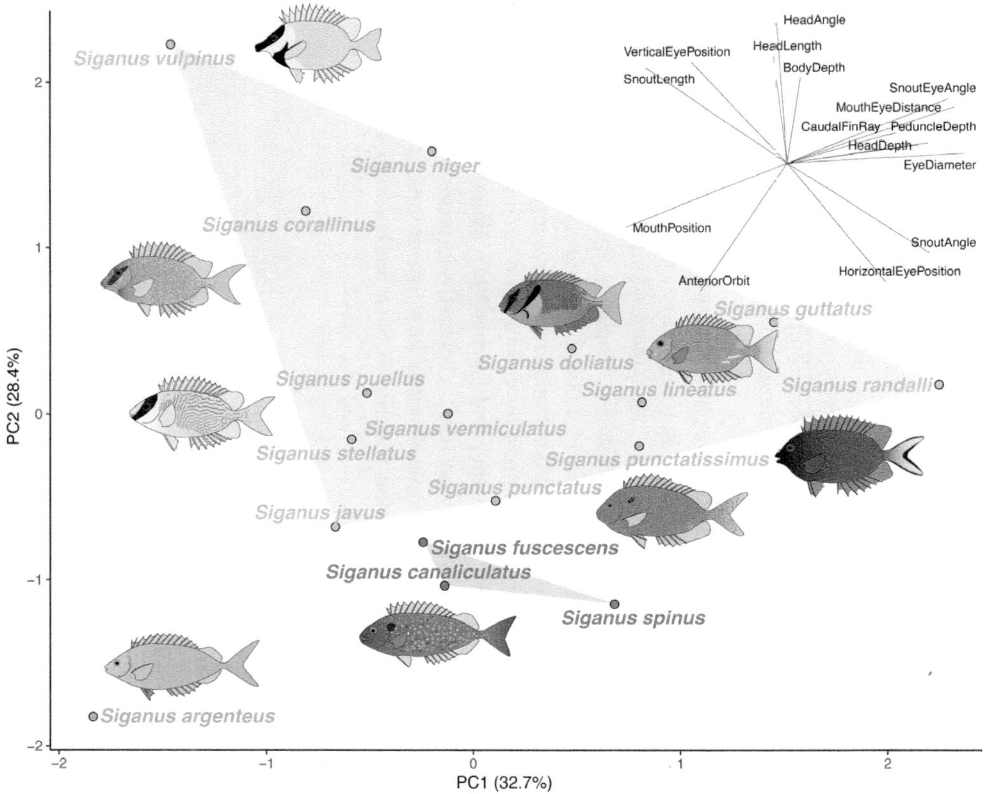

Fig. 3. Morphological differentiation in rabbitfishes. The plot depicts the ordination from a Principal Components Analysis based on 17 standardized morphological measurements from planar, lateral photographs of rabbitfishes (two of the morphological traits had loading values < 0.5 on PC1 or PC2 and were removed for clarity). Morphological traits were standardized by calculating each species' averaged residual in a simple regression with standard length. Measurements correspond to those in Bellwood et al. (2014). Colors in points and convex hull polygons reflect the three clades.

high burst speeds (as indicated by the thick caudal peduncle) (Fisher and Hogan 2007). Instead, these species appear more adapted to perform longer, slow-paced movement paths, which matches the observation that species such as *S. lineatus*, *S. guttatus*, or *S. randalli* often dwell in deeper, low-energy reef zones during the day and migrate into off-reef habitats such as sheltered mangroves at night (Woodland 1990, Fox and Bellwood 2011, Woodland and Anderson 2014, Moleana et al. 2016), which may make them important for the transport of inorganic material and nutrients to coral reefs. The latter function is often assumed to be performed predominantly by mobile predators (e.g., Haemulidae, Lutjanidae) (Clark et al. 2009, Berkström et al. 2012, McMahon et al. 2012, Khan et al. 2017), but it appears that some siganids may substantially contribute to this process in the Indo-Pacific (discussed in more detail below). Finally, some of the species for which we have the least information on their functional niches (such as *S. javus*, *S. punctatus*, and *S. punctatissimus*) were all located at the morphological break between the pairing and schooling cluster, raising the possibility that these occupy a unique functional role that has yet to be fully elucidated.

Thus, overall, both categorical and morphological trait-based analyses suggest differences among rabbitfish species regarding their functional roles on reefs. While there is clear separation between the main clades, the fundamental niches of species within each clade also diverge. In contrast, sister species appear to be practically interchangeable, based on broad categorical traits and existing ecological knowledge. This supports the notion that geographic boundaries between sister species in rabbitfishes are, at least in part, maintained by competitive exclusion based on the principle of limiting similarity (MacArthur and Levins 1967, Woodland 1997). Yet, to fully grasp the role of rabbitfishes on reefs, especially in terms of their role as consumers of benthic organisms relative to other key fish groups, it is necessary to examine their trait space occupation relative to other herbivorous families and to gauge their realized functional niches under the constraint of biotic and abiotic drivers in the wild.

Rabbitfishes in the Coral Reef Herbivore Trait-Space

Relative to other herbivores, rabbitfishes appear to fall between the two other major reef fish lineages (Acanthuridae and scarine Labridae) in both categorical and morphological trait space. The limited resolution of categorical traits predictably results in substantial overlap among herbivores at this high level of comparison, which may also suggest limited phylogenetic conservatism in trait evolution (Floeter et al. 2018). However, there are relatively few trait entities (TEs) that rabbitfishes share with other families and TEs with high 'functional redundancy' (i.e., multiple points in the same entity; cf. Mouillot et al. 2014) in rabbitfishes are sparse compared to surgeon- and parrotfishes (Fig. 4A) (drive, in part, by lower species richness in the family). Furthermore, several rabbitfish species occupy a completely unique area of the trait space. On the negative end of PCoA2, this space is dilated most strongly by the crepuscular/nocturnal group, including *S. lineatus*, *S. guttatus*, and *S. vermiculatus*, highlighting the distinctive role that these species may play on coral reefs, including their capacity to move between reefs and nearby habitats at night

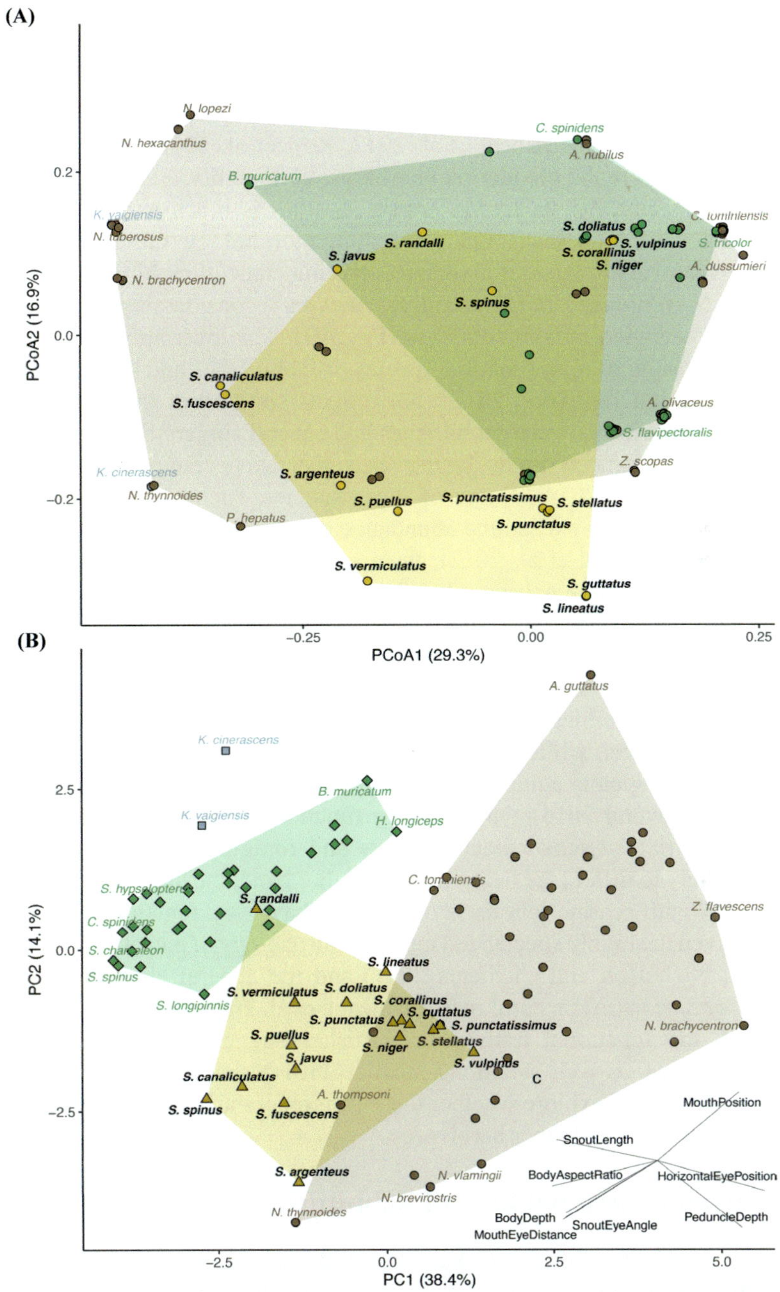

Fig. 4. Trait-space occupation of rabbitfish species relative to the Acanthuridae (brown), scarine Labridae (green), and Kyphosidae (blue). **(A)** Categorical trait space occupation based on a Principal Co-ordinates Analysis of six categorical traits (Size, Sociality, Diet, Activity, Mobility), revealing that while rabbitfishes are completely nested among other species on PCoA1, several species occupy unique trait space on PCoA2. **(B)** Morphological trait space occupation based on a suite of linear morphometrics (bottom right), showing that most rabbitfish species occupy a unique section of the herbivore morphospace. All rabbitfish species names are shown, and only the outer vertices for the other families are named.

(Fox et al. 2009, Fox and Bellwood 2011). Other species in this trait space include *S. punctatus*, *S. stellatus* (Forsskål 1775), *S. punctatissimus* and *S. puellus*. Except for *S. puellus* (Hoey et al. 2013, Brandl and Bellwood 2015), relatively little is known about the functional ecology of these species and although they generally occur in relatively low abundances (Cheal et al. 2012, Roff et al. 2019), they may occupy niches that differ from the broader reef herbivore community.

Morphologically, the separation among families is much stronger, suggesting strong phylogenetic conservatism in overarching morphology (Ghilardi et al. 2021). Nevertheless, especially the reef-associated, pairing clade of rabbitfishes (including, for instance, *S. vulpinus*, *S. doliatus*, and *S. corallinus*) considerably extends into the morphospace occupied by surgeonfishes (Fig. 4B). This overlap was not apparent when a smaller suite of representative species of rabbitfishes and surgeonfishes was examined (Fox and Bellwood 2013), but reflects similarities in the morphology, sociality and behavior between rabbitfishes and several surgeonfish species that are characterized by elongated snouts, large eyes, and concave cranial structures (Brandl and Bellwood 2013a, 2014). Some of these species, such as *Zebrasoma scopas* (Cuvier 1829), typically exceed the abundance of all pairing rabbitfishes combined on offshore reefs (Cheal et al. 2012), thus potentially raising questions about the functional impact of these pairing rabbitfish species. However, *Zebrasoma* spp. are generally rare on inshore reefs, and pairing rabbitfishes, such as *S. doliatus*, can far outnumber them on these reefs (Cheal et al. 2012). This highlights the potential for rabbitfishes to play important roles in inshore reef environments, which often host depauperate assemblages of parrotfishes and surgeonfishes (Russ 1984b, Cheal et al. 2012). Moreover, while a considerable suite of rabbit- and surgeonfishes are probably morphologically equipped to exploit small crevices and interstices on reefs (Brandl and Bellwood 2016) – the primary function linked to the aforementioned morphological traits – actual dietary targets and foraging behavior (i.e., realized niches) are likely to differ (see below). On the other hand, the morphological overlap of *S. randalli* with scarine labrids is unlikely to have any functional significance, especially given the highly specialized internal morphology of parrotfishes (Clements et al. 2017, Nicholson and Clements 2021) and the general elusiveness and low abundance of *S. randalli* on coral reefs (Moleana et al. 2016).

Overall, the functional traits of rabbitfishes suggest significant differences within rabbitfishes, as well as among most rabbitfish species and other coral reef herbivores. Below, we explore whether these differences are mirrored in the realized niches of rabbitfishes and other herbivores.

The Realized Functional Role of Rabbitfishes on Reefs

From a realized functional perspective, rabbitfishes primarily shape sessile organismal communities via consumptive effects. Indeed, when feeding, rabbitfishes have been found to remove and assimilate a variety of benthic resources (Figs. 5 and 6) (Hoey et al. 2013, Ebrahim et al. 2020), with the potential to play major roles in shaping benthic community composition through this top-down control. Given the fierce competition for space in most shallow, benthic tropical marine ecosystems and the association of increased algal cover with ecosystem

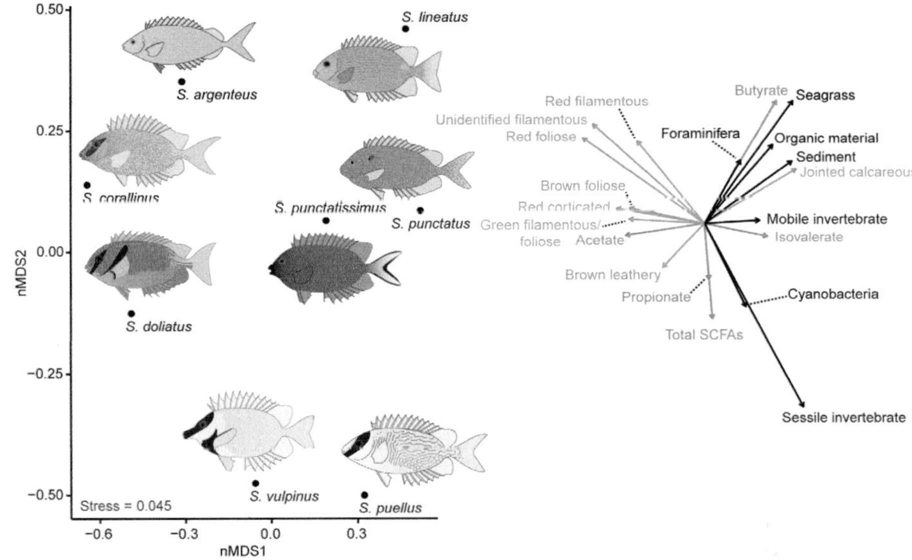

Fig. 5. Non-metric multidimensional scaling ordination based on gut content and short chain fatty acid (SCFA) profile data to provide an insight into the material being removed, ingested and utilized by rabbitfishes on coral reefs. The vectors indicate how these data are arranged in multivariate space, based on different algal groups, non-algal nutritional items, and short-chain fatty acids. The ordination was based on a Bray Curtis similarity matrix of range standardized data. Methods for the ordination follow those outlined in Tebbett et al. (2022a). Gut content data were sourced from Hoey et al. (2013), while SCFA profile data were sourced from Clements and Choat (1995).

change (Bellwood et al. 2004, Graham et al. 2006), research into rabbitfish feeding effects has largely focused on how they shape macroalgal and algal turf communities (e.g., Bennett et al. 2015, Humphries et al. 2015, Loffler et al. 2015, Plass-Johnson et al. 2015). However, various species of rabbitfishes may also be playing roles in the removal of particulate material (detritus and sediments), cyanobacteria, as well as other organisms including sponges and seagrass (Fox et al. 2009, Hoey et al. 2013). Below we explore the realized role of rabbitfishes in the removal of these groups of benthic organisms/material, consider the relative importance of rabbitfishes in these roles under different contexts, and discuss the overall functional impact of rabbitfishes on benthic communities.

Realized Functional Roles of Rabbitfishes: Dietary Targets

Macroalgae Removal

While upright, tall (> 2 cm), fleshy macroalgae are a prominent component of many tropical marine ecosystems (Fulton et al. 2019), they are frequently associated with 'phase shifts' on coral reefs into macroalgae-dominated, degraded states (Hughes et al. 2007, Roff et al. 2015). Therefore, the importance of fish-based consumption of macroalgae and the potential for this consumption to limit–or reverse–macroalgal phase shifts has received substantial attention (e.g., Rasher and Hay 2010,

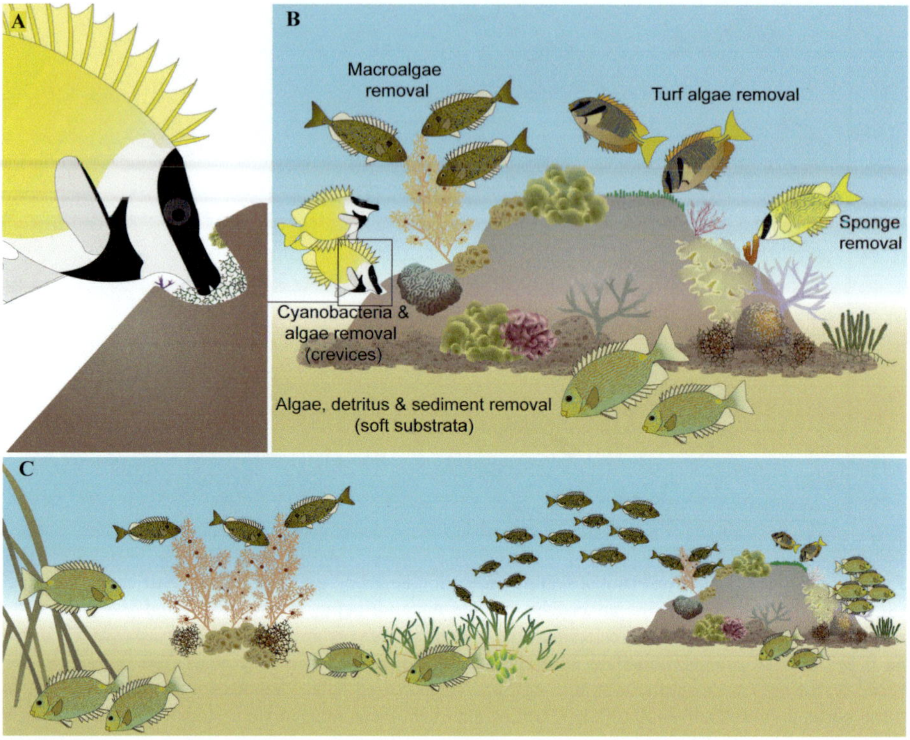

Fig. 6. The ecosystem roles of rabbitfishes in tropical ecosystems across different spatial scales. **(A)** Feeding in microtopographic refuges by pair-forming rabbitfish species, such as *Siganus vulpinus*, can be important for keeping these essential coral settlement areas free of algae and cyanobacteria. **(B)** The removal of key nutritional resources and material by rabbitfishes around a coral reef ecosystems. **(C)** Rabbitfishes acting as mobile cross-habitat links in tropical seascapes composed of mangrove, macroalgae, seagrass and coral reef habitats. Algae, plant and coral vector images: C. Donovan, J. Woerner, T. Saxby, C. Collier, A. Jones, D. Tracey, and J. Thomas.

Hoey and Bellwood 2011, Loffler et al. 2015). Given that increased macroalgae cover has been found to suppress key functions on reefs such as coral recruitment (Diaz-Pulido et al. 2010, Johns et al. 2018) and coral growth (Vega Thurber et al. 2012, Clements et al. 2018), macroalgae removal by fishes is viewed as a critical ecosystem function in some contexts (Bellwood et al. 2004, 2019, Brandl et al. 2019).

In terms of macroalgae consumption in tropical marine systems, rabbitfishes can play an important role (e.g., Fox and Bellwood 2008, Rasher et al. 2013, Chong-Seng et al. 2014, Humphries et al. 2015, Plass-Johnson et al. 2015). Indeed, out of the five most important macroalgae removing fish species in the Pacific Ocean, three are rabbitfishes (*Siganus doliatus*, *S. canaliculatus*, and *S. virgatus*), while the remaining two species are the kyphosid *Kyphosus vaigiensis* (Quoy and Gaimard 1825) and the surgeonfish *Naso unicornis* (Forsskål 1775) (Topor et al. 2019). However, while rabbitfishes are frequently involved in macroalgal removal, their feeding mechanics differs from other key fishes such as *N. unicornis* and *K. vaigiensis* (Streit et al. 2015). Specifically, the latter two species frequently remove entire macroalgae thallae, while the smaller body size and different morphology of rabbitfishes limits them to nibbling on leaves without removing holdfasts or stalks (Streit et al. 2015). This means that macroalgal removal by rabbitfishes is often less extensive in magnitude than other key macroalgae removing fishes, with macroalgae potentially recovering faster following feeding by rabbitfishes.

Despite the comparatively limited capacity for individual rabbitfishes to remove macroalgae, several macroalgae-removing species form large schools (Woodland 1990, Pitt 1997) that can extensively remove macroalgae and potentially act as a mobile link between different coastal tropical habitats (see below). For example, (Fox and Bellwood 2008) found that the schooling rabbitfish *S. canaliculatus* was responsible for 46 percent of all bites delivered to macroalgal assays at Orpheus Island on the inner-shelf GBR, contributing to substantial macroalgal loss from assays. Interestingly, *S. canaliculatus* had not been observed in the study location prior to this experiment, suggesting it generally occurred in habitats that were not commonly surveyed (i.e., non-coral reef habitats) and/or is particularly sensitive to 'diver effects' during surveys. Effects of schooling species on macroalgal removal is also supported by the feeding-effects of schooling *S. fuscescens* on nearshore reefs in the Spermonde Archipelago, Indonesia (Plass-Johnson et al. 2015), in Moreton Bay, Australia (Gilby et al. 2017), as well as *S. sutor* on lagoonal back reef sites in Kenya (Humphries et al. 2015). Moreover, the capacity for schooling rabbitfishes to denude macroalgae is perhaps best highlighted by the impact of rabbitfishes as invaders in temperate coastal marine ecosystems. In these ecosystems rabbitfishes have been found to reshape benthic communities via the direct removal of habitat-forming macroalgae and/or the maintenance of low-complexity algal turf-dominated systems (see below). However, all the examples above are from coastal or inshore reef habitats, suggesting that the feeding impacts of rabbitfishes on macroalgae are largely restricted to such environments, rather than offshore reefs or oceanic atolls. Importantly, this inference also aligns with the distribution patterns of schooling rabbitfishes, which are generally only abundant on inshore reefs (Cheal et al. 2012, Hoey et al. 2013). Thus, some species of rabbitfishes–especially from the schooling clade–clearly shape benthic competition by consuming erect macroalgae on inshore reefs, even if per capita impacts on macroalgal thalli are limited.

Algal Turf Removal

Algal turfs are generally considered short (~ < 2 cm) multispecies assemblages of algae that cover most hard substrata in shallow marine ecosystems (Connell et al. 2014, Tebbett and Bellwood 2019). Algal turfs frequently occupy the majority of benthic space in these ecosystems and can play a key role in benthic primary productivity (Hatcher 1988, Klumpp and McKinnon 1989). Accordingly, most herbivory on tropical reefs is enacted on these short algal communities rather than tall macroalgae (Steneck 1988, Choat and Clements 1998). While settlement and recovery of other benthic organisms (e.g., corals) is possible in turf-dominated ecosystems, the configuration of the algal turf community may be key to this (Speare et al. 2019, Evans et al. 2020). Indeed, we now recognize that algal turfs range from short productive algal turfs to long sediment-laden algal turfs (Goatley et al. 2016, Tebbett and Bellwood 2019), with the latter impeding ecosystem functions such as the recruitment and survival of habitat-forming organisms (Alestra et al. 2014,

Speare et al. 2019, Evans et al. 2020), benthic productivity (Tebbett et al. 2018), and fish feeding (Bellwood and Fulton 2008, Goatley et al. 2016). The process of herbivory plays a key role in shaping the nature of algal turfs, and maintaining them in a short, clean, productive state.

Herbivory rates on algal turfs are often intense in many shallow marine ecosystems (Randall 1961, Bennett et al. 2015, Longo et al. 2019), matching the high growth rates of algal turf communities (Carpenter 1986, Russ and McCook 1999, Kelly et al. 2017). Across the Indo-Pacific, rabbitfishes form a core-component of the broader assemblage of grazing fishes that feed on these turfs (Cheal et al. 2012, Heenan et al. 2016, Robinson et al. 2020). However, in terms of total feeding rates on algal turfs, rabbitfishes appear to play a relatively minor role compared to other key groups such as surgeonfishes and parrotfishes, predominantly due to the low abundances of turf-feeding rabbitfish species and their slow bite rates (Russ 1984b, Cheal et al. 2012). Indeed, where feeding rates on tropical Indo-Pacific reefs have been quantified, rabbitfishes have generally been found to make far smaller contributions than surgeonfishes or parrotfishes (e.g., Bellwood and Fulton 2008, Goatley and Bellwood 2012, Goatley et al. 2016, Pessarrodona et al. 2022). This relatively small contribution of rabbitfishes is particularly apparent on offshore and oceanic reefs, where rabbitfishes generally occur in far lower abundances when compared to parrotfishes and surgeonfishes (Russ 1984b, Cheal et al. 2012, Hoey et al. 2013). Therefore, while rabbitfishes contribute to algal turf removal, their relative impact on turf algal growth and cover is likely limited.

The functional role of rabbitfishes in algal turf removal is further limited by their feeding mode. Turf-feeding rabbitfishes feed using bicuspid teeth with rapid, 'nibbling' bite rates, suggesting they are 'cropping' the apical portion of algal turf filaments similar to cropping surgeonfishes (Tebbett et al. 2017). This contrasts strongly to the removal of epi- and endolithic organisms by scraping and excavating parrotfishes, which essentially reset turf algal succession dynamics (Clements et al. 2017). However, the exact nature of feeding by rabbitfishes, especially their individual bite-level impact on algal turfs is not currently resolved. Detailed morphological and aquaria-based studies would be necessary to understand the exact nature of feeding and how this does or does not compare to other groups such as cropping surgeonfishes. As algal turfs are poised to remain an abundant component of tropical reefs in the future (Tebbett et al. 2023), a more comprehensive understanding of rabbitfish interactions with these algae could be interesting.

Detritus/Sediment Removal

In marine ecosystems, detrital (i.e., non-living particulate organic matter containing microbes) food chains represent a major, but often underappreciated, energetic trophic pathway (Wilson et al. 2003). The sources of detritus are variable [including algal breakdown (Wilson et al. 2001), coral mucus (Wild et al. 2004), or fish feces (Zarco-Perello et al. 2019)], but the end product is highly nutritious (Crossman et al. 2001). Yet, accumulating benthic detritus in algal turfs also becomes laden with inorganic sediments, which dilute these nutritional qualities when ingested by fishes (Bellwood 1996, Goatley and Bellwood 2010, Bowden et al. 2022). The ingestion and movement of both detritus and sediments by fishes can, in-turn, represent an important ecosystem function (Bellwood 1996, Goatley and Bellwood 2010).

The importance of both fish-based detritivory (Wilson et al. 2003) and sediment removal/transport (Tebbett and Bellwood 2019, Perry et al. 2022) has been increasingly recognized in marine systems. However, within this research space, rabbitfishes have received limited attention with most research focusing on parrotfishes (e.g., Bellwood 1996, Schlaefer et al. 2022) and surgeonfishes (e.g., Goatley and Bellwood 2010, Krone et al. 2011). Nevertheless, gut contents analyses suggest that some rabbitfishes, such as *S. lineatus* and *S. javus*, ingest substantial quantities of organic and inorganic particulate material (Fig. 5; Pitt 1997, Fox et al. 2009, Hoey et al. 2013). For example, organic particulate material accounted for 40 percent of the stomach contents of *S. lineatus* individuals at Lizard Island, in the northern GBR (Fox et al. 2009). However, according to nocturnal movement data, *S. lineatus* may, in fact, be feeding over the sand aprons that surround coral reefs, while resting on the reef during the day (Fig. 6B) (Fox and Bellwood 2011), suggesting that this species actually promotes the net deposition of sediments onto reefs rather than off-reef transport. Furthermore, *S. lineatus* has been observed to graze heavily on seagrass blades in Western Australia (Vanderklift et al. 2021), further supporting the notion that the species tends to feed mostly in off reef environments and is not specialized on ingesting (and assimilating) detrital material. Therefore, with the possible exception of *S. javus*, on-reef detritivory and sediment movement largely appears to be the domain of a suite of specialized surgeonfishes (Tebbett et al. 2022a) and parrotfishes (Bonaldo et al. 2014), rather than rabbitfishes. Indeed, rabbitfishes likely lack the morphological and behavioral traits to efficiently remove particulates from hard substrata.

Cyanobacteria Removal

Cyanobacteria are found on and within virtually all surfaces in shallow marine ecosystems (Charpy et al. 2012). These organisms are highly productive and can rapidly capitalize on changing conditions (Bakker et al. 2017, Ford et al. 2018, Tebbett et al. 2022b), especially those associated with nutrient fluxes (Albert et al. 2005, Haan et al. 2016). Consequently, cyanobacteria are also being increasingly recognized as a valuable nutritional resource for fishes, with a growing body of evidence documenting fish consumption of cyanobacteria (e.g., (Clements et al. 2017, Cissell et al. 2019, Nicholson and Clements 2020)).

Most recent research on cyanobacterial consumption by fishes has focused on parrotfishes (Clements et al. 2017, Nicholson and Clements 2020), which successfully exploit this resource from the upper endo- and epilithic layers of exposed reef surfaces (Clements et al. 2017), and often represent a dominant component of herbivorous fish assemblages on reefs (Russ 1984b, Cheal et al. 2012). Among rabbitfishes, the gut contents of *S. vulpinus* (but not its sister species, *S. unimaculatus*; (Nanami 2018)) were found to be dominated by cyanobacteria filaments, comprising 46.6–60.8 percent of ingested material (Hoey et al. 2013), supporting earlier observations in Palau (Robertson 1982). *S. vulpinus* is also typified by higher levels of short chain fatty acids (SCFAs) in their hindguts compared to many other rabbitfish species (Fig. 5). As the amount of SCFAs in the hindgut is indicative of fermentation of nutritional resources, these elevated levels in *S. vulpinus* could relate to how these fishes process cyanobacteria, which are often rich in secondary metabolites (Nagle and Paul 1998, Capper et al. 2006). In addition to *S. vulpinus*, *S. lineatus*, *S. punctatissimus* and *S. punctatus* were found to ingest notable (11.3–26.1 percent of gut contents) amounts of cyanobacteria on reefs in the northern GBR (Hoey et al. 2013), corresponding with higher levels of the SCFA isovalerate in their hindguts relative to species which feed predominantly on algae (Fig. 5). This could be indicative of ingestion and assimilation of microbial trophic resources as isovalerate is a metabolite of leucine catabolism in all organisms (including bacteria) (Choat and Clements 1998). Together, this evidence suggests that cyanobacteria could be a major dietary target for some species of rabbitfishes. While parrotfishes dominate the removal of cyanobacteria from exposed surfaces, rabbitfishes may be key in removing filamentous cyanobacteria from inaccessible locations such as microtopographic refuges (discussed in more detail below).

Sponge Removal and Feeding on other Nutritional Resources

Rabbitfishes have been documented to feed on an array of other potential nutritional resources beyond those outlined above, and there appears to be a high degree of dietary opportunism. For example, the gut contents of *S. puellus* were found to be dominated by sessile invertebrates (sponges) in the northern GBR (Hoey et al. 2013) and in Okinawa, Japan (Nanami 2018). This is mirrored by its sister species *S. puelloides* in the Indian Ocean, which has been documented to feed on tunicates (Woodland and Randall 1979). Similar to *S. vulpinus*, *S. puellus* also has high total SCFA levels in its hindguts relative to other rabbitfishes (Fig. 5), possibly related to a diverse array of secondary metabolites in sponges (Proksch 1994). Alternatively, the similar SCFA profiles in *S. vulpinus* and *S. puellus* could indicate that *S. puellus* is, at least in part, targeting and processing the cyanobacteria and other microbial photoautotrophs that often associate with sponges (Charpy et al. 2012), rather than the sponge itself. It is also interesting to note that both *S. puellus* and *S. vulpinus*, which feed on sponges and cyanobacteria, respectively, generally have low population densities on reefs (Cheal et al. 2012, Hoey et al. 2013). There is increasing evidence that coral reef herbivores *sensu lato* may be limited by resource availability (Russ et al. 2015, Tootell and Steele 2016). Exploring the potential mechanisms behind these limited population sizes may be interesting in future research.

In addition to *S. puellus* and *S. puelloides* feeding on sponges and tunicates, respectively, there are several reports of rabbitfish species feeding opportunistically on other benthic organisms in tropical systems. For example, schools of *S. fuscescens* have been documented feeding heavily on soft corals on reefs around Taiwan (Kuo et al. 2015). The same is the case for *S. vulpinus*, which has been observed to feed vigorously on *Sinularia* soft corals, while *S. punctatus* has been observed biting the apical tips of branching acroporid corals (Brandl, pers. obs.). Meanwhile, *S. rivulatus* has been reported to feed on ctenophores and scyphozoans in the Red Sea (Bos et al. 2017), while *S. lineatus* (Vanderklift et al. 2021), *S. fuscescens* (Chelsky Budarf et al. 2011), and *S. rivulatus* (Forsskål and Niebuhr 1775) (Lundberg and Golani 1995) have been documented feeding heavily on seagrass in various tropical systems (the functional implications of this, in terms of cross-habitat linkages, are addressed below). However, our understanding of just how widespread or important the ingestion and assimilation of alternative nutritional resources may be in rabbitfishes, and whether they are targeting them directly or the epiphytic nutritional resources on them (e.g., cyanobacteria), is limited. Given the potential for alternative benthic organisms such as sponges and soft corals to increase following disturbances in some marine ecosystems (Norström et al. 2009), understanding the potential for rabbitfishes to consume these organisms may be important in the future.

Realized Functional Roles of Rabbitfishes: Spatial Impact

Microtopographic Refuges

On hard benthic substrata in marine ecosystems, microtopographic complexity represents a crucial driver of broader ecosystem dynamics by governing available surface area and grazing pressure (Brandl and Bellwood 2016, Mallela 2018). Specifically, on coral reefs, micro-topographic crevices offer grazing refuges in which important habitat-forming organisms, such as corals (Whalan et al. 2015), can settle and develop under protection from incidental or intentional predation (O'Leary et al. 2013, Trapon et al. 2013). However, algae and sediments also accumulate in these refuges which can negatively impact the settlement and survival of corals (Brandl et al. 2014, Ricardo et al. 2017, Evensen et al. 2021). Therefore, nominally herbivorous fishes that can feed in these refuges may play an important role in broader ecosystem dynamics (Brandl and Bellwood 2016). Rabbitfishes are particularly important crevice-feeding fishes on reefs (Fox and Bellwood 2013, Brandl and Bellwood 2014). These species, such as *S. vulpinus* (Fig. 6A, B), are generally characterized by specialized behavioral (e.g., pair-forming) and morphological (e.g., elongated snouts) adaptations which are relatively limited in other herbivorous fish groups on reefs. Indeed, the only other group of nominally herbivorous fishes adapted to feed in crevices are surgeonfishes in the genus *Zebrasoma* (Brandl et al. 2015), highlighting the relative importance of rabbitfishes in this role.

A key example of the importance of this crevice-feeding function on reefs was provided by Brandl et al. 2014, who demonstrated that early life stages of both corals and a macroalgae competitor (*Turbinaria* sp.) were more abundant in concealed, compared to open, microhabitats on exposed reefs in the northern GBR. In this system, the abundance of early life-stage corals was positively related to the abundance of pair forming rabbitfishes and *Zebrasoma* spp. surgeonfishes, while the abundance of *Turbinaria* showed opposite patterns (Brandl et al. 2014). Importantly, the results of this correlative work were later confirmed with experimental work, which also highlighted the importance of pairing rabbitfishes in controlling algae in microtopographic refuges (Brandl and Bellwood 2016). Thus, top-down control of algae in refuges may govern the capacity of coral reef ecosystems to recover from disturbances, as competition between early life-stage algae and corals is likely to be a key factor determining benthic dominance and appears to play out predominantly in small crevices. Indeed, given that crevice-feeding herbivorous fishes are completely lacking in the Western Atlantic (Siqueira et al. 2019), this could be a potential factor which pre-disposes reefs in this region to increased macroalgae cover when corals are lost (Roff and Mumby 2012, Tebbett et al. 2023). Overall, the unique spatial impact of rabbitfishes at the scale of millimeters or centimeters may differentiate several rabbitfish species from many other nominal herbivores and could make them particularly important for broader reef ecosystem dynamics.

Reef-Scale and Cross-Habitat Movements

Increasingly, shallow tropical marine ecosystems are being viewed as a diverse mosaic of interconnected habitats (Sambrook et al. 2019, Fulton et al. 2020). One of the ways in which these habitats are connected is via the assimilation, movement, and release of nutrients (Meyer and Schultz 1985a, Davis et al. 2014, Francis and Côté 2018). Indeed, nutrient transport and release across habitats, especially through fishes, may be important in shaping processes such as herbivory rates (Shantz et al. 2015) aggregating behavior through time, they can create nutrient hotspots where rates of biogeochemical activity are higher than those found in the surrounding environment. In turn, these hotspots can influence ecosystem processes and community structure. We examined the potential for reef fishes from the family Haemulidae (grunts and coral growth rates (Meyer and Schultz 1985b). While many fish species can play important roles in these cross-habitat trophic linkages, several species of rabbitfishes appear to contribute to these processes (Fig. 6C).

A number of rabbitfish species occupy multiple tropical coastal habitats beyond coral reefs, including estuaries, mangroves, seagrass beds and macroalgal reefs (Gell and Whittington 2002, Sambrook et al. 2019, Hemingson and Bellwood 2020). For example, a review of non-reef habitat use by coral reef fishes, (Sambrook et al. 2019) found that 85 percent of rabbitfish species utilized seagrass habitats, while 54 percent utilized mangrove habitats. Importantly, rabbitfishes are not just occupying these other tropical habitats, with studies based on feeding assays (Henderson et al. 2019) and stable isotopes (Davis et al. 2014) suggesting that rabbitfishes can feed in, and derive nutritional resources from, a range of tropical habitats. Not all rabbitfish species, however, are likely to be equally important in driving cross-habitat linkages. In this respect, wide-ranging, schooling species, such as *S. canaliculatus*, *S. fuscescens*, and *S. sutor*, are frequently observed in multiple habitats and have the capacity to move over large distances (Olds et al. 2014, Henderson et al. 2017, Eggertsen et al. 2019). For example, *S. sutor* individuals have been documented travelling distances of 30 km in as little as 33 days (Kaunda-Arara and Rose 2004), while an acoustic telemetry study suggested their average home ranges were in the order of 4.5 hectares and covered multiple habitat types (Ebrahim et al. 2020). By contrast, many pair-forming rabbitfish species, such as *S. doliatus*, *S. corallinus*, and *S. vulpinus*, are often more reef associated (Pitt 1997, Hemingson and Bellwood 2020, Sambrook et al. 2020) and generally occupy relatively small and stable home ranges (in the order of 1.5 hectares) (Brandl and Bellwood 2013b, Semmler et al. 2021). It is also important to note that, unlike reef-associated, pairing rabbitfishes, the large number of individuals in schooling rabbitfish groups (> 200 in some cases (Zarco-Perello et al. 2020)), would lead to substantial cross-habitat nutrient movement at localized scales. Thus, while schooling rabbitfish species could represent key cross-habitat trophic links in tropical Indo-Pacific seascapes, pairing reef-associated species likely have a concentrated, spatially-limited footprint on coral reefs.

The capacity for certain rabbitfishes to occupy and move among multiple tropical habitats is also important for other reasons beyond trophic linkages. For example, by acting as 'mobile links' among habitats certain rabbitfishes may help bolster the resilience of ecosystems, such as coral reefs, after disturbance (Nyström and Folke 2001). Specifically, the movement of herbivorous fishes, such as rabbitfishes, onto disturbed coral reefs may increase grazing pressure and limit macroalgal development, enhancing the resilience of the system (Welsh and Bellwood 2015). Furthermore, the capacity of certain rabbitfishes to occupy multiple habitats may mean they are less sensitive to disturbances in a given habitat, which may help support the ecosystem services on which people depend, even in heavily disturbed ecosystems. For example, following coral bleaching in the Seychelles, substantial coral loss led to the occurrence of phase-shifts to macroalgae dominated systems in several areas (Graham et al. 2006). While the abundance of many coral-dependent fishes were negatively impacted by these changes (Robinson et al. 2019b), rabbitfish populations responded positively (Robinson et al. 2019a). This increase in rabbitfish populations not only helped sustain, but also increase, the productivity of the fishery in this location, with higher catch per unit effort after the phase-shifts occurred (Robinson et al. 2019a). Therefore, the capacity of rabbitfishes to utilize multiple habitats means they may cope well with marked alterations in habitat condition, which may become increasingly important as ecosystems continue to change into the future.

Regional Range Expansions in Rabbitfishes

As our oceans warm, temperate and subtropical reef ecosystems–which are often dominated by benthic macroalgae–are faced with changing ecological dynamics due to the increasingly common presence of tropical herbivores (Wernberg et al. 2016, Vergés et al. 2019). Rabbitfishes are among the most influential fishes involved in the topicalization and transformation of subtropical and temperate macrophyte communities (Vergés et al. 2014b). Indeed, warming ocean waters have facilitated notable range extensions and subsequent increases in the abundance of *S. fuscescens* off Japan (Yamaguchi 2010) as well as along the east and west coasts of Australia (Bennett et al. 2015, Vergés et al. 2016). Moreover, the opening of the Suez Canal, in addition to warming, allowed the invasion and spread of *S. luridus* (Rüppell 1829) and *S. rivulatus* into the Mediterranean Sea (Rilov and Galil 2009). In these subtropical and temperate geographic areas where rabbitfishes have spread, they are now often a major component of herbivorous fish assemblages (Vergés et al. 2016, Zarco-Perello et al. 2021). For example, in warmer eastern regions of the Mediterranean, rabbitfishes can now account for ~ 64–84 percent of all herbivorous fish biomass (Vergés et al. 2014a). This abundance of rabbitfishes in the ecosystems they have invaded means that they are now having considerable impacts on how these subtropical/temperate ecosystems function via their feeding effects. Notably, only schooling rabbitfishes, which are known to have more extensive spatial movements, occur in greater abundances, and generally feed on upright macroalgae, have thus far appeared to successfully invade temperate and subtropical systems.

The spread of rabbitfishes into temperate geographic areas is frequently associated with shifts in benthic community composition (Vergés et al. 2014b), characterized by the decline of upright fleshy macrophyte communities such as kelp and the increase in short algal turf communities (Wernberg et al. 2016, Pessarrodona et al. 2021). However, the role of rabbitfishes in these shifts, via feeding effects, can be multifaceted. On one hand, stressors such as marine heatwaves may initially drive extensive loss of upright macrophytes, with rabbitfishes playing little direct role (Bennett et al. 2015, Wernberg et al. 2016). Once the macrophytes are lost, however, the intense grazing pressure of rabbitfishes on the substratum can maintain a benthic condition typified by algal turf dominance, as the grazing impedes the recovery of upright macrophytes (Bennett et al. 2015, Zarco-Perello et al. 2021). On the other hand, rabbitfishes have also been documented to feed on the upright macrophytes directly, with intense feeding pressure leading to the overgrazing of these algae and subsequent expansion of algal turfs, even without marine heatwaves (Vergés et al. 2016). In this case, expansion of tropical herbivores, especially S. fuscescens, appeared to be the primary driver leading to upright macrophyte decline (Vergés et al. 2016). However, in either case, rabbitfishes play a key role in the transformation of subtropical and temperate reef ecosystems via pronounced consumptive effects.

Conclusion

Overall, rabbitfishes provide an intriguing family for the study of functional differentiation and adaptive trait diversification. While a deep split divides rabbitfishes in their functional roles, there appears to be strong niche differentiation among sympatric species within each clade, with traceable connections between morphology, behavior, physiology, and life-history. In turn, allopatrically distributed sister species are likely to be functionally equivalent. While the extent of divergent niches appears to be more pronounced in the reef-associated, deep-bodied clade, schooling species have rarely been assessed through a lens of ecological differentiation. Furthermore, several species of rabbitfishes remain poorly studied—such as *S. javus*, *S. punctatus*, or *S. punctatissimus*—with very little ecological information available. We suggest that the Siganidae may be a promising group to further examine the eco-evolutionary linkages between morphology, physiology, and behavior to better understand how biodiversity affects the functioning of marine ecosystems.

References

Albert, S., J.M. O'Neil, J.W. Udy, K.S. Ahern, C.M. O'Sullivan and W.C. Dennison. 2005. Blooms of the cyanobacterium *Lyngbya majuscula* in coastal Queensland, Australia: Disparate sites, common factors. Mar. Pollut. Bull. 51: 428–437.

Alestra, T., L.W. Tait and D.R. Schiel. 2014. Effects of algal turfs and sediment accumulation on replenishment and primary productivity of fucoid assemblages. Mar. Ecol. Prog. Ser. 511: 59–70.

Bakker, D.M. de, F.C. van Duyl, R.P.M. Bak, M.M. Nugues, G. Nieuwland and E.H. Meesters. 2017. 40 Years of benthic community change on the Caribbean reefs of Curaçao and Bonaire: the rise of slimy cyanobacterial mats. Coral Reefs 36: 355–367.

Bejarano, S., J.-B. Jouffray, I. Chollett, R. Allen, G. Roff, A. Marshell et al. 2017. The shape of success in a turbulent world: wave exposure filtering of coral reef herbivory. Funct. Ecol. 31: 1312–1324.

Bellwood, D.R. 1996. Production and reworking of sediment by parrotfishes (family Scaridae) on the Great Barrier Reef, Australia. Mar. Biol. 125: 795–800.

Bellwood, D.R., T.P. Hughes, C. Folke and M. Nyström. 2004. Confronting the coral reef crisis. Nature 429: 827–833.

Bellwood, D.R. and C.J. Fulton. 2008. Sediment-mediated suppression of herbivory on coral reefs: decreasing resilience to rising sea levels and climate change? Limnol. Oceanogr. 53: 2695–2701.

Bellwood, D., C. Goatley, S. Brandl and O. Bellwood. 2014. Fifty million years of herbivory on coral reefs: fossils, fish and functional innovations. Proc. R. Soc. Lond. B Biol. Sci. 281: 20133046.

Bellwood, D.R., C.H. Goatley, J.A. Khan and S.B. Tebbett. 2016. Site fidelity and homing in juvenile rabbitfishes (Siganidae). Coral Reefs 35: 1151–1155.

Bellwood, D.R., R.P. Streit, S.J. Brandl and S.B. Tebbett. 2019. The meaning of the term 'function' in ecology: a coral reef perspective. Funct. Ecol. 33: 948–961.

Bennett, S., T. Wernberg, E.S. Harvey, J. Santana-Garcon and B.J. Saunders. 2015. Tropical herbivores provide resilience to a climate-mediated phase shift on temperate reefs. Ecol. Lett. 18: 714–723.

Berkström, C., M. Gullström, R. Lindborg, A.W. Mwandya, S.A. Yahya, N. Kautsky et al. 2012. Exploring 'knowns' and 'unknowns' in tropical seascape connectivity with insights from East African coral reefs. Estuar. Coast. Shelf Sci. 107: 1–21.

Bonaldo, R.M., A.S. Hoey and D.R. Bellwood. 2014. The ecosystem roles of parrotfishes on tropical reefs. Oceanogr. Mar. Biol. Annu. Rev. 52: 81–132.

Borsa, P., S. Lemer and D. Aurelle. 2007. Patterns of lineage diversification in rabbitfishes. Mol. Phylogenet. Evol. 44: 427–435.

Bos, A.R., E. Cruz-Rivera and A.M. Sanad. 2017. Herbivorous fishes *Siganus rivulatus* (Siganidae) and *Zebrasoma desjardinii* (Acanthuridae) feed on Ctenophora and Scyphozoa in the Red Sea. Mar. Biodivers. 47: 243–246.

Bowden, C.L., R.P. Streit, D.R. Bellwood and S.B. Tebbett. 2022. A 3D perspective on sediment turnover and feeding selectivity in blennies. Mar. Pollut. Bull. 180: 113799.

Brandl, S. and D. Bellwood. 2013a. Morphology, sociality, and ecology: can morphology predict pairing behavior in coral reef fishes? Coral Reefs 32: 835–846.

Brandl, S. and D. Bellwood. 2013b. Pair formation in the herbivorous rabbitfish *Siganus doliatus*. J. Fish Biol. 82: 2031–2044.

Brandl, S.J. and D.R. Bellwood. 2014. Individual-based analyses reveal limited functional overlap in a coral reef fish community. J. Anim. Ecol. 83: 661–670.

Brandl, S.J., A.S. Hoey and D.R. Bellwood. 2014. Micro-topography mediates interactions between corals, algae, and herbivorous fishes on coral reefs. Coral Reefs 33: 421–430.

Brandl, S.J. and D.R. Bellwood. 2015. Coordinated vigilance provides evidence for direct reciprocity in coral reef fishes. Sci. Rep. 5: 14556.

Brandl, S.J., W.D. Robbins and D.R. Bellwood. 2015. Exploring the nature of ecological specialization in a coral reef fish community: morphology, diet and foraging microhabitat use. Proc. R. Soc. Lond. B Biol. Sci. 282: 20151147.

Brandl, S.J. and D.R. Bellwood. 2016. Microtopographic refuges shape consumer-producer dynamics by mediating consumer functional diversity. Oecologia 182: 203–217.

Brandl, S.J., M.J. Emslie and D.M. Ceccarelli. 2016. Habitat degradation increases functional originality in highly diverse coral reef fish assemblages. Ecosphere 7: e01557.

Brandl, S.J., D.B. Rasher, I.M. Côté, J.M. Casey, E.S. Darling, J.S. Lefcheck et al. 2019. Coral reef ecosystem functioning: eight core processes and the role of biodiversity. Front. Ecol. Environ. 17: 445–454.

Brandl, S.J., J.S. Lefcheck, A.E. Bates, D.B. Rasher and T. Norin. 2022. Can metabolic traits explain animal community assembly and functioning? Biol. Rev. 98: 1–18.

Burkepile, D.E. and M.E. Hay. 2008. Herbivore species richness and feeding complementarity affect community structure and function on a coral reef. Proc. Natl. Acad. Sci. USA 105: 16201–16206.

Burkepile, D.E. and M.E. Hay. 2010. Impact of herbivore identity on algal succession and coral growth on a Caribbean reef. PLoS ONE 5: e8963.

Capper, A., I.R. Tibbetts, J.M. O'Neil and G.R. Shaw. 2006. Feeding preference and deterrence in rabbitfish *Siganus fuscescens* for the cyanobacterium *Lyngbya majuscula* in Moreton Bay, south-east Queensland, Australia. J. Fish Biol. 68: 1589–1609.

Carpenter, R.C. 1986. Partitioning herbivory and its effects on coral reef algal communities. Ecol. Monogr. 56: 345–364.

Charpy, L., B.E. Casareto, M.J. Langlade and Y. Suzuki. 2012. Cyanobacteria in coral reef ecosystems: a review. J. Mar. Biol. 2012: 259571.

Cheal, A., M. Emslie, I. Miller and H. Sweatman. 2012. The distribution of herbivorous fishes on the Great Barrier Reef. Mar. Biol. 159: 1143–1154.

Chelsky Budarf, A., D.D. Burfeind, W.K.W. Loh and I.R. Tibbetts. 2011. Identification of seagrasses in the gut of a marine herbivorous fish using DNA barcoding and visual inspection techniques. J. Fish Biol. 79: 112–121.

Choat, J.H. and K.D. Clements. 1998. Vertebrate herbivores in marine and terrestrial environments: a nutritional ecology perspective. Annu. Rev. Ecol. Evol. Syst. 29: 375–403.

Choat, J., K. Clements and W. Robbins. 2002. The trophic status of herbivorous fishes on coral reefs. Mar. Biol. 140: 613–623.

Chong-Seng, K., K. Nash, D. Bellwood and N. Graham. 2014. Macroalgal herbivory on recovering versus degrading coral reefs. Coral Reefs 33: 409–419.

Cissell, E.C., J.C. Manning and S.J. McCoy. 2019. Consumption of benthic cyanobacterial mats on a Caribbean coral reef. Sci. Rep. 9: 12693.

Clark, R.D., S. Pittman, C. Caldow, J. Christensen, B. Roque, R.S. Appeldoorn et al. 2009. Nocturnal fish movement and trophic flow across habitat boundaries in a coral reef ecosystem (SW Puerto Rico). Caribb. J. Sci. 45: 282–303.

Clements, C.S., D.B. Rasher, A.S. Hoey, V.E. Bonito and M.E. Hay. 2018 Spatial and temporal limits of coral-macroalgal competition: the negative impacts of macroalgal density, proximity, and history of contact. Mar. Ecol. Prog. Ser. 586: 11–20.

Clements, K.D. and J.H. Choat. 1995. Fermentation in tropical marine herbivorous fishes. Physiol. Zool. 68: 355–378.

Clements, K.D., D.P. German, J. Piché, A.D. Tribollet and J.H. Choat. 2017. Integrating ecological roles and trophic resources on coral reefs: multiple lines of evidence identify parrotfishes as microphages. Biol. J. Linn. Soc. 120: 729–751.

Connell, S.D., M.S. Foster and L. Airoldi. 2014. What are algal turfs? Towards a better description of turfs. Mar. Ecol. Prog. Ser. 495: 299–307.

Crossman, D.J., J.H. Choat, K.D. Clements, T. Hardy and J. McConochie. 2001. Detritus as food for grazing fishes on coral reefs. Limnol. Oceanogr. 46: 1596–1605.

Cvitanovic, C. and D. Bellwood. 2009. Local variation in herbivore feeding activity on an inshore reef of the Great Barrier Reef. Coral Reefs 28: 127–133.

D'agata, S., D. Mouillot, M. Kulbicki, S. Andréfouët, D.R. Bellwood, J.E. Cinner et al. 2014. Human-mediated loss of phylogenetic and functional diversity in coral reef fishes. Curr. Biol. 24: 555–560.

Davis, J.P., K.A. Pitt, B. Fry, A.D. Olds and R.M. Connolly. 2014. Seascape-scale trophic links for fish on inshore coral reefs. Coral Reefs 33: 897–907.

Devictor, V., J. Clavel, R. Julliard, S. Lavergne, D. Mouillot, W. Thuiller et al. 2010. Defining and measuring ecological specialization. J. Appl. Ecol. 47: 15–25.

Diaz-Pulido, G., S. Harii, L.J. McCook and O. Hoegh-Guldberg. 2010. The impact of benthic algae on the settlement of a reef-building coral. Coral Reefs 29: 203–208.

Ebrahim, A., T.S. Martin, P. Mumby, A.D. Olds and I. Tibbetts. 2020. Differences in diet and foraging behaviour of commercially important rabbitfish species on coral reefs in the Indian Ocean. Coral Reefs 39: 977–988.

Eggertsen, M., D.H. Chancin, C. Åkerlund, C. Halling and C. Berkström. 2019. Contrasting distribution and foraging patterns of herbivorous and detritivorous fishes across multiple habitats in a tropical seascape. Mar. Biol. 166: 51.

Evans, R.D., S.K. Wilson, R. Fisher, N.M. Ryan, R. Babcock, D. Blakeway et al. 2020. Early recovery dynamics of turbid coral reefs after recurring bleaching events. J. Environ. Manage. 268: 110666.

Evensen, N.R., I. Vanwonterghem, C. Doropoulos, M. Gouezo, E.S. Botté, N.S. Webster et al. 2021. Benthic micro- and macro-community succession and coral recruitment under overfishing and nutrient enrichment. Ecology 102: e03536.

Fisher, R. and J.D. Hogan. 2007. Morphological predictors of swimming speed: a case study of pre-settlement juvenile coral reef fishes. J. Exp. Biol. 210: 2436–2443.

Floeter, S.R., M.G. Bender, A.C. Siqueira and P.F. Cowman. 2018. Phylogenetic perspectives on reef fish functional traits. Biol. Rev. 93: 131–151.

Ford, A.K., S. Bejarano, M.M. Nugues, P.M. Visser, S. Albert and S.C.A. Ferse. 2018. Reefs under siege—the rise, putative drivers, and consequences of benthic cyanobacterial mats. Front. Mar. Sci. 5: 18.

Fox, R.J. and D.R. Bellwood. 2007. Quantifying herbivory across a coral reef depth gradient. Mar. Ecol. Prog. Ser. 339: 49–59.

Fox, R. and D. Bellwood. 2008. Remote video bioassays reveal the potential feeding impact of the rabbitfish *Siganus canaliculatus* (f: Siganidae) on an inner-shelf reef of the Great Barrier Reef. Coral Reefs 27: 605–615.

Fox, R.J., T.L. Sunderland, A.S. Hoey and D.R. Bellwood. 2009. Estimating ecosystem function: contrasting roles of closely related herbivorous rabbitfishes (Siganidae) on coral reefs. Mar. Ecol. Prog. Ser. 385: 261–269.

Fox, R.J. and D.R. Bellwood. 2011. Unconstrained by the clock? Plasticity of diel activity rhythm in a tropical reef fish, *Siganus lineatus*. Funct. Ecol. 25: 1096–1105.

Fox, R. and D. Bellwood. 2013. Niche partitioning of feeding microhabitats produces a unique function for herbivorous rabbitfishes (Perciformes, Siganidae) on coral reefs. Coral Reefs 32: 13–23.

Francis, F.T. and I.M. Côté. 2018. Fish movement drives spatial and temporal patterns of nutrient provisioning on coral reef patches. Ecosphere 9: e02225.

Friedman, S., S. Price, K. Corn, O. Larouche, C. Martinez and P. Wainwright. 2020. Body shape diversification along the benthic–pelagic axis in marine fishes. Proc. R. Soc. Lond. B Biol. Sci. 287: 20201053.

Fulton, C.J., R.A. Abesamis, C. Berkström, M. Depczynski, N.A.J. Graham, T.H. Holmes et al. 2019. Form and function of tropical macroalgal reefs in the Anthropocene. Funct. Ecol. 33: 989–999.

Fulton, C.J., C. Berkström, S.K. Wilson, R.A. Abesamis, M. Bradley, C. Åkerlund et al. 2020. Macroalgal meadow habitats support fish and fisheries in diverse tropical seascapes. Fish Fish. 21: 700–717.

Gagic, V., I. Bartomeus, T. Jonsson, A. Taylor, C. Winqvist, C. Fischer et al. 2015. Functional identity and diversity of animals predict ecosystem functioning better than species-based indices. Proc. R. Soc. Lond. B Biol. Sci. 282: 20142620.

Gell, F.R. and M.W. Whittington. 2002. Diversity of fishes in seagrass beds in the Quirimba Archipelago, northern Mozambique. Mar. Freshwater Res. 53: 115–121.

Ghilardi, M., N.M. Schiettekatte, J.M. Casey, S.J. Brandl, S. Degregori, A. Mercière et al. 2021. Phylogeny, body morphology, and trophic level shape intestinal traits in coral reef fishes. Ecol. Evol. 11: 13218–13231.

Gilby, B.L., I.R. Tibbetts and T. Stevens. 2017. Low functional redundancy and high variability in Sargassum browsing fish populations in a subtropical reef system. Mar. Freshwater Res. 68: 331–341.

Gill, A. 2003. The dynamics of prey choice in fish: the importance of prey size and satiation. J. Fish Biol. 63: 105–116.

Goatley, C. and D.R. Bellwood. 2009. Morphological structure in a reef fish assemblage. Coral Reefs 28: 449–457.

Goatley, C.H.R. and D.R. Bellwood. 2010. Biologically mediated sediment fluxes on coral reefs: sediment removal and off-reef transportation by the surgeonfish *Ctenochaetus striatus*. Mar. Ecol. Prog. Ser. 415: 237–245.

Goatley, C.H.R. and D.R. Bellwood. 2012. Sediment suppresses herbivory across a coral reef depth gradient. Biol. Lett. 8: 1016–1018.

Goatley, C.H.R., R.M. Bonaldo, R.J. Fox and D.R. Bellwood. 2016. Sediments and herbivory as sensitive indicators of coral reef degradation. Ecol. Soc. 21: 29.

Graham, N.A.J., S.K. Wilson, S. Jennings, N.V.C. Polunin, J.P. Bijoux and J. Robinson. 2006. Dynamic fragility of oceanic coral reef ecosystems. Proc. Natl. Acad. Sci. USA 103: 8425–8429.

Gundermann, N., D. Popper and T. Lichatowich. 1983. Biology and life cycle of S*iganus vermiculatus* (Siganidae, Pisces). Pac. Sci. 37: 165–180.

Haan, J. den, J. Huisman, H.J. Brocke, H. Goehlich, K.R.W. Latijnhouwers, S. van Heeringen et al. 2016. Nitrogen and phosphorus uptake rates of different species from a coral reef community after a nutrient pulse. Sci Rep. 6: 28821.

Hadj-Hammou, J., D. Mouillot and N.A. Graham. 2021. Response and effect traits of coral reef fish. Front. Mar. Sci. 8: 640619.

Hatcher, B.G. 1988. Coral reef primary productivity: a beggar's banquet. Trends Ecol. Evol. 3: 106–111.

Healy, K., T.H. Ezard, O.R. Jones, R. Salguero-Gómez and Y.M. Buckley. 2019. Animal life history is shaped by the pace of life and the distribution of age-specific mortality and reproduction. Nat. Ecol. Evol. 3: 1217–1224.

Heenan, A., A.S. Hoey, G.J. Williams and I.D. Williams. 2016. Natural bounds on herbivorous coral reef fishes. Proc. R. Soc. Lond. B Biol. Sci. 283: 20161716.

Hemingson, C.R. and D.R. Bellwood. 2020. Greater multihabitat use in Caribbean fishes when compared to their Great Barrier Reef counterparts. Estuar. Coast. Shelf Sci. 239: 106748.

Henderson, C.J., A.D. Olds, S.Y. Lee, B.L. Gilby, P.S. Maxwell, R.M. Connolly et al. 2017. Marine reserves and seascape context shape fish assemblages in seagrass ecosystems. Mar. Ecol. Prog. Ser. 566: 135–144.

Henderson, C.J., T. Stevens, S.Y. Lee, B.L. Gilby, T.A. Schlacher, R.M. Connolly et al. 2019. Optimising seagrass conservation for ecological functions. Ecosyst. 22: 1368–1380.

Hoey, A.S. and D.R. Bellwood. 2011. Suppression of herbivory by macroalgal density: a critical feedback on coral reefs? Ecol. Lett. 14: 267–273.

Hoey, A., S. Brandl and D. Bellwood. 2013. Diet and cross-shelf distribution of rabbitfishes (f. Siganidae) on the northern Great Barrier Reef: implications for ecosystem function. Coral Reefs 32: 973–984.

Hughes, T.P., M.J. Rodrigues, D.R. Bellwood, D. Ceccarelli, O. Hoegh-Guldberg, L. McCook et al. 2007. Phase shifts, herbivory, and the resilience of coral reefs to climate change. Curr. Biol. 17: 360–365.

Humphries, A.T., C.D. McQuaid and T.R. McClanahan. 2015. Context-dependent diversity-effects of seaweed consumption on coral reefs in Kenya. PLoS ONE 10: e0144204.

Hutchinson, G.E. 1957. Cold spring harbor symposium on quantitative biology. Concluding Remarks 22: 415–427.

Johns, K.A., M.J. Emslie, A.S. Hoey, K. Osborne, M.J. Jonker and A.J. Cheal. 2018. Macroalgal feedbacks and substrate properties maintain a coral reef regime shift. Ecosphere 9: e02349.

Kaunda-Arara, B. and G.A. Rose. 2004. Long-distance movements of coral reef fishes. Coral Reefs 23: 410–412.

Kelly, E.L., Y. Eynaud, I.D. Williams, R.T. Sparks, M.L. Dailer, S.A. Sandin et al. 2017. A budget of algal production and consumption by herbivorous fish in an herbivore fisheries management area, Maui, Hawaii. Ecosphere 8: e01899.

Khan, J.A., C.H. Goatley, S.J. Brandl, S.B. Tebbett and D.R. Bellwood. 2017. Shelter use by large reef fishes: long-term occupancy and the impacts of disturbance. Coral Reefs 36: 1123–1132.

Klumpp, D.W. and A.D. McKinnon. 1989. Temporal and spatial patterns in primary production of a coral-reef epilithic algal community. J. Exp. Mar. Biol. Ecol. 131: 1–22.

Kraft, N.J., R. Valencia and D.D. Ackerly. 2008. Functional traits and niche-based tree community assembly in an Amazonian forest. Science 322: 580–582.

Krone, R., M. Paster and H. Schuhmacher. 2011. Effect of the surgeonfish *Ctenochaetus striatus* (Acanthuridae) on the processes of sediment transport and deposition on a coral reef in the Red Sea. Facies 57: 215–221.

Kuo, F.-W., C.-Y. Kuo, T.-Y. Fan, M.-C. Liu and C.A. Chen. 2015. Hidden ecosystem function of rabbitfishes? *Siganus fuscescens* feeds on the soft coral, *Sarcophyton* sp. Coral Reefs 34: 57.

Lemer, S., D. Aurelle, L. Vigliola, J.-D. Durand and P. Borsa. 2007. Cytochrome b barcoding, molecular systematics and geographic differentiation in rabbitfishes (Siganidae). C. R. Biol. 330: 86–94.

Loffler, Z., D.R. Bellwood and A.S. Hoey. 2015. Among-habitat algal selectivity by browsing herbivores on an inshore coral reef. Coral Reefs 34: 597–605.

Longo, G.O., M.E. Hay, C.E.L. Ferreira and S.R. Floeter. 2019. Trophic interactions across 61 degrees of latitude in the Western Atlantic. Glob. Ecol. Biogeogr. 28: 107–117.

Lundberg, B. and D. Golani. 1995. Diet adaptations of Lessepsian migrant rabbitfishes, *Siganus luridus* and *S. rivulatus*, to the algal resources of the Mediterranean coast of Israel. Mar. Ecol. 16: 73–89.

MacArthur, R. and R. Levins. 1967. The limiting similarity, convergence, and divergence of coexisting species. Am. Nat. 101: 377–385.

Mallela, J. 2018. The influence of micro-topography and external bioerosion on coral-reef-building organisms: recruitment, community composition and carbonate production over time. Coral Reefs 37: 227–237.

McGill, B.J., B.J. Enquist, E. Weiher and M. Westoby. 2006. Rebuilding community ecology from functional traits. Trends Ecol. Evol. 21: 178–185.

McMahon, K.W., M.L. Berumen and S.R. Thorrold. 2012. Linking habitat mosaics and connectivity in a coral reef seascape. Proc. Natl. Acad. Sci. USA 109: 15372–15376.

Meyer, J.L. and E.T. Schultz. 1985a. Migrating haemulid fishes as a source of organic matter on coral reefs. Limnol. Oceanogr. 30: 146–156.

Meyer, J.L. and E.T. Schultz. 1985b. Tissue condition and growth rate of corals associated with schooling fish. Limnol. Oceanogr. 30: 157–166.

Mihalitsis, M. and D.R. Bellwood. 2017. A morphological and functional basis for maximum prey size in piscivorous fishes. PLoS ONE 12: e0184679.

Mirbach, C.E. and S.J. Brandl. 2016. Ontogenetic shifts in the social behaviour of pairing coral reef rabbitfishes (Siganidae). Mar. Biol. Res. 12: 874–880.

Moleana, T., L. Della Patrona, T. Meziane and Y. Letourneur. 2016. First record of *Siganus randalli* (Teleost, Siganidae) in New Caledonia, and comments on its diet. Mar. Biodivers. Rec. 9: 1–5.

Mouillot, D., N.A. Graham, S. Villéger, N.W. Mason and D.R. Bellwood. 2013. A functional approach reveals community responses to disturbances. Trends Ecol. Evol. 28: 167–177.

Müller, M., C.F. Staab, L.D. Puk, E.M. Schoenig, S.C. Ferse and C. Wild. 2021. The rabbitfish *Siganus virgatus* as key macroalgae browser in coral reefs of the Gulf of Thailand. Diversity 13: 123.

Nagle, D.G. and V.J. Paul. 1998. Chemical defense of a marine cyanobacterial bloom. J. Exp. Mar. Bio. Ecol. 225: 29–38.

Nanami, A. 2015. Pair formation, home range, and spatial variation in density, size and social status in blotched foxface *Siganus unimaculatus* on an Okinawan coral reef. PeerJ 3: e1280.

Nanami, A. 2018. Spatial distributions, feeding ecologies, and behavioral interactions of four rabbitfish species (*Siganus unimaculatus*, *S. virgatus*, *S. corallinus*, and *S. puellus*). PeerJ 6: e6145.

Nicholson, G.M. and K.D. Clements. 2020. Resolving resource partitioning in parrotfishes (Scarini) using microhistology of feeding substrata. Coral Reefs 39: 1313–1327.

Nicholson, G.M. and K.D. Clements. 2021. Ecomorphological divergence and trophic resource partitioning in 15 syntopic Indo-Pacific parrotfishes (Labridae: Scarini). Biol. J. Linn. Soc. 132: 590–611.

Norström, A.V., M. Nyström, J. Lokrantz and C. Folke. 2009. Alternative states on coral reefs: beyond coral-macroalgal phase shifts. Mar. Ecol. Prog. Ser. 376: 293–306.

Nyström, M. and C. Folke. 2001. Spatial resilience of coral reefs. Ecosyst. 4: 406–417.

O'Leary, J., D. Potts, K. Schoenrock and T. McClahanan. 2013. Fish and sea urchin grazing opens settlement space equally but urchins reduce survival of coral recruits. Mar. Ecol. Prog. Ser. 493: 165–177.

Olds, A.D., R.M. Connolly, K.A. Pitt, P.S. Maxwell, S. Aswani and S. Albert. 2014. Incorporating surrogate species and seascape connectivity to improve marine conservation outcomes. Conserv. Biol. 28: 982–991.

Orpwood, J.E., A.E. Magurran, J.D. Armstrong and S.W. Griffiths. 2008. Minnows and the selfish herd: effects of predation risk on shoaling behaviour are dependent on habitat complexity. Anim. Behav. 76: 143–152.

Perry, C.T., M.A. Salter, I.D. Lange, D.P. Kochan, A.R. Harborne and N.A.J. Graham. 2022. Geo-ecological functions provided by coral reef fishes vary among regions and impact reef carbonate cycling regimes. Ecosphere 13: e4288.

Pessarrodona, A., K. Filbee-Dexter, T. Alcoverro, J. Boada, C.J. Feehan, S. Fredriksen et al. 2021. Homogenization and miniaturization of habitat structure in temperate marine forests. Glob. Change Biol. 27: 5262–5275.

Pessarrodona, A., S.B. Tebbett, N.E. Bosch, D.R. Bellwood and T. Wernberg. 2022. High herbivory despite high sediment loads on a fringing coral reef. Coral Reefs 41: 161–173.

Pickholtz, R.S., M. Kiflawi, A.M. Friedlander and J. Belmaker. 2018. Habitat utilization by an invasive herbivorous fish (*Siganus rivulatus*) in its native and invaded range. Biol. Invasions 20: 3499–3512.

Pitt, J.M. 1997. The feeding ecology of rabbitfish (Siganidae) at Green Island reef: ontogenetic and interspecific differences in diet, morphology and habitat utilisation. Ph.D. Thesis, James Cook University, Townsville, AUS.

Plass-Johnson, J.G., S.C.A. Ferse, J. Jompa, C. Wild and M. Teichberg. 2015. Fish herbivory as key ecological function in a heavily degraded coral reef system. Limnol. Oceanogr. 60: 1382–1391.

Pratchett, M.S., O. Pradjakusuma and G.P. Jones. 2006. Is there a reproductive basis to solitary living versus pair-formation in coral reef fishes? Coral Reefs 25: 85–92.

Price, S.A., R. Holzman, T.J. Near and P.C. Wainwright. 2011. Coral reefs promote the evolution of morphological diversity and ecological novelty in labrid fishes. Ecol. Lett. 14: 462–469.

Proksch, P. 1994. Defensive roles for secondary metabolites from marine sponges and sponge-feeding nudibranchs. Toxicon 32: 639–655.

Randall, J.E. 1961. Overgrazing of algae by herbivorous marine fishes. Ecology 42: 812.

Randall, J.E. and M. Kulbicki. 2005. *Siganus woodlandi*, new species of rabbitfish (Siganidae) from New Caledonia. Cybium 29: 185–189.

Rasher, D.B. and M.E. Hay. 2010 Chemically rich seaweeds poison corals when not controlled by herbivores. Proc. Natl. Acad. Sci. USA 107: 9683–9688.

Rasher, D.B., A.S. Hoey and M.E. Hay. 2013. Consumer diversity interacts with prey defenses to drive ecosystem function. Ecology 94: 1347–1358.

Ricardo, G.F., R.J. Jones, M. Nordborg and A.P. Negri. 2017. Settlement patterns of the coral *Acropora millepora* on sediment-laden surfaces. Sci. Total Environ. 609: 277–288.

Richardson, L.E., N.A. Graham, M.S. Pratchett and A.S. Hoey. 2017. Structural complexity mediates functional structure of reef fish assemblages among coral habitats. Environ. Biol. Fishes 100: 193–207.

Rilov, G. and B.S. Galil. 2009. Marine bioinvasions in the Mediterranean Sea - History, distribution and ecology. pp. 549–575. *In*: Rilov, G. and J.A. Crooks [eds.]. Biological Invasions in Marine Ecosystems. Springer, Berlin.

Rizzari, J.R., A.J. Frisch, A.S. Hoey and M.I. McCormick. 2014. Not worth the risk: apex predators suppress herbivory on coral reefs. Oikos 123: 829–836.

Roberts, C.M. and R.F. Ormond. 1992. Butterflyfish social behaviour, with special reference to the incidence of territoriality: a review. Environ. Biol. Fishes 34: 79–93.

Robertson, D. 1982. Fish feces as fish food on a Pacific coral reef. Mar. Ecol. Prog. Ser. 7: 253–265.

Robinson, J.P.W., S.K. Wilson, J. Robinson, C. Gerry, J. Lucas, C. Assan et al. 2019a. Productive instability of coral reef fisheries after climate-driven regime shifts. Nat. Ecol. Evol. 3: 183–190.

Robinson, J.P.W., S.K. Wilson, S. Jennings and N.A.J. Graham. 2019b. Thermal stress induces persistently altered coral reef fish assemblages. Glob. Change Biol. 25: 2739–2750.

Robinson, J.P.W., J.M. McDevitt-Irwin, J.C. Dajka, J. Hadj-Hammou, S. Howlett, A. Graba-Landry et al. 2020. Habitat and fishing control grazing potential on coral reefs. Funct. Ecol. 34: 240–251.

Roff, G. and P.J. Mumby. 2012. Global disparity in the resilience of coral reefs. Trends Ecol. Evol. 27: 404–413.

Roff, G., I. Chollett, C. Doropoulos, Y. Golbuu, R.S. Steneck, A.L. Isechal et al. 2015. Exposure-driven macroalgal phase shift following catastrophic disturbance on coral reefs. Coral Reefs 34: 715–725.

Roff, G., S. Bejarano, M. Priest, A. Marshell, I. Chollett, R.S. Steneck et al. 2019. Seascapes as drivers of herbivore assemblages in coral reef ecosystems. Ecol. Monogr. 89: e01336.

Russ, G. 1984a. Distribution and abundance of herbivorous grazing fishes in the central Great Barrier Reef. II. Patterns of zonation of mid-shelf and outershelf reefs. Mar. Ecol. Prog. Ser. 20: 35–44.

Russ, G.R. 1984b. Distribution and abundance of herbivorous grazing fishes in the central Great Barrier Reef. I. Levels of variability across the entire continental shelf. Mar. Ecol. Prog. Ser. 20: 23–34.

Russ, G.R. and L.J. McCook. 1999. Potential effects of a cyclone on benthic algal production and yield to grazers on coral reefs across the central Great Barrier Reef. J. Exp. Mar. Bio. Ecol. 235: 237–254.

Russ, G.R., S.-L.A. Questel, J.R. Rizzari and A.C. Alcala. 2015. The parrotfish–coral relationship: refuting the ubiquity of a prevailing paradigm. Mar. Biol. 162: 2029–2045.

Sambrook, K., A.S. Hoey, S. Andréfouët, G.S. Cumming, S. Duce and M.C. Bonin. 2019. Beyond the reef: the widespread use of non-reef habitats by coral reef fishes. Fish Fish. 20: 903–920.

Sambrook, K., M.C. Bonin, M. Bradley, G.S. Cumming, S. Duce, S. Andréfouët et al. 2020. Broadening our horizons: seascape use by coral reef-associated fishes in Kavieng, Papua New Guinea, is common and diverse. Coral Reefs 39: 1187–1197.

Schlaefer, J.A., S.B. Tebbett, C.L. Bowden, W.P. Collins, S. Duce, C.R. Hemingson et al. 2022. A snapshot of sediment dynamics on an inshore coral reef. Mar. Environ. Res. 181: 105763.

Schmitz, L. and P.C. Wainwright. 2011. Nocturnality constrains morphological and functional diversity in the eyes of reef fishes. BMC Evol. Biol. 11: 338.

Semmler, R.F., S.J. Brandl, S.A. Keith and D.R. Bellwood. 2021. Fine-scale foraging behavior reveals differences in the functional roles of herbivorous reef fishes. Ecol. Evol. 11: 4898–4908.

Shantz, A.A., M.C. Ladd, E. Schrack and D.E. Burkepile. 2015. Fish-derived nutrient hotspots shape coral reef benthic communities. Ecol. Appl. 25: 2142–2152.

Siqueira, A.C., D.R. Bellwood and P.F. Cowman. 2019. The evolution of traits and functions in herbivorous coral reef fishes through space and time. Proc. R. Soc. Lond. B Biol. Sci. 286: 20182672.

Speare, K.E., A. Duran, M.W. Miller and D.E. Burkepile. 2019. Sediment associated with algal turfs inhibits the settlement of two endangered coral species. Mar. Pollut. Bull. 144: 189–195.

Steneck, R.S. 1988. Herbivory on coral reefs: a synthesis. Proc. 6th Intl. Coral Reef Symp. 1: 37–49.

Streit, R.P., A.S. Hoey and D.R. Bellwood. 2015. Feeding characteristics reveal functional distinctions among browsing herbivorous fishes on coral reefs. Coral Reefs 34: 1037–1047.

Streit, R.P. and D.R. Bellwood. 2022. To harness traits for ecology, let's abandon 'functionality'. Trends Ecol. Evol. 38: 402–411.

Stuart-Smith, R.D., A.E. Bates, J.S. Lefcheck, J.E. Duffy, S.C. Baker, R.J. Thomson et al. 2013. Integrating abundance and functional traits reveals new global hotspots of fish diversity. Nature 501: 539–542.

Tebbett, S.B., C.H.R. Goatley and D.R. Bellwood. 2017. Clarifying functional roles: algal removal by the surgeonfishes *Ctenochaetus striatus* and *Acanthurus nigrofuscus* on coral reefs. Coral Reefs 36: 803–813.

Tebbett, S.B., D.R. Bellwood and S.W. Purcell. 2018. Sediment addition drives declines in algal turf yield to herbivorous coral reef fishes: implications for reefs and reef fisheries. Coral Reefs 37: 929–937.

Tebbett, S.B. and D.R. Bellwood. 2019. Algal turf sediments on coral reefs: what's known and what's next. Mar. Pollut. Bull. 149: 110542.

Tebbett, S.B., A.C. Siqueira and D.R. Bellwood. 2022a. The functional roles of surgeonfishes on coral reefs: past, present and future. Rev. Fish Biol. Fish. 32: 387–439.

Tebbett, S.B., R.P. Streit, J. Morais, J.A. Schlaefer, S. Swan and D.R. Bellwood. 2022b. Benthic cyanobacterial mat formation during severe coral bleaching at Lizard Island: the mediating role of water currents. Mar. Environ. Res. 181: 105752.

Tebbett, S.B., S.R. Connolly and D.R. Bellwood. 2023. Benthic composition changes on coral reefs at global scales. Nat. Ecol. Evol. 7: 71–81.

Tootell, J.S. and M.A. Steele. 2016. Distribution, behavior, and condition of herbivorous fishes on coral reefs track algal resources. Oecologia 181: 13–24.

Topor, Z.M., D.B. Rasher, J.E. Duffy and S.J. Brandl. 2019. Marine protected areas enhance coral reef functioning by promoting fish biodiversity. Conserv. Lett. 12: e12638.

Trapon, M.L., M.S. Pratchett, A.S. Hoey and A.H. Baird. 2013. Influence of fish grazing and sedimentation on the early post-settlement survival of the tabular coral *Acropora cytherea*. Coral Reefs 32: 1051–1059.

Vanderklift, M., R. Pillans, M. Hutton, L. De Wever, G. Kendrick, A. Zavala-Perez et al. 2021. High rates of herbivory in remote northwest Australian seagrass meadows by rabbitfish and green turtles. Mar. Ecol. Prog. Ser. 665: 63–73.

Vega Thurber, R., D.E. Burkepile, A.M.S. Correa, A.R. Thurber, A.A. Shantz, R. Welsh et al. 2012. Macroalgae decrease growth and alter microbial community structure of the reef-building coral, *Porites astreoides*. PLoS ONE 7: e44246.

Vergés, A., F. Tomas, E. Cebrian, E. Ballesteros, Z. Kizilkaya, P. Dendrinos et al. 2014a. Tropical rabbitfish and the deforestation of a warming temperate sea. J. Ecol. 102: 1518–1527.

Vergés, A., P.D. Steinberg, M.E. Hay, A.G.B. Poore, A.H. Campbell, E. Ballesteros et al. 2014b. The tropicalization of temperate marine ecosystems: climate-mediated changes in herbivory and community phase shifts. Proc. R. Soc. Lond. B Biol. Sci. 281: 20140846.

Vergés, A., C. Doropoulos, H.A. Malcolm, M. Skye, M. Garcia-Pizá, E.M. Marzinelli et al. 2016. Long-term empirical evidence of ocean warming leading to tropicalization of fish communities, increased herbivory, and loss of kelp. Proc. Natl. Acad. Sci. USA 113: 13791–13796.

Vergés, A., E. McCosker, M. Mayer-Pinto, M.A. Coleman, T. Wernberg, T. Ainsworth et al. 2019. Tropicalisation of temperate reefs: Implications for ecosystem functions and management actions. Funct. Ecol. 33: 1000–1013.

Welsh, J.Q. and D.R. Bellwood. 2015. Simulated macro-algal outbreak triggers a large-scale response on coral reefs. PLoS ONE 10: e0132895.

Wernberg, T., S. Bennett, R.C. Babcock, T. De Bettignies, K. Cure, M. Depczynski et al. 2016. Climate-driven regime shift of a temperate marine ecosystem. Science 353: 169–172.

Whalan, S., M.A. Abdul Wahab, S. Sprungala, A.J. Poole and R. De Nys. 2015. Larval settlement: the role of surface topography for sessile coral reef invertebrates. PLoS ONE 10: e0117675.

Wild, C., M. Huettel, A. Klueter, S.G. Kremb, M.Y.M. Rasheed and B.B. Jørgensen. 2004. Coral mucus functions as an energy carrier and particle trap in the reef ecosystem. Nature 428: 66–70.

Wilson, S.K., K. Burns and S. Codi. 2001. Identifying sources of organic matter in sediments from a detritivorous coral reef fish territory. Org. Geochem. 32: 1257–1269.

Wilson, S.K., D.R. Bellwood, J.H. Choat and M.J. Furnas. 2003. Detritus in the epilithic algal matrix and its use by coral reef fishes. Oceanogr. Mar. Biol. Annu. Rev. 41: 279–309.

Woodland, D.J. 1990. Revision of the fish family Siganidae with descriptions of two new species and comments on distribution and biology. Indo-Pacific Fishes 19: 1–136.

Woodland, D. 1997. An examination of the effect of ecological factors, especially competitive exclusion, on the distributions of species of an inshore, tropical, marine family of Indo-Pacific fishes (Siganidae). Proc. 5th Indo-Pacific Fish Conf., Nouméa, New Caledonia: 3–8.

Woodland, D.J. and R.C. Anderson. 2014. Description of a new species of rabbitfish (Perciformes: Siganidae) from southern India, Sri Lanka and the Maldives. Zootaxa 3811: 129–136.

Yamaguchi, A. 2010. Biological aspects of herbivorous fishes in the coastal areas of western Japan. Bull. Fish. Res. Agen. 32: 89–94.

Zarco-Perello, S., T.J. Langlois, T. Holmes, M.A. Vanderklift and T. Wernberg. 2019. Overwintering tropical herbivores accelerate detritus production on temperate reefs. Proc. R. Soc. Lond. B Biol. Sci. 286: 20192046.

Zarco-Perello, S., G. Carroll, M. Vanderklift, T. Holmes, T.J. Langlois and T. Wernberg. 2020. Range-extending tropical herbivores increase diversity, intensity and extent of herbivory functions in temperate marine ecosystems. Funct. Ecol. 34: 2411–2421.

Zarco-Perello, S., N.E. Bosch, S. Bennett, M.A. Vanderklift and T. Wernberg. 2021. Persistence of tropical herbivores in temperate reefs constrains kelp resilience to cryptic habitats. J. Ecol. 109: 2081–2094.

7

Vulnerability of Rabbitfishes to Environmental Change

Andrew S. Hoey,[1,*] *Alexia Dubuc,*[2] *Jennifer M. Donelson,*[1]
Eva C. McClure[1] and *Morgan S. Pratchett*[1]

Introduction

Variation in environmental conditions has long been recognised as a major driver of ecological communities across spatial and temporal scales (Tittensor et al. 2010, Boyd et al. 2016). For example, differences in environmental conditions with altitude, depth, latitude, and distance from the coast have been related to changes in community composition across a range of terrestrial and marine ecosystems (e.g., Gentry 1988, Hoey and Bellwood 2008, Rath et al. 2017, Kenitz et al. 2010). Such changes in community composition along environmental gradients often reflect the physiological tolerances of individual species, with species entering and leaving communities based on their physiological tolerances (Franklin 1995, Sunday et al. 2019), and/or changes in biotic interactions. Similarly, habitats or ecosystems that experience considerable variation in environmental conditions over short temporal scales (e.g., seasonal, diurnal, tidal) favour species with broad ranges in physiological tolerances (Tewksbury et al. 2008, Gutiérrez-Pesquera et al. 2016, Blewett et al. 2022).

Increasingly, however, anthropogenic activities are altering the physical, chemical, and/or biological environments of many of the world's ecosystems and are becoming a major driver of ecological communities, and the good and services they provide (Walther 2010, Grimm et al. 2013, Pecl et al. 2017, Hughes et al. 2017). Indeed, the effects of local anthropogenic activities (e.g., land clearing, harvesting,

[1] College of Science and Engineering, James Cook University, Townsville, Australia.
Emails: jennifer.donelson@jcu.edu.au; eva.mcclure@jcu.edu.au; morgan.pratchett@jcu.edu.au
[2] School of Life Sciences, University of Essex, Colchester, United Kingdom.
Email: alexia.dubuc1@my.jcu.edu.au
* Corresponding author: andrew.hoey1@jcu.edu.au

pollution) are being greatly compounded by the effects of climate change, leading to shifts in the structure and functioning of ecological communities (Hoegh-Guldberg and Bruno 2010, Newbold et al. 2015, Diaz et al. 2019). Moreover, projected increases in the frequency, intensity, diversity and spatial extent of disturbances paints a bleak picture for the future of many ecosystems (e.g., Halpbern et al. 2015, Breitberg et al. 2018 Hughes et al. 2018a). Understanding the differential vulnerability of populations, species and communities to current and future disturbances is key to the conservation and management of ecosystems.

The effects of increasing anthropogenic disturbance are perhaps greatest in coastal tropical marine ecosystems where the combined effects of terrestrial run-off (i.e., eutrophication, sedimentation, pollution), coastal development, high human population densities and exploitation, and climate change are causing the loss and degradation of habitat-forming taxa (Selig et al. 2019). For example, upwards of 38 percent of the world's mangrove forests have been lost since the 1990's (Valiela et al. 2001, Goldberg et al. 2020), and 19 to 29 percent of the areal extent of the world's seagrass meadows have been lost since 1880 (Waycott et al. 2009, Dunic et al. 2021). Similarly, there have been widespread declines in the cover of scleractinian (hard) corals across tropical reefs globally (Gardner et al. 2003, Hughes et al. 2017, Souter et al. 2021). While historically the major cause/s of loss and degradation of coral reefs has varied geographically (e.g., Pandolfi et al. 2003, Bruno and Selig 2007), increasing seawater temperature due to climate change is now the foremost threat to coral reefs globally (Hoegh-Guldberg et al. 2007; Hughes et al. 2003, 2017). Indeed, modelled increases in the incidence of bleaching and the subsequent mortality of corals based on projected climate change scenarios have led to suggestions that there will be significant declines (*ca.* 70 percent), if not near complete loss of coral reefs before the end of this century (Hoegh-Guldberg et al. 2018). Moreover, tropical coastal ecosystems are some of the most physiologically challenging environments, with human activities posing a high risk of exacerbating these environmental conditions, making them unsuitable for most organisms (Dubuc et al. 2024). Species inhabiting these coastal marine ecosystems must not only deal with the direct effects of changes in their physical and chemical environment, but also the indirect effects related to the reductions in habitat-forming organisms, or ecosystem engineers (i.e., corals, seagrass, mangroves, macroalgae), and the structural complexity they provide.

Changing environmental conditions and the degradation of shallow coastal marine ecosystems is of considerable concern given the ecosystem goods and services they provide. Globally, up to one billion people live within 100 km of the coast across 117 tropical nations (many within Small Island Developing States), and their well-being and security is already at risk due to the degradation of tropical seascapes (Sing Wong et al. 2022). Notably, the degradation of tropical seascapes is leading to declines in the abundance, biomass and diversity, and shifts in the composition of shallow water fish communities (e.g., Pratchett et al. 2014, Malik et al. 2017, Richardson et al. 2018), and ultimately reducing the productivity of associated fisheries (Lam et al. 2020) and/or ecosystem functions (Graham et al. 2015, Richardson et al. 2020). There is, however, considerable variation among fish species in their susceptibility to habitat degradation, and their contribution to fisheries production and ecological processes.

Rabbitfishes (f. Siganidae) are a conspicuous component of the fish fauna in shallow tropical and subtropical marine and estuarine habitats throughout the Indo-west Pacific region (except the Hawaiian Islands and Easter Island; Woodland 1990), and more recently the Mediterranean Sea (Zarco-Perello et al. – Chapter 8). Despite broadly similar morphologies (Huertas and Hemmingson – Chapter 2) and generally considered to be herbivorous, there is considerable variation in diet among rabbitfish species (Hoey et al. 2013, Clements – Chapter 4). Rabbitfishes are most abundant in coastal marine and estuarine habitats (e.g., Woodland 1990; Cheal et al. 2012; Hoey et al. 2013; Dubuc et al. 2019, Pombo-Ayora et al. 2020), where they are common targets of commercial and artisanal fishers throughout much of their range (e.g., Lam 1974, Dalzell et al. 1996, Grandcourt et al. 2007, Kamukuru 2009, Olds et al. 2012, Taylor et al. – Chapter 5). As such, any environmentally-driven changes in rabbitfish populations could affect the livelihoods and food security of local communities.

The objective of this chapter is to explore species-specific vulnerability of rabbitfishes to environmental change, considering both the direct effects of changes in their physical and chemical environment and the indirect effects of the degradation of biogenic habitats. Specifically, we describe the tolerance of rabbitfishes to environmental conditions (temperature, salinity, dissolved oxygen), and resource requirements (diet, habitat), and then explore the species-specific vulnerability of rabbitfishes to habitat degradation using a database on changes in the abundance of fishes following acute disturbances on coral reefs, and a case study from the Great Barrier Reef.

Resource Use and Plasticity

The vulnerability of species to changing environmental and habitat conditions is directly related to their requirements for environmental conditions and resources, with consideration for both their optimal or preferred range, and their tolerance range for a particular environmental variable or resource. Such relationships may be visualised using species performance curves (Fig. 1). The left- and right-hand bounds of a species performance curve are defined by the critical minimum (CR_{min}) and maximum (CR_{max}) of the environmental variable, respectively. This represents the species tolerance range, beyond which individual survival is compromised. Within a species' tolerance range the relationship between the environmental variable and performance is typically represented by a bell-shaped curve (e.g., thermal performance curve, Fig. 1). As the availability or concentration of the environmental variable increases from the critical minimum, performance often increases exponentially before slowing as it approaches the optimum (R_{opt}): the availability or concentration of the environmental variable at which the highest performance is achieved. At values above R_{opt}, performance declines until it reaches the critical maximum. While these performance curves are typically applied to environmental variables (e.g., temperature) they could equally be applied to resource availability (e.g., food or habitat).

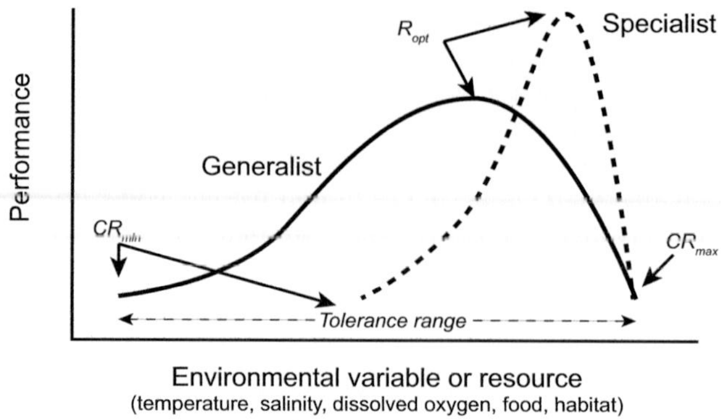

Fig. 1. A hypothetical performance curve showing the relationship between the level or concentration of an environmental variable and organism performance for a specialist (*dashed line*) and generalist (*solid line*). The critical environmental variable minimum (CR_{min}) and maximum (CR_{max}) bound the tolerance range, beyond which survival is compromised. Note the broader tolerance range of the generalist versus the specialist.

The response of an individual to changing environmental conditions will depend on the shape of their performance curve and the position of their local environment along the performance curve. For example, the temperature performance curve of tropical ectotherms is often shown to be more compressed than that of temperate ectotherms, rendering tropical species more vulnerable than temperate species to similar increases in temperature (e.g., Tewksbury et al. 2008, Sunday et al. 2011). Such relationships can be translated to other resource requirements, whereby species with compressed or narrow performance curves (resource specialists) are more constrained by resource availability and more vulnerable to changes in environmental conditions and resource availability than species with broad performance curves (resource generalists). It is, however, important to note that in natural environments some species may use only restricted range of preferred resources when available but use a wider range of resources when their preferred resources are depleted or absent (e.g., Feary et al. 2018). Manipulative experiments that test the complete range of resources under controlled conditions are required to differentiate resource use from requirements, and fully understand the resource requirements and plasticity of individual species; such experiments are, however, rare.

Environmental Conditions

Temperature

Increasing environmental temperature due to climate change is having a pervasive and immediate effect on marine ecosystems (e.g., Hoey et al. 2016a, Hughes et al. 2017, Pinsky et al. 2019), largely due to the predominance of ectothermic species in these ecosystems. Ectotherms are particularly sensitive to changes in temperature, although the mechanistic understanding of resulting responses is still debated (e.g., Pörtner 2010, Deutsch et al. 2015, Jutfelt et al. 2018, Pauly 2021). Despite the controversy, there is a consensus in that warming induces profound effects on biochemical and cellular processes ultimately influencing fitness-related functions such as metabolism, growth, development and reproduction (Huey and Stevenson 1979, Brown et al. 2004, Schulte 2015). Ultimately, temperature-induced changes in these functions can influence species distributions, population dynamics, and biotic interactions (Dell et al. 2011, Pecl et al. 2017). Although tropical species are generally considered to be more sensitive to increasing temperature than their subtropical or temperate counterparts having evolved in relatively stable thermal environments (Tewksbury et al. 2008), a growing body of research is showing that the responses of organisms to increasing temperature varies considerably among populations, species, and the traits considered (e.g., Johansen et al. 2014, LaMonica et al. 2021).

Despite a rapidly growing body of research investigating the effects of increasing temperature on tropical fish, in particular coral reef fish (e.g., Zarco-Perelló et al. 2012, Johansen et al. 2015, Motson and Donelson 2017, Nay et al. 2020, 2021), there has been very limited investigation on the thermal tolerances or effect of increasing temperature on the performance of rabbitfishes (see Graba-Landry 2020, LaMonica et al. 2021 for exceptions). While this somewhat limits our ability to project the direct impacts of environmental change, insight into the thermal tolerance of rabbitfishes can, however, be drawn from their geographic distributions and investigation of the optimal conditions for growth in aquaculture.

The broad latitudinal distribution of some rabbitfish species is indicative of a wide thermal tolerance range. Notably, the distribution of *Siganus fuscescens* spans over 70° of latitude, from northern Japan to southern Australia, equating to > 15°C range in seawater temperature (e.g., 15°C in winter on the southern New South Wales coast to 31°C in summer in Sulawesi; Reef Life Survey, 2024; Fig. 2). Numerous tropical and subtropical fish species, including *S. fuscescens*, have been reported to have undergone recent poleward range extensions due to strengthening of western boundary currents carrying warm waters to higher latitudes along the Australian and Japanese coasts (e.g., Feary et al. 2014, Verges et al. 2014, 2015, Pecl et al. 2017). While these range expansions may be relatively recent for some species, early records indicate that *S. fuscescens* has historically occurred at higher latitudes, and its current distribution is not a result of recent range expansion. For example, Quoy and Gaimaird collected two rabbitfish in Sydney Harbour in 1825, and from their original descriptions, Woodland (1990) concluded they could only be *S. fuscescens*. Several other studies refer to this species along the subtropical and temperate coast of eastern and western Australia as *Siganus nebulosus* (e.g., Malcolm et al. 2007, Vanderklift et al. 2019), however, *S. nebulosus* is almost certainly a junior synonym of *S. fuscescens*. Similarly, the distribution of *Siganus sutor* spans at least 46° of latitude, from Knysna, South Africa (34°S; mean monthly seawater temperature range: 17.3–20.8°C) to the Gulf of Aden (12.5°N; 25.8–30.9°C; Woodland 1990, DiBattista et al. 2016), with a recent report of two individuals collected in southern Iraq (30°N; 15.7–32.2°C; Jawad et al. 2023). The sister species *S. punctatus* and *S. stellatus*, while most abundant on tropical reefs, also span from equatorial waters to as far south as 32°S (Woodland 1990). While the extensive latitudinal distributions of these rabbitfish suggest they may be thermal generalists, these broad distributions could equally have been driven by their ability to occupy soft sediment habitats (e.g., seagrass). Unlike other tropical piscine herbivores, namely parrotfishes, surgeonfishes, and herbivorous damselfishes, many rabbitfish species can exploit soft benthic and estuarine habitats (Johannes 1978, Gundermann et al. 1983, Woodland 1990, Dubuc et al. 2021, Vanderklift et al. 2021). This ability to use soft sediment habitats may have allowed rabbitfishes to extend beyond tropical habitats more effectively than the other herbivores.

It should also be noted that not all rabbitfish species have a broad latitudinal range. Many rabbitfish species have a restricted tropical distribution, which creates the potential for thermal specialisation and enhanced performance at warmer ocean temperatures. For example, the distributions of *Siganus magnificus, Siganus labyrinthodes and Siganus randalli* do not extend beyond 15 degrees latitude (N or S). While direct whole organism performance measurements are not available, a narrow thermal range is expected due to a maximal seasonal temperature range of around 4°C and latitudinal thermal range from 25.5 to 29.7°C (15.5°S–5.5°S/N; Lough 2012). For tropical rabbitfish species such as *Siganus doliatus, Siganus lineatus, Siganus corallinus, Siganus vulpinus, Siganus punctatissimus,* and *Siganus puellus* that encompass the full breadth of the tropical oceans, but do not extend beyond 30 degrees latitude (N or S), temperatures experienced range from 20.2 to 31.4°C (*S. doliatus* and *S. javus*, respectively) giving a season thermal range per species of 6.9 to 10.9°C (*S. punctatissimus* and *S. doliatus*, respectively; Reef Life

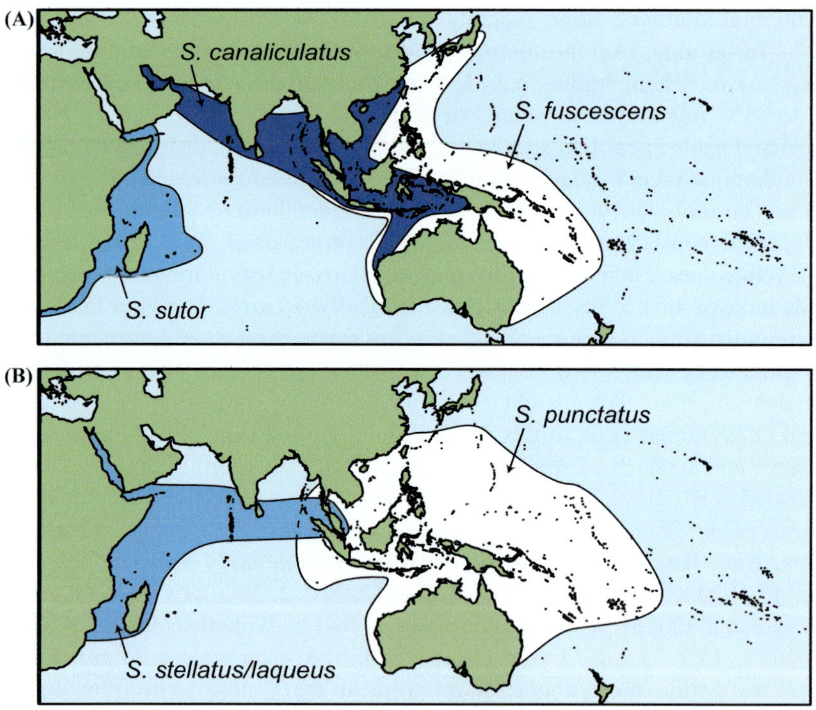

Fig. 2. The geographic distributions of five species of rabbitfish showing the occupancy of a broad range of thermal environments. **(A)** broad latitudinal extent of *Siganus fuscescens* from Japan to southern Australia, and *Siganus sutor* from the Gulf of Aden to South Africa; and the presence of *S. canaliculatus* in the thermally variable environment of the Arabian Gulf. **(B)** broad latitudinal extent of *Siganus punctatus* in the central Indo-Pacific, and the sister taxa *Siganus stellatus* and *Siganus laqueus* in the western Indian Ocean.

Survey, 2024). Considering this reasonably broad thermal range for these rabbitfish species there is perhaps less concern of thermal sensitivity to warming, however, it is unknown exactly how close species currently live to their upper thermal limits especially for sensitive processes like reproduction (Pankhurst and Munday 2011).

In addition to these broad latitudinal distributions, several rabbitfish species occupy thermally variable environments. For example, *S. canaliculatus* and *S. javus* are common and/or abundant within the extreme thermal environment of the southern Arabian Gulf (Burt et al. 2013; Fig. 2), where water temperatures vary from *ca.* 14 to 36°C annually (Hoey et al. 2018, Brandl et al. 2020, Burt et al. 2020). Further, several rabbitfish species inhabit estuarine environments either as juveniles (*S. guttatus, S. insomnis, S. javus, S. lineatus, S. randalli,* and *S. vermiculatus*) and/or adults (*S. vermiculatus*; Johannes 1978, Gundermann et al. 1983, Woodland 1990, Woodland and Anderson 2014, Moleana et al. 2016, Dubuc et al. 2021, Vanderklift et al. 2021) where they experience large fluctuations in water temperature (4–6°C) over short (diurnal) temporal scales (Dubuc et al. 2019). Some individuals are likely to be exposed to even greater variation in temperature and other environmental conditions. For example, juvenile *S. lineatus* have been collected from tidal wetland pools in the upper reaches of Ross River, Townsville, north Queensland (Dubuc et al. 2019), where daily variation in water temperature can exceed 10°C (Dubuc et al. 2017). This diurnal variability may benefit species in response to ocean warming even when latitudinal distributions result in similar upper thermal limit expectations of ~ 31.2°C (Reef Life Survey 2024). This was observed experimentally with maintained thermal performance of juvenile *S. lineatus* compared to *S. doliatus* (reef associated; Table 1) at 31.5°C (LaMonica et al. 2021).

The broad latitudinal distribution of some rabbitfish species, coupled with the thermally variable habitats occupied by other rabbitfish species, suggest most species have a high degree of flexibility in their temperature tolerance (i.e., thermal generalists). Indeed, the broad thermal tolerance of rabbitfishes has likely contributed to the successful invasion of the Mediterranean Sea by *S. rivulatus* and *S. luridus* following the opening of the Suez canal (Stergiou 1988, Rilov and Galil 2009, Zarco-Perello et al. – Chapter 8), and increases in abundance of *S. fuscescens* along the eastern and western coasts of Australia (Verges et al. 2016; Zarco-Perello et al. 2021) and northern Japan (Kumagai et al. 2018).

The flexibility in the thermal tolerance of rabbitfishes has also been cited as one of the characteristics that make them amenable to aquaculture (Camus et al. – Chapter 9). For example, juvenile *S. rivulatus* can withstand temperatures as low as 14°C and as high as 36°C, with no difference in survival of juveniles reared in temperatures from 17-32°C (Saoud et al. 2008). Similarly, juvenile *S. canaliculatus* have a thermal tolerance range of 12.1-32.5°C under culture conditions (Huang et al. 2024), and juvenile *S. vermiculatus* have been reported to survive variable thermal environments (25-35°C) in a tank culture trial in Fiji (Popper et al. 1976). While direct comparisons with other fish species are difficult due to methodological differences among studies, the thermal tolerance of rabbitfishes is generally greater than those of other co-occurring species in both temperate (Marras et al. 2015) and tropical reef ecosystems (Rajaguru 2002, Zarco-Perllo et al. 2012, Messmer et al. 2017).

Table 1. Habitat associations of recently-settled and juvenile rabbitfishes. The vast majority of juvenile rabbitfish associate with vegetated habitats (i.e., macroalgae, seagrass, and mangroves) and appear to show a high degree of plasticity in their habitat associations. * indicates habitat association of sub-adults.

Species	Habitat	Location	Source
S. argenteus	Shallow reef flat	Guam	Tsuda and Bryan 1973, Tobias (1976 in Woodland (1990)
	Seagrass and Macroalgae	New Caledonia	Mellin et al. (2007)
S. canaliculatus	Macroalgae	Ishigaki Island, Japan	Sugiyama and Tomori (1990)
	Macroalgae - *Sargassum*	Orpheus Island, GBR	Bellwood et al. 2006, Waldeland (2017), Tang et al (2020), Martin et al. (2024)
		Turtle Group, GBR	Hoey et al. (2013)
	Seagrass	Qatar	El-Sayad (1994)
		Arabian Gulf	Lin et al. (2019)
		Thailand	Sichum et al. (2013), Du et al. (2020)
		Magnetic Island, GBR	Kwak et al. (2015)
	Mangrove estuary	Thailand	Ikejima et al. (2003)
S. corallinus	Coral-rich habitat	Okinawa, Japan	Nanami (2018)
	Coral – branching *Acropora*	not specified	Woodland (1990)
	Coral rubble	Lizard Island, GBR	Bellwood et al. (2016)
	Seagrass	not specified	Woodland (1990)
S. doliatus	Macroalgae – *Sargassum*	Orpheus Island, GBR	Tang et al. (2020), LaMonica et al. (2021), Webber et al. (2024), Martin et al. (2024)
	Macroalgae – *Sargassum*, and seagrass	Green Island, GBR	Pitt (1997)
	Coral rubble	Lizard Island, GBR	Bellwood et al. (2016)
S. fuscescens	Macroalgae - high canopy	Montebello and Barron Island, Western Australia	Evans et al. (2014)
	Seagrass	Port Gregory, Western Australia	Zarco-Perello et al. (2021)
		Ryukyu Isl, Japan	Nakamura and Tsuchiya (2008)
		Palau	Woodland (1990)
	Seagrass and macroalgae	Green Island, GBR	Pitt (1997)
		New Caledonia	Mellin et al. (2007)
	Mangrove	Philippines	Honda et al. (2013)

Table 1 contd. ...

...Table 1 contd.

Species	Habitat	Location	Source
S. guttatus	Macroalgae	Ishigaki Island, Japan	Sugiyama and Tomori (1990)
	Mangrove	Philippines	Honda et al. (2013)
	Seagrass	Lagonoy Gulf, Philippines	Soliman et al. (2009)
	Seagrass and mangrove	Gulf of Thailand	Sichum et al. (2013)
S. insomnis	estuarine habitat	Sri Lanka	Woodland and Anderson (2014)
	seagrass	Puttalam lagoon, Sri Lanka	Jones et al. (2018)
S. javus	Mangrove estuary	Thailand	Ikejima et al. (2003)
	Mangrove and seagrass	Negombo estuary, Sri Lanka	Pinto and Punchihewa (1996)
	Seagrass	Qatar	El-Sayad (1994)
		Puttalam lagoon, Sri Lanka	Jones et al. (2018)
	Seagrass, mangrove and mudflat	Gulf of Thailand	Sichum et al. (2013)
S. lineatus	Estuarine habitat	Ross River, North Queensland	LaMonica et al. (2021)
	Mangrove	New Caeldonia	Dubuc et al. (2019)
	Macroalgae – *Sargassum*, *Padina*	Orpheus Island, GBR	Waldeland (2017), Tang et al. (2020)
	Macroalgae – *Sargassum*, and seagrass	Green Island, GBR	Pitt (1997)
S. luridus	Seagrass	eastern Mediterranean	Kalogirou et al. (2010)
	Shallow reef flat	not specified	Woodland (1990)
	Macroalgae*	Zanzibar, Tanzania	Tano et al. (2017)
S. puellus	Coral-rich habitat	Okinawa, Japan	Nanami (2018)
	Seagrass	Ryukyu Islands, Japan	Nakamura and Tsuchiya (2008)
S. punctatus	Macroalgae - *Sargassum*	Orpheus Island, GBR	Hoey pers. obs.
		Ishigaki Island, Japan	Sugiyama and Tomori (1990)
	Macroalgae – *Sargassum*, and seagrass	Green Island, GBR	Pitt (1997)
S. rivulatus	Macroalgae	Thuwal, Saudi Arabia	Pombo-Ayora et al. (2020)
	Mangrove	southern Egypt	El-Regal and Ibrahim (2014)
	Seagrass	eastern Mediterranean	Kalogirou et al. (2010)
		not specified	Woodland (1990)

Table 1 contd. ...

...Table 1 contd.

Species	Habitat	Location	Source
S. spinus	Shallow (algal) reef flat	Guam	Tsuda and Bryan (1973), Thacker et al. (1997), Priest et al. (2016)
	Macroalgae	Ishigaki Island, Japan	Sugiyama and Tomori (1990)
		Orpheus Island, GBR	Martin et al. (2024)
		multiple	Fulton et al. (2020)
	Seagrass	Ryukyu Islands, Japan	Nakamura and Tsuchiya (2008)
S. stellatus	Mangrove lined estuary	Tanzania	Mziray pers comms in Woodland (1990)
S. sutor	Mangrove	Zanzibar, Tanzania	Lugendo et al. (2005)
	Seagrass	Tanzania	Kimirei et al. (2011)
		Mozambique	Gell and Whittington (2002)
	Seagrass and mangrove	Madagascar	Jaonalison et al. (2022)
	Seagrass and macroalgae	Zanzibar, Tanzania	Tano et al. (2017)
S. unimaculatus	Coral-rich habitat	Okinawa, Japan	Nanami (2018)
S. vermiculatus	Mangrove roots	Vitu Levu, Fiji	Gundermann et al. (1983)
S. virgatus	Coral-rich habitat	Okinawa, Japan	Nanami (2018)
	Live coral	Montebello and Barron Island, Western Australia	Evans et al. (2014)
	Seagrass	Trat, Thailand	Du et al. (2020)
S. vulpinus	Live coral - branching	Kimbe Bay, PNG	Giffin et al. (2019)
		not specified	Woodland (1983)

More recently, however, experimental studies have begun to investigate the effects of increasing temperature on rabbitfishes from an ecological perspective. For example, Graba-Landry (2020) reared juvenile *S. doliatus* at four temperatures (summer minimum 26°C, summer mean 28°C, summer maximum 30°C, future maximum 32°C) for eight weeks, and found that feeding rate, feeding selectivity, oxygen uptake rate (a proxy for metabolic rate), growth rate, and hepatosomatic condition were relatively robust to changes in temperature. Similarly, LaMonica et al. (2021) found that temperature treatment (28°C, 30°C, 31.5°C) had no effect on the growth, condition, or immune response of juvenile *S. lineatus* over 14 weeks, and small, but significant, effects on the immune response and hepatosomatic condition of juvenile *S. doliatus*. The responses were, however, relatively small compared to other small-bodied tropical fish (e.g., Zarco-Perelló et al. 2012, Motson and Donelson 2017). Importantly, most studies to date have quantified the effect of temperature on growth, survival and/or physiological condition of rabbitfish. The effects of temperature on reproduction and population sustainability of rabbitfishes are largely unknown, and represents a key area for future research.

Salinity and Dissolved Oxygen

Together with their broad temperature tolerance, at least some rabbitfish species can withstand a wide range of salinities, dissolved oxygen levels, and turbidity in their natural habitat. As noted above, several closely related species typically spend some (*S. guttatus, S. insomnis, S. javus, S. lineatus, S. randalli,* and *S. vermiculatus*) or all (*S. vermiculatus*) of their life cycle in tropical estuarine and brackish environments where salinity can vary widely and rapidly if in close proximity to freshwater catchments (Johannes 1978, Gundermann et al. 1983, Woodland 1990, Woodland and Anderson 2014, Moleana et al. 2016, Dubuc et al. 2019, Vanderklift et al. 2021). Such habitats associations indicate that these species not only tolerate, but may even prefer, areas of low salinity. Experimental studies have also shown that the tolerance of a broad range of salinities extends beyond these rabbitfish species that inhabit tropical estuaries. For example, Saoud et al. (2007) found that the survival and growth of juvenile *S. rivulatus* was similar across nine salinities (10–50‰) over six weeks, although there was some evidence of decreased growth in the lowest salinity treatment (i.e., 10‰) after 8-weeks. Similarly, juvenile *S. canaliculatus* have been reported to tolerate salinities as low as 3.5‰ and as high as 75‰, with a suitable range for culturing of 8–59‰ (Huang et al. 2024).

Tolerance to a broad range of salinity means that rabbitfishes have effective physiological plasticity to maintain ion and water homeostasis. However, these mechanisms can be energetically costly with studies suggesting that up to 50 percent of standard metabolic rate could be allocated to these functions when individuals are transferred from freshwater to saltwater (Ern et al. 2014). Few studies have properly assessed salinity tolerance of rabbitfishes and its consequences on fitness traits. Considering the relatively broad habitat range of rabbitfishes, it would be interesting to compare different populations of the same species living in brackish and seawater environments to assess the impact of salinity on fitness traits. This is especially relevant as climate change and anthropogenic activities are predicted to amplify salinity variations in costal ecosystems as extreme weather events such as cyclones and heavy monsoon, sea level rise, altered terrestrial runoff and desalinisation processes become more frequent (Röthig et al. 2023). Species occupying coastal ecosystems will most likely encounter even greater salinity fluctuations, potentially further increasing energetic cost of ionoregulation and osmoregulation with repercussion on fitness-related functions.

Low dissolved oxygen (DO) events are common (almost daily) in some tidally influenced coastal ecosystems such as estuaries, wetlands and mangroves (Dubuc et al. 2017, 2019). Fish are aerobic organisms, meaning that they depend on oxygen to produce energy to fuel all vital functions. A lack of oxygen has therefore important adverse effects at the individual level leading to changes in physiology and behaviour, affecting growth, reproduction, feeding and eventually survival (Claireaux and Chabot 2019). Several species of rabbitfishes are known to occupy some of these ecosystems that regularly experience hypoxia. To cope with the detrimental effects of hypoxia, they can rely on rapid and reversible phenotypic plasticity to adjust their physiology to successfully use these productive yet challenging environments. In fact, several rabbitfish species appear to have a high tolerance to low levels of

DO, presumably linked to their pattern of use of estuarine habitats (Babikian et al. 2017, Dubuc et al. 2021). The frequency of occurrence (a proxy of abundance) of *S. lineatus* for instance, was largely insensitive to changes in DO (40–110 percent saturation) in natural tropical estuaries (Dubuc et al. 2019). The corresponding critical oxygen level (O_{2crit}) of *S. lineatus* was 13 percent saturation, lower than that of other fish species that use these environments but less commonly (*Heniochus acuminatus*: 18.7 percent saturation; *Acanthopagrus pacificus*: 20 percent saturation; *Chaetodon vagabundus*: 20.2 percent saturation; Dubuc et al. 2021), emphasising that being tolerant to low DO is an advantageous physiological trait for this species. Moreover, the tolerance of low DO levels (ca. 5 percent oxygen saturation) has been linked to the use of mangrove swamps by *S. vermiculatus* (Gundermann et al. 1983), and led to suggestions that *S. rivulatus*, and perhaps rabbitfishes more broadly, must have evolved in environments prone to hypoxia making hypoxia tolerance a widespread physiological trait of this family (Babikian et al. 2016).

Ocean deoxygenation is one of the most important threats stemming from climate change and urbanisation, primarily driven by two main pathways: rising temperatures, which reduce oxygen solubility, and nutrient enrichment, which promotes eutrophication (Breitburg et al. 2018). Coastal ecosystems are particularly susceptible to eutrophication, which would exacerbate natural hypoxic events, placing additional pressure on organisms like rabbitfishes already coping with challenging conditions (Dubuc et al. 2024). Additional stress could push them beyond their capacity to acclimate and adapt, ultimately excluding them from vital habitats with major consequences for coastal ecosystems functioning and services.

Multiple Stressor Environments

Rabbitfishes exhibit a remarkable ability to thrive in diverse environmental conditions, including dynamically fluctuating environments. They likely employ a variety of rapid and reversible mechanisms to maintain essential functions while displaying high tolerance to temperature, hypoxia, and salinity. Field studies suggest that rabbitfishes frequently encounter these stressors concurrently in the wild, indicating their capacity to cope with multiple co-occurring environmental challenges. However, our understanding of the underlying physiological mechanisms and potential fitness trade-offs resulting from this phenotypic plasticity remains limited. With climate change and other anthropogenic activities projected to intensify these stressors, there is concern that rabbitfishes may struggle to withstand further environmental pressure, potentially leading to detrimental consequences for their populations.

Diet and Habitat Requirements

Rabbitfishes are most abundant and diverse in coastal marine and estuarine habitats in tropical and subtropical regions of the Indo-Pacific, where they are found in a range of habitats, including coral reefs, macroalgae beds, seagrass meadows, mangrove forests, and/or estuaries (e.g., Cheal et al. 2012, Hoey et al. 2013). This distribution contrasts with that of other groups of tropical herbivorous fishes (i.e., Labridae-Scarini, Acanthuridae, Kyphosidae) that are typically most abundant and diverse on shallow coral reefs, and particularly exposed seaward reef crests (e.g., Hoey and Bellwood 2008, Cheal et al. 2013, Tebbett et al. 2022). Rabbitfishes, like parrotfishes and surgeonfishes, exhibit clear distinctions in habitat preferences across both small (e.g., among sites and habitats within individual reefs; Cvitanovic and Bellwood 2009, Hoey et al. 2013) and large spatial scales (e.g., inshore vs. offshore reefs, among latitudes; Cheal et al. 2012, Hoey et al. 2013). For example, on Australia's Great Barrier Reef (GBR), *S. canaliculatus* are largely restricted to inner-shelf reefs where they associate with shallow macroalgal beds and seagrass meadows (Pitt 1997, Hoey et al. 2013), yet are rare or absent on mid- and outer shelf reefs (Cheal et al. 2012, Hoey and Bellwood 2009, 2010a). In contrast, *S. argenteus* is most abundant on mid- and outer-shelf reefs of the GBR (Hoey et al. 2013), with its distribution extending beyond the GBR and onto the oceanic reefs of the Coral Sea (Galbraith et al. 2024). These spatial patterns in the distribution and abundance of rabbitfishes on the GBR and elsewhere (e.g., Red Sea: Pombo-Ayora et al. 2020) may be attributable to their tolerance to environmental conditions, ability to use soft-sediment habitats, and/or spatial variation in the availability of preferred habitat and/or dietary resources among life stages.

Diet

A comprehensive assessment of the nutritional ecology and diet of rabbitfishes is provided elsewhere in this volume (Clements – Chapter 4), however it is worth providing a brief overview of the specificity of rabbitfishes diet here. Despite limited variation in the morphology of the feeding apparatus among rabbitfish species, there is considerable variation in diet both among and within species (Hoey et al. 2013, Clements – Chapter 4). For example, examination of the stomach content of eleven co-occurring rabbitfish species on the northern GBR revealed marked differences in ingested material among species, and some intra-specific variation (albeit smaller relative to the inter-specific differences) between inner- and mid-shelf reefs (Hoey et al. 2013). Based on this data the authors proposed four dietary categories: browsers of leathery brown macroalgae (*S. canaliculatus*, *S. javus*), croppers of red and green algal turfs (*S. argenteus, S. corallinus*, *S. doliatus*, *S. spinus*), mixed feeders of diverse algal material, cyanobacteria, detritus and sediment (*S. lineatus*, *S. punctatissimus*, *S. punctatus*, *S. vulpinus*), and spongivores (*S. puellus*). While these groups provide a useful construct to examine broader patterns, they mask variation both among and within species. As highlighted by Clements (Chapter 4), assignments of rabbitfish species to these groups may reflect spatial and temporal variation in resource availability rather than trophic diversification. Indeed, a recent comparison of the diets of closely related species pairs *S. doliatus* – *S. virgatus* and *S. corallinus* – *S. trispilos* across three locations in the central Indo-Pacific (Japan, GBR, Ningaloo Reef) revealed greater intra-specific variation (i.e., among locations) than inter-specific variation within each location (Zarco-Perello et al. 2024). Similarly, Brentebraten et al. (unpublished data) found differences in the stomach content of *S. doliatus* between inner- and mid-shelf reefs on the northern GBR. This intra-specific variation indicates that the diets of these species are flexible and likely driven by local availability of dietary resources. Similar intra-specific variation in the diet of rabbitfish is evident across spatial and temporal scales from comparisons of gut content studies, direct feeding observations and video assays of tethered macroalgae (see Clements – Chapter 4 for detailed discussion).

Habitat Use

Juvenile Rabbitfish

Settlement is a critical phase in the life history of demersal fish, marking the transition from a pelagic larval phase to a benthic or demersal phase. Mortality of reef fish at, and in the days to weeks following, settlement is extremely high as recently-settled fish are exposed to demersal predators for the first time (Fakan et al. 2024). For coral reef fish, upwards of 57 percent die within the first 2-days post-settlement (e.g., McCormick and Hoey 2004, Almany and Webster 2006), with even fewer surviving to maturity (e.g., Jones 1990). As such, recently-settled and juvenile fishes often select habitats that reduce predation risk and/or optimise growth, before moving to adult habitats as they grow (McCormick and Makey 1997, Kimirei et al. 2013). For example, many coral reef fish species settle in live branching corals, yet rarely associate with corals as adults (Wilson et al. 2010, Coker et al. 2014), while others species settle in macrophyte habitats (i.e., seagrass, macroalgae, mangroves) before moving to coral reefs as sub-adults (Sambrook et al. 2019).

While many rabbitfish species are considered to be coral reef fish or reef-associated fish as adults (e.g., Woodland 1990, Siqueira and Cowman – Chapter 3), very few are reported to preferentially associate with live corals or coral rich habitats at settlement or as juveniles (Table 1). Rather, the vast majority of rabbitfishes appear to settle to and/or associate with various macrophyte habitats (i.e., macroalgae, mangroves, seagrass) during their early life stages (Table 1). For example, Woodland (1990) describes *S. doliatus* are being strongly corallophilic, yet recently-settled and juvenile fishes appear to associate with canopy-forming macroalgae, and in particular *Sargassum*, on inshore reefs of the GBR (e.g., Tang et al. 2020), and coral rubble on mid-shelf reefs (Bellwood et al. 2016) where canopy-forming algae is rare (Wismer et al. 2009, Hoey and Bellwood 2009, 2010b). Moreover, juveniles of other 'reef-associated' rabbitfishes appear to either always associate with macrophyte habitats (*S. guttatus, S insomnis, S. javus, S. lineatus, S. puntatus, S. stellatus*) or under certain conditions and/or at some locations (*S. corallinus, S. puellus, S. virgatus*). The only exceptions to this appear to be those species with a long tubular snout (subgenus *Lo*), whose juveniles appear to associate exclusively with live branching corals (Table 1). This lack of association with live corals in recently-settled and juvenile fish is counter to that of many other reef-associated fishes where juveniles show a strong association with live coral even though adult conspecifics rarely associate with live coral (Jones et al. 2004).

Together with the limited use of live corals by recently-settled and juvenile rabbitfish, there appears to be considerable intra-species variability in juvenile habitat use. For example, juvenile *S. fuscescens* and *S. rivulatus* have been reported to associate with seagrass, macroalgae and mangrove habitats, and juvenile *S. javus* with seagrass, mangrove, mudflat and estuarine habitats (Table 1). This association with macrophytes has been suggested to be related to the physical structure, and hence refugia from predation, they provide (Tano et al. 2017) and/or the provision of dietary resources. Early qualitative observations describe juvenile rabbitfish browsing on the "aufwuchs" or epibiota on the blades of seagrass and macroalgae (Drew 1973, Woodland 1990). These early observations are supported by recent quantitative studies linking various structural attributes of macroalgae (biomass, density, and intra- vs inter-thallus space) with the abundance of juvenile rabbitfish (Tang et al. 2020, Webber et al. 2024), and the predominance of epiphytic filamentous algae in their stomach content (Martin et al. 2024). Irrespective of the driver/s for associating with macrophyte habitats, the limited reliance on live coral and the flexibility in juvenile habitat suggest that rabbitfishes may be more resilient to habitat degradation and coral loss than other reef-associated fish taxa.

Adult Rabbitfish

Rabbitfishes are widely distributed throughout the Indo-west Pacific. While some species, namely *S. argenteus* and *S. woodlandi*, are common on oceanic reefs (e.g., Hoey et al. 2020), and have been recorded in relatively deep waters (*S. argenteus*: 58 m, *S. woodlandi*: 47 m; Galbraith et al. 2024), rabbitfish assemblages generally attain their greatest abundance, biomass and richness on shallow inshore reefs and seagrass habitat (e.g., Woodland 1990, Cheal et al. 2012, Hoey et al. 2013, Pombo-Ayora et al. 2020; Fig. 3). This is in direct contrast to other groups of herbivorous coral reef fish that attain their greatest abundance, biomass and/or richness on offshore reefs (e.g., parrotfishes: Hoey and Bellwood 2008, Hoey et al. 2016b; surgeonfishes: McClure et al. 2019, Tebbett et al. 2022). Moreover, the evolutionary history of rabbitfishes differs from that of other tropical piscine herbivores (Siqueira and Cowman – Chapter 3). Although present throughout the entire Cenozoic, rabbitfishes appear to have never undergone the episodes of diversification that are evident in the parrotfishes, surgeonfishes and damselfishes.

Phylogenies of rabbitfishes have generally identified three clades (Woodland 1990, Borsa et al. 2007, Kuriiwa et al. 2007, Siqueira et al. 2019); the first clade (Clade A in Fig. 3.1) consists of two predominantly schooling, habitat generalist species with deeply forked caudal fins (*S. argenteus, S. woodlandi*). The second clade (Clade B in Fig. 3.1) consisting of six species (*S. canaliculatus*, *S. fuscescens*, *S. luridus*, *S. rivulatus*, *S. spinus*, *S. sutor*) are also schooling habitat generalist species, associating with coral reefs, seagrass, macroalgal beds, estuaries, rocky reefs and temperate kelp forests, and have a truncated caudal fin. The remaining 23 species make up the third clade (Clade C in Fig. 3.1) and is often described as being composed of predominantly 'pairing reef-associated species' (Woodland 1990). Except for members of the subgenus *Lo* (i.e., *S. magnificus, S. niger S. unimaculatus, S. uspi* and *S. vulpinus*) that appear to be restricted to coral reef habitats, the majority of rabbitfish species within this 'pairing reef-associated' clade also display a high degree of plasticity in their habitat associations, being commonly observed across multiple habitats in tropical seascapes (Sambrook et al. 2019). Several closely related species within the 'reef-associated clade' spend some (*S. guttatus, S. insomnis, S. lineatus, S. randalli*) or all (*S. vermiculatus*) of their life cycle in tropical estuarine and brackish environments, while the remaining species are also regularly observed in shallow seagrass and macroalgal habitats (e.g., Fox and Bellwood 2011,

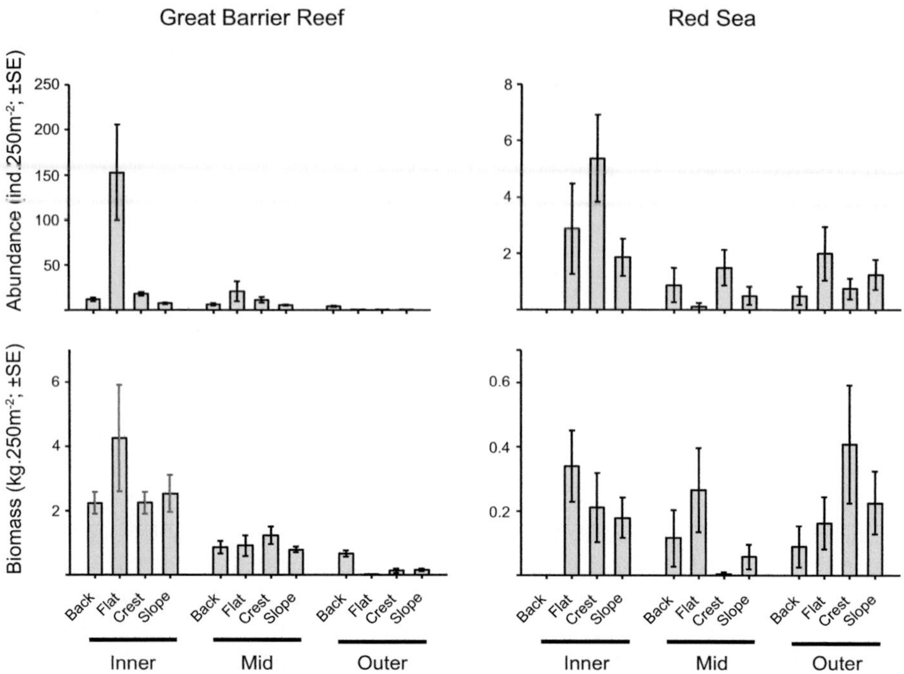

Fig. 3. Cross-shelf variation in the abundance and biomass of rabbitfish assemblages in the northern Great Barrier Reef and central Red Sea. Data are based on four replicate visual census within each of four habitats, at each of six reefs in each region (two inner-, two mid-, and two outer-shelf reefs). See Hoey et al. (2013) for description of habitats.

Bauman et al. 2017). This apparent flexibility in habitat associations has led to suggestions that rabbitfishes may provide mobile links among habitats (Fox and Bellwood 2011, Ebrahim et al. 2020), and provide some resilience to the degradation and/or loss of some habitats.

Vulnerability to Habitat Degradation and Loss

The increasing frequency, intensity, and/or diversity of climate-induced and local anthropogenic disturbances is having a major impact on many of the world's ecosystems. The effects of climate change, and in particular increasing environmental temperatures, are most pronounced in tropical marine ecosystems due to the narrow thermal tolerances of tropical species, and the proximity of their current environments to upper thermal limits (e.g., Pörtner and Farrell 2008, Pinsky et al. 2019). Scleractinian (hard) corals, the foundational species of coral reefs, often bleach and die in response to relatively small increases (≥ 1°C) in summer seawater temperatures, especially if these elevated temperatures are experienced over periods of weeks to months (Jokiel and Coles 1990, Hughes et al. 2018b). The widespread mortality of corals not only leads to declines in live coral cover and shifts in the composition of coral and benthic assemblages (Graham et al. 2015, Hughes et al. 2018b), but also shifts in the taxonomic and functional composition of reef fishes (Wilson et al. 2006, Pratchett et al. 2008, 2011, Richardson et al. 2018, Robinson et al. 2019). Numerous empirical studies and reviews have shown that fish species that are directly dependent on live corals for food and/or habitat are generally the first and most adversely affected by coral loss, however, other species that appear to have no direct association with coral are sensitive to changes in coral cover (e.g., Wilson et al. 2006, Pratchett et al. 2015).

Data on the responses of rabbitfishes to disturbances that caused loss of live coral cover (i.e., thermally-induced bleaching events, cyclones, or population irruptions of crown-of-thorns starfish) were compiled from 105 published studies, and the standardised response of each species to reductions in coral cover calculated (following Wilson et al. 2006, Pratchett et al. 2008, 2011, 2014, 2018). Briefly, the standardised response is the proportional change in the abundance of each rabbitfish species relative to the proportional reduction in coral cover. These standardised responses were then averaged across all studies for each fish species.

Of the eleven rabbitfish species considered in this analysis, the majority (eight species) experienced consistent declines in abundance, and only one species (*S. stellatus*) experienced an increase in abundance following episodes of acute coral loss (Fig. 4). The remaining two species (*S. argenteus* and *S. doliatus*) showed no consistent change in abundance. With the exception of *S. unimaculatus* and *S. vulpinus*, none of the remaining six species that experienced significant declines (*S. javus, S. lineatus, S. sutor, S. corallinus, S. spinus*, and *S. puelloides*) appear to be directly reliant on live coral cover (see Table 1 and the discussion above). The predominance of declines in abundance following coral loss among rabbitfish species contrasts with that of another group of nominally herbivorous reef fish, the surgeonfishes in which only four of the 22 species (for which data was available) declined following coral loss (Fig. 4B). Moreover, the declines in abundance of rabbitfish does not align with previously identified feeding groups, phylogenetic clades, or social structures (i.e., pairing versus schooling) among species. The results of this, and similar, meta-analyses should be interpreted with caution due to differences in the nature and intensity of disturbances, survey methods used, and the temporal scale post-disturbance among individual studies.

Given the lack of association with live coral habitats, the predominance of algal material in the diet of most species, and the general flexibility in resource requirements, it would seem reasonable to predict rabbitfishes to be relatively insensitive to changes in coral cover. To investigate the response of rabbitfish populations to changes in benthic cover using standardised methods, we use a temporal dataset that spans multiple acute disturbances (two severe [category 4] tropical cyclones: TC Ita 2014, TC Nathan 2015, and two mass coral-bleaching events in 2016 and 2017) from six reefs spanning the continental shelf in the northern Great Barrier Reef (see McClure et al. 2019 for reef descriptions). The density of rabbitfish displayed no significant change from pre- (2009), immediately post- (2016, i.e., after two cyclones and the first of two bleaching events), and five years post- (2022) the most recent disturbance (Fig. 5A,B). This lack of change in rabbitfish density was evident despite 27–44 percent declines in absolute hard coral cover across the three shelf positions, and a 26 percent increase in absolute macroalgae cover on the inner-shelf reefs from 2009 to 2016, and a 13–30 percent increase in absolute hard coral cover across the three shelf positions, and a 5 percent decrease in absolute macroalgae cover

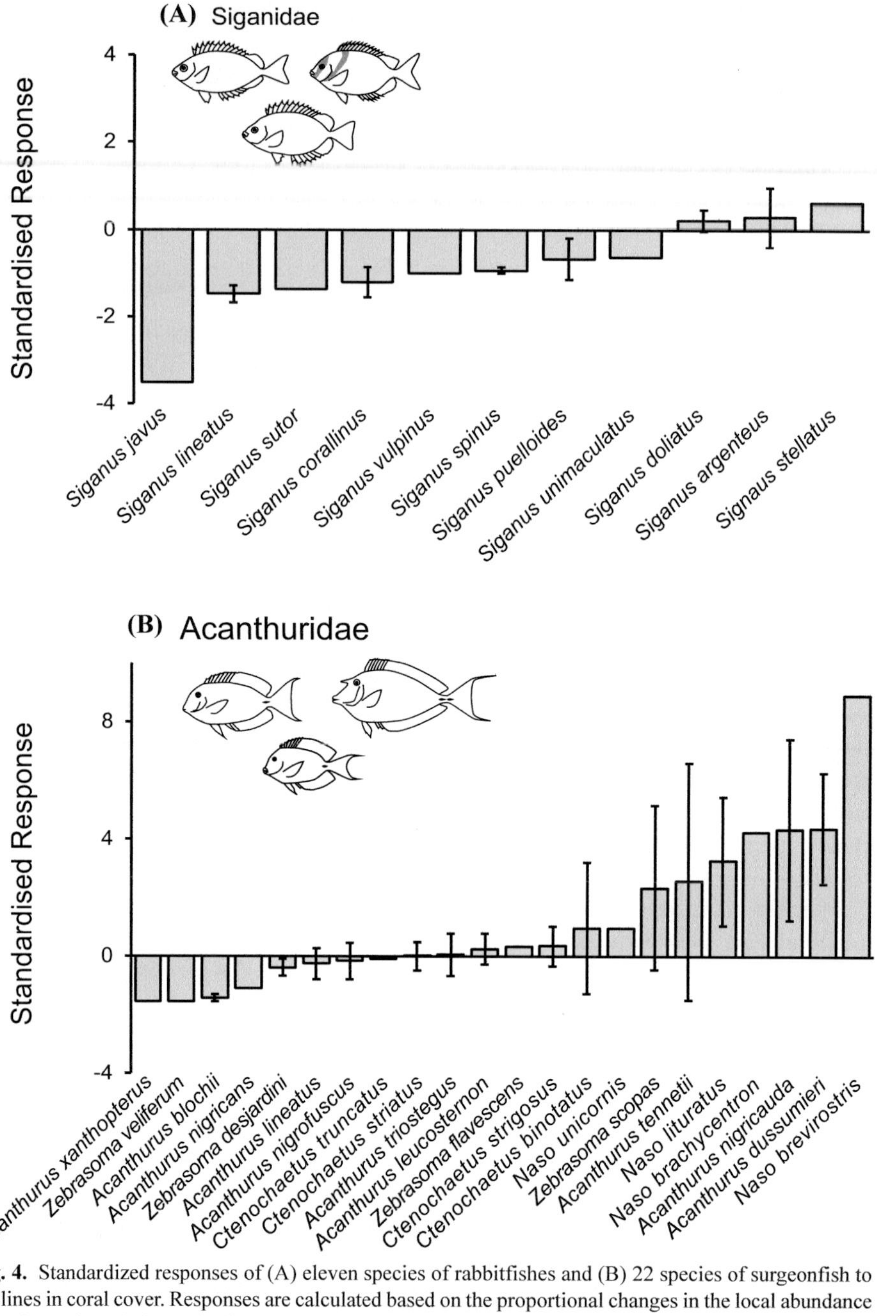

Fig. 4. Standardized responses of (A) eleven species of rabbitfishes and (B) 22 species of surgeonfish to declines in coral cover. Responses are calculated based on the proportional changes in the local abundance of each species relative to declines in total live coral cover, following Wilson et al. (2006). Where multiple studies had recorded the change in abundance of a species, the standard error was calculated to show variation in responses recorded among different studies.

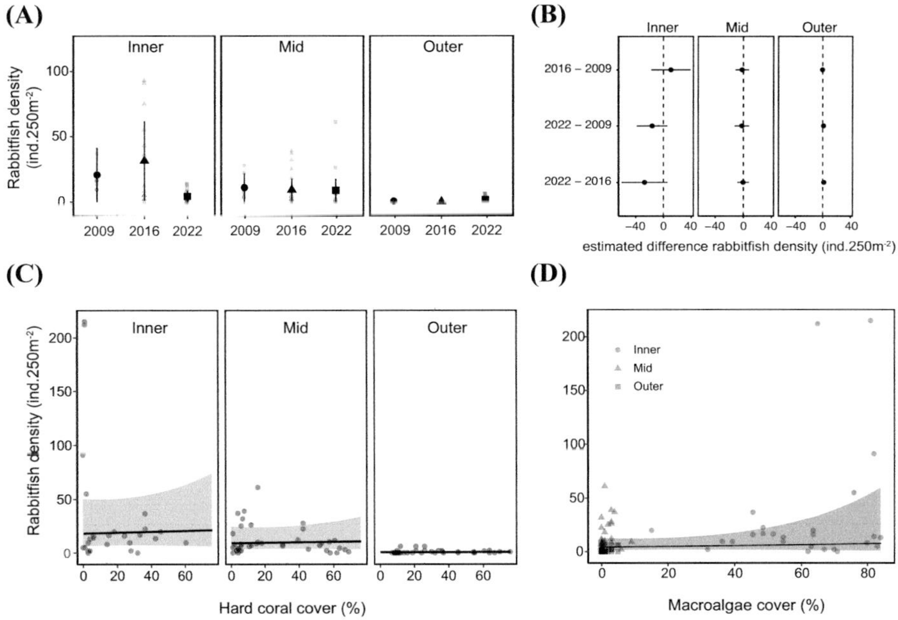

Fig. 5. The response of rabbitfish assemblages to multiple acute disturbances. (A) temporal changes in the estimated mean density of rabbitfishes across inner-, mid- and outer-shelf reefs in the northern Great Barrier Reef; (B) estimated differences in the mean density of rabbitfishes among years across inner-, mid- and outer-shelf reefs; relationship between rabbitfish density and (C) live coral cover at each shelf position, and (D) macroalgae cover. Solid lines are modelled relationships and shaded areas 95 percent confidence intervals. Plots are based on the estimated modelled (A) means, (B) mean differences and (C, D) relationships, with 95 percent confidence limits, fit with Generalized Linear Mixed Models (GLMMs). All models were fit against a negative binomial distribution (log-link) with transect nested in site as a random effect. Light grey points represent raw data. Data are based on three to four replicate visual surveys conducted at two reef crest sites on each of six reefs (two inner-, 2 mid- and 2 outer-shelf) in 2009, 2016, and 2022. These reefs were exposed to multiple acute disturbances from 2014–2017 (two tropical cyclones: TC Ita 2014, TC Nathan 2015, and two mass coral-bleaching events in 2016 and 2017).

on inner-shelf reefs from 2016 to 2022. Indeed, the density of rabbitfish displayed no relationship with the cover of live corals or macroalgae across these spatial and temporal scales (Fig. 5C,D). Similarly, changes in the biomass of rabbitfishes did not correspond with changes in benthic communities on recovering versus regime-shifted (i.e., macroalgal-dominated) reefs following the 1998 coral bleaching event in the Seychelles (Robinson et al. 2019). The biomass of *S. stellatus* and *S. puelloides* declined, and the biomass of *S. sutor* increased across all reefs, irrespective of whether coral cover had returned to pre-bleaching levels (i.e., recovering reef) or had undergone a regime shift to macroalgae (*ca.* 20 percent cover; Robinson et al. 2019). Collectively, these findings suggest that fluctuations of rabbitfish populations are largely insensitive to habitat changes, and may be influenced by other factors (e.g., stochastic recruitment, fishing, predation).

Conclusion

Rabbitfishes appear to display a high degree of flexibility in terms of environmental conditions, diet and habitat requirements. This flexibility likely influences their capacity to establish in new habitats (e.g., Mediterranean Sea), increase in abundance in sub-tropical and temperate habitats, and their inhabitation of a broad range of habitats (i.e., generalist species). While the majority of rabbitfish species are often referred to as being 'reef-associated', it appears that coral reefs are only one of many habitats that rabbitfishes use across tropical seascapes (Sambrook et al. 2019). The only exceptions to this appear to be the long-snouted members of the subgenus *Lo*, that settle to branching corals and remain associated with coral habitats throughout ontogeny. The apparent insensitivity of rabbitfishes to habitat change and degradation, coupled with their broad tolerances for physico-chemical conditions and life history characteristics (rapid growth, short time to maturity and high fecundity; Taylor et al. – Chapter 5) may allow rabbitfish populations to flourish in future tropical seascapes.

References

Almany, G.R. and M.S. Webster. 2006. The predation gauntlet: early post-settlement mortality in reef fishes. Coral Reefs 25: 19–22.

Babikian, J., N. Nasser, S. Monzer and I.P. Saoud. 2017. Survival and respiration of marbled rabbitfish (*Siganus rivulatus*) fingerlings at various oxygen tensions. Aquaculture Res. 48: 4219–4227.

Bauman, A.G., A.S. Hoey, G. Dunshea, D. Feary, J. Low and P.A. Todd. 2017. Macroalgal browsing on a heavily degraded, urbanized equatorial reef system. Sci. Repo. 7: 8352.

Bellwood, D.R., T.P. Hughes and A.S. Hoey. 2006. Sleeping functional group drives coral-reef recovery. Current Biology 16: 2434–2439.

Bellwood, D.R., C.H. Goatley, J.A. Khan and S.B. Tebbett. 2016. Site fidelity and homing in juvenile rabbitfishes (Siganidae). Coral Reefs 35: 1151–1155.

Blewett, T.A., S.A. Binning, A.M. Weinrauch, C.M. Ivy, G.S. Rossi, B.G. Borowiec et al. 2022. Physiological and behavioural strategies of aquatic animals living in fluctuating environments. J. Exp. Biol. 225: jeb242503.

Borsa, P., S. Lemer and D. Aurelle. 2007. Patterns of lineage diversification in rabbitfishes. Mol. Phylogen. Evol. 44: 427–435.

Boyd, P.W., C.E. Cornwall, A. Davison, S.C. Doney, M. Fourquez, C.L. Hurd et al. 2016. Biological responses to environmental heterogeneity under future ocean conditions. Glob. Chang. Biol. 22: 2633–2650.

Brandl, S.J., J.L. Johansen, J.M. Casey, L. Tornabene, R.A. Morais and J.A. Burt. 2020. Extreme environmental conditions reduce coral reef fish biodiversity and productivity. Nat. Comm. 11: 3832.

Breitburg, D., L.A. Levin, A. Oschlies, M. Grégoire, F.P. Chavez, D.J. Conley et al. 2018. Declining oxygen in the global ocean and coastal waters. Science 359: eam7240.

Bruno, J.F. and E.R. Selig. 2007. Regional decline of coral cover in the Indo-Pacific: timing, extent, and subregional comparisons. PLoS One 2: e711.

Burt, J.A., D.A. Feary, G. Cavalcante, A.G. Bauman and P. Usseglio. 2013. Urban breakwaters as reef fish habitat in the Persian Gulf. Mar. Poll. Bull. 72: 342–350.

Burt, J.A., E.F. Camp, I.C. Enochs, J.L. Johansen, K.M. Morgan, B. Riegl et al. 2020. Insights from extreme coral reefs in a changing world. Coral Reefs 39: 495–507.

Cheal, A., M. Emslie, I. Miller and H. Sweatman. 2012. The distribution of herbivorous fishes on the Great Barrier Reef. Mar. Biol. 159: 1143–1154.

Claireaux, G. and D. Chabot. 2019. The significance of ocean deoxygenation for the physiology of marine organisms. pp. 461–484. *In*: Laffoley, D. and J.M. Baxter [eds.]. Ocean Deoxygenation: Everyone's Problem. IUCN, Gland, Switzerland.

Coker, D.J., S.K. Wilson and M.S. Pratchett. 2014. Importance of live coral habitat for reef fishes. Rev. Fish Biol. Fisher. 24: 89–126.

Costanza, R., R. d'Arge, R. de Groot, S. Farber, M. Grasso, B. Hannon et al. 1997. The value of the world's ecosystem services and natural capital. Nature 387: 253–260.

Cvitanovic, C. and D.R. Bellwood. 2009. Local variation in herbivore feeding activity on an inshore reef of the Great Barrier Reef. Coral Reefs 28: 127–133.

Dalzell, P., T.J.H. Adams and N.V.C. Polunin. 1996. Coastal fisheries in the Pacific Islands. Oceanogr. Mar. Biol.: Ann. Rev. 34: 395–531.

Dell, A.I., S. Pawar and V.M. Savage. 2011. Systematic variation in the temperature dependence of physiological and ecological traits. Proc. Natl Acad. Sci. USA 108: 10591–10596.

Deutsch, C., A. Ferrel, B.A. Seibel, H.-O. Pörtner and R.B. Huey. 2015. Climate change tightens a metabolic constraint on marine habitats. Science 348: 1132–1135.

Díaz, S., J. Settele, E.S. Brondízio, H.T. Ngo, J. Agard, A. Arneth et al. 2019. Pervasive human-driven decline of life on Earth points to the need for transformative change. Science 366: eaax3100.

DiBattista, J.D., J.H. Choat, M.R. Gaither, J.P.A. Hobbs, D.F. Lozano-Cortés, R.F. Myers et al. 2016. On the origin of endemic species in the Red Sea. J. Biogeogr. 43: 13–30.

Drew, A.W. 1973. The comparative ecology of *Siganus lineatus* and *Siganus canaliculatus*: two fishes of the Palau Islands (Pisces: Siganidae). MSc thesis, Northern Illinois University, 46pp.

Dubuc, A., N. Waltham, M. Malerba and M. Sheaves. 2017. Extreme dissolved oxygen variability in urbanised tropical wetlands: The need for detailed monitoring to protect nursery ground values. Estuar. Coastal Shelf Sci. 198: 163–171.

Dubuc, A., R. Baker, C. Marchand, N.J. Waltham and M. Sheaves. 2019. Hypoxia in mangroves: occurrence and impact on valuable tropical fish habitat. Biogeosciences 16: 3959–3976.

Dubuc, A., G.M. Collins, L. Coleman, N.J. Waltham, J.L. Rummer and M. Sheaves 2021. Association between physiological performance and short temporal changes in habitat utilisation modulated by environmental factors. Mar. Environ. Res. 170: 105448.

Dubuc, A., J.L. Rummer, L. Vigliola and H. Lemonnier. 2024. Coping with environmental degradation: physiological and morphological adjustments of wild mangrove fish to decades of aquaculture-induced nutrient enrichment. Mar. Poll. Bull. 205: 116599.

Dunic, J.C., C.J. Brown, R.M. Connolly, M.P. Turschwel and I.M. Côté. 2021. Long-term declines and recovery of meadow area across the world's seagrass bioregions. Glob. Change Biol. 27: 4096–4109.

Ebrahim, A., J.P. Bijoux, P.J. Mumby and I.R. Tibbetts. 2020. The commercially important shoemaker spinefoot, *Siganus sutor*, connects coral reefs to neighbouring seagrass meadows. J. Fish Biol. 96: 1034–1044.

El-Regal, M.A.A. and N.K. Ibrahim. 2014. Role of mangroves as a nursery ground for juvenile reef fishes in the southern Egyptian Red Sea. Egyptian J. Aquat. Res. 40: 71–78.

El-Sayed, A.M. 1994. Feeding habits of rabbitfishes, *Siganus canaliculatus* and *Siganus javus* fingerlings from the Arabian Gulf waters off Qatar. Indian J. Mar. Sci. 23: 112–114.

Emslie, M.J. and M.S. Pratchett. 2018. Differential vulnerabilities of parrotfishes to habitat degradation. pp. 355–382. *In*: Hoey, A.S. and R.M. Bonaldo [eds.]. Biology of Parrotfishes. CRC Press, Boca Raton.

Ern, R., D.T.T. Huong, N.V. Cong, M. Bayley and T. Wang. 2014. Effect of salinity on oxygen consumption in fishes: A review. J. Fish Biol. 84: 1210–1220.

Evans, R.D., S.K. Wilson, S.N. Field and J.A.Y. Moore. 2014. Importance of macroalgal fields as coral reef fish nursery habitat in north-west Australia. Marine Biology 161: 599–607.

Fakan, E.P., M.I. McCormick, G.P. Jones and A.S. Hoey. 2024. Habitat and morphological characteristics affect juvenile mortality in five coral reef damselfishes. Coral Reefs 43: 171–183.

Feary, D.A., A.G. Bauman, J. Guest and A.S. Hoey. 2018. Trophic plasticity in an obligate corallivorous butterflyfish. Mar. Ecol. Progr. Ser. 605: 165–171.

Feary, D.A., M.S. Pratchett, M. Emslie, A.M. Fowler, W.F. Figueira, O.J. Luiz et al. 2014. Latitudinal shifts in coral reef fishes: why some species do and others do not shift. Fish Fisher. 15: 593–615.

Fox, R.J. and D.R. Bellwood. 2011. Unconstrained by the clock? Plasticity of diel activity rhythm in a tropical reef fish, *Siganus lineatus*. Func. Ecol. 25: 1096–1105.

Franklin, J. 1995. Predictive vegetation mapping: geographic modelling of biospatial patterns in relation to environmental gradients. Progr. Physical Geography 19: 474–499.

Fulton, C.J., C. Berkström, S.K. Wilson, R.A. Abesamis, M. Bradley, C. Åkerlund et al. 2020. Macroalgal meadow habitats support fish and fisheries in diverse tropical seascapes. Fish Fisher. 21: 700–717.

Galbraith, G.F., B.J. Cresswell, E.C. McClure and A.S. Hoey. 2024. Tropical seamounts as stepping-stones for coral reef fishes: range extensions and new regional distributions from mesophotic ecosystems in the Coral Sea, Australia. Mar. Biodiv. 54: 17.

Gardner, T.A., I.M. Côté, J.A. Gill, A. Grant and A.R. Watkinson. 2003. Long-term region-wide declines in Caribbean corals. Science 301: 958–960.

Gell, F.R. and M.W. Whittington. 2002. Diversity of fishes in seagrass beds in the Quirimba Archipelago, northern Mozambique. Mar. Freshw. Res 53: 115–121.

Gentry, A.H. 1988. Changes in plant community diversity and floristic composition on environmental and geographical gradients. Annals Missouri Botanical Garden 75: 1–34.

Goldberg, L., D. Lagomasino, N. Thomas and T. Fatoyinbo. 2020. Global declines in human-driven mangrove loss. Glob. Change Biol. 26: 5844–5855.

Graba-Landry, A. 2020. The effect of increasing temperature on algae-fish interactions on coral reefs. Doctoral dissertation, James Cook University, Townsville.

Graham, N.A., S. Jennings, M.A. MacNeil, D. Mouillot and S.K. Wilson. 2015. Predicting climate-driven regime shifts versus rebound potential in coral reefs. Nature 518: 94–97.

Grandcourt, E., T. Al Abdessalaam, F. Francis and A. Al Shamsi. 2007. Population biology and assessment of the white-spotted spinefoot, Siganus canaliculatus (Park, 1797), in the southern Arabian Gulf. J. Appl. Ichthyol. 23: 53–59.

Grimm, N.B., F.S. Chapin III, B. Bierwagen, P. Gonzalez, P.M. Groffman, Y. Luo et al. 2013. The impacts of climate change on ecosystem structure and function. Frontiers Ecol. Environ. 11: 474–482.

Gundermann, N., D.M. Popper and T. Lichatowich. 1983. Biology and life cycle of *Siganus vermiculatus* (Siganidae, Pisces). Pac. Sci. 37: 165–180.

Gutiérrez-Pesquera, L.M., M. Tejedo, M.Á. Olalla-Tárraga, H. Duarte, A. Nicieza and M. Solé. 2016. Testing the climate variability hypothesis in thermal tolerance limits of tropical and temperate tadpoles. J. Biogeogr. 43: 1166–1178.

Halpern, B.S., M. Frazier, J. Potapenko, K.S. Casey, K. Koenig, C. Longo et al. 2015. Spatial and temporal changes in cumulative human impacts on the world's ocean. Nat. Comm. 6: 1–7.

Hoegh-Guldberg, O. and J.F. Bruno. 2010. The impact of climate change on the world's marine ecosystems. Science 328: 1523–1528.

Hoegh-Guldberg, O., D. Jacob, M. Taylor, M. Bindi, S. Brown, I. Camilloni et al. 2018. Impacts of 1.5C Global Warming on Natural and Human Systems. pp. 175–312. *In*: Global Warming of 1.5C. An IPCC Special Report on the impacts of global warming of 1.5C above pre-industrial levels and related global greenhouse gas emission pathways, in the context of strengthening the global response to the threat of climate change, sustainable development, and efforts to eradicate poverty. Cambridge University Press, Cambridge, UK and New York, NY, USA.

Hoegh-Guldberg, O., P.J. Mumby, A.J. Hooten, R.S. Steneck, P. Greenfield, E. Gomez et al. 2007. Coral reefs under rapid climate change and ocean acidification. Science 318: 1737–1742.

Hoey, A.S., H. Harrison and M. Pratchett. 2020. Coral Reef Health in the Coral Sea Marine Park – Surveys 2018–2020. Report prepared for Parks Australia.

Hoey, A.S. and D.R. Bellwood. 2008. Cross-shelf variation in the role of parrotfishes on the Great Barrier Reef. Coral Reefs 27: 37–47.

Hoey, A.S. and D.R. Bellwood. 2009. Limited functional redundancy in a high diversity system: single species dominates key ecological process on coral reefs. Ecosystems 12: 1316–1328.

Hoey, A.S. and D.R. Bellwood. 2010a. Among-habitat variation in herbivory on *Sargassum* spp. on a mid-shelf reef in the northern Great Barrier Reef. Mar. Biol. 157: 189–200.

Hoey, A.S. and D.R. Bellwood. 2010b. Damselfish territories as a refuge for macroalgae on coral reefs. Coral Reefs 29: 107–118.

Hoey, A.S., S.J. Brandl and D.R. Bellwood. 2013. Diet and cross-shelf distribution of rabbitfishes (f. Siganidae) on the northern Great Barrier Reef: implications for ecosystem function. Coral Reefs 32: 973–984.

Hoey, A.S., E. Howells, J.L. Johansen, J.P.A. Hobbs, V. Messmer, D.M. McCowan et al. 2016a. Recent advances in understanding the effects of climate change on coral reefs. Diversity 8: 12.

Hoey, A.S., D.A. Feary, J.A. Burt, G. Vaughan, M.S. Pratchett and M.L. Berumen. 2016b. Regional variation in the structure and function of parrotfishes on Arabian reefs. Mar. Poll. Bull. 105: 524–531.

Hoey, A.S. 2018. Feeding in parrotfishes: the influence of species, body size, and temperature. pp. 119–133. *In*: Hoey, A.S. and R.M. Bonaldo [eds.]. Biology of Parrotfishes. CRC Press, Boca Raton.

Honda, K., Y. Nakamura, M. Nakaoka, W.H. Uy and M.D. Fortes. 2013. Habitat use by fishes in coral reefs, seagrass beds and mangrove habitats in the Philippines. PLoS ONE 8: e65735.

Huang, X., Z. Lao, Y. Yang, T. Li, Z. Huang, W. You et al. 2024. Research on tolerance of juvenile *Siganus oramin* to water temperature, salinity and dissolved oxygen. South China Fisheries Science doi: 10.12131/20230143.

Huey, R.B. and R.D. Stevenson. 1979. Integrating thermal physiology and ecology of ectotherms: a discussion of approaches. Am. Zool. 19: 357–366.

Hughes, T.P., A.H. Baird, D.R. Bellwood, M. Card, S.R. Connolly, C. Folke et al. 2003. Climate change, human impacts, and the resilience of coral reefs. Science 301: 929–933.

Hughes, T.P., J.T. Kerry, M. Álvarez-Noriega, J.G. Álvarez-Romero, K.D. Anderson, A.H. Baird et al. 2017. Global warming and recurrent mass bleaching of corals. Nature 543: 373–377.

Hughes, T.P., K.D. Anderson, S.R. Connolly, S.F. Heron, J.T. Kerry, J.M. Lough et al. 2018a. Spatial and temporal patterns of mass bleaching of corals in the Anthropocene. Science 359: 80–83.

Hughes, T.P., J.T. Kerry, A.H. Baird, S.R. Connolly, A. Dietzel, C.M. Eakin et al. 2018b. Global warming transforms coral reef assemblages. Nature 556: 492–496.

Ikejima, K., P. Tongnunui, T. Medej and T. Taniuchi. 2003. Juvenile and small fishes in a mangrove estuary in Trang province, Thailand: seasonal and habitat differences. Estuar. Coastal Shelf Sci. 56: 447–457.

Jaonalison, H., H. Ravelohasina, J. Mahafina, L. Ranivoarivelo, G. Lepoint, R. Rasolofonirina et al. 2022. The potential nursery areas and recruitment seasons of *Siganus sutor* in Madagascar. 12th WIOMSA Scientific Symposium.

Jawad, L., J. Abed, A.K. Resen, R. Al-Yassein and A.M. Kulw. 2023. The shoemaker spinefoot *Siganus sutor* (Valenciennes, 1835) in the marine waters of Iraq: second appearance and the northernmost record in the northern Indian Ocean. Oceanolog. Hydrobiol. Stud. 52: 166–171.

Johannes, R.E. 1978. Reproductive strategies of coastal marine fishes in the tropics. Environ. Biol. Fishes 3: 65–84.

Johansen, J.L., V. Messmer, D.J. Coker, A.S. Hoey and M.S. Pratchett. 2014. Increasing ocean temperatures reduce activity patterns of a large commercially important coral reef fish. Glob. Change Biol. 20: 1067–1074.

Johansen, J.L., M.S. Pratchett, V. Messmer, D.J. Coker, A.J. Tobin and A.S. Hoey. 2015. Large predatory coral trout species unlikely to meet increasing energetic demands in a warming ocean. Sci. Rep. 5: 13830.

Jokiel, P.L. and S.L. Coles. 1990. Response of Hawaiian and other Indo-Pacific reef corals to elevated temperature. Coral Reefs 8: 155–162.

Jones, G.P. 1990. The importance of recruitment to the dynamics of a coral reef fish population. Ecology 71: 1691–1698

Jones, B.L., R.K. Unsworth, S. Udagedara and L.C. Cullen-Unsworth. 2018. Conservation concerns of small-scale fisheries: by-catch impacts of a shrimp and finfish fishery in a Sri Lankan lagoon. Frontiers Mar. Sci. 5: 52.

Jones, G.P., M.I. McCormick, M. Srinivasan and J.V. Eagle. 2004. Coral decline threatens fish biodiversity in marine reserves. Proc. Natl Acad. Sci. USA. 101: 8251–8253.

Jutfelt, F., T. Norin, R. Ern, J. Overgaard, T. Wang, D.J. McKenzie et al. 2018. Oxygen- and capacity-limited thermal tolerance: Blurring ecology and physiology. J. Exp. Biol. 221: 2016–2019.

Kalogirou, S., M. Corsini-Foka, A. Sioulas, H. Wennhage and L. Pihl. 2010. Diversity, structure and function of fish assemblages associated with Posidonia oceanica beds in an area of the eastern Mediterranean Sea and the role of non-indigenous species. J. Fish Biol. 77: 2338–2357.

Kamukuru, A.T. 2009. Trap fishery and reproductive biology of the whitespotted rabbitfish *Siganus sutor* (Siganidae), within the Dar es Salaam Marine Reserves, Tanzania. West. Indian Ocean J. Mar. Sci. 8: 75–86.

Kenitz, K.M., A.W. Visser, M.D. Ohman, M.R. Landry and K.H. Andersen., 2019. Community trait distribution across environmental gradients. Ecosystems 22: 968–980.

Kimirei, I.A., I. Nagelkerken, M. Trommelen, P. Blankers, N. Van Hoytema, D. Hoeijmakers et al. 2013. What drives ontogenetic niche shifts of fishes in coral reef ecosystems? Ecosystems 16: 783–796.

Kumagai, N.H., J. García Molinos, H. Yamano, S.Takao, M. Fujii and Y. Yamanaka. 2018. Ocean currents and herbivory drive macroalgae-to-coral community shift under climate warming. Proc. Natl. Acad. Sci. USA 115: 8990–8995.

Kuriiwa, K., N. Hanzawa, T. Yoshino, S. Kimura and M. Nishida. 2007. Phylogenetic relationships and natural hybridization in rabbitfishes (Teleostei: Siganidae) inferred from mitochondrial and nuclear DNA analyses. Mol. Phylogen. Evol. 45: 69–80.

Kwak, S.N., D.W. Klumpp and J.M. Park. 2015. Feeding relationships among juveniles of abundant fish species inhabiting tropical seagrass beds in Cockle Bay, North Queensland, Australia. NZ J. Mar. Freshw. Res.

Lam, T.J. 1974. Siganids: Their biology and mariculture potential. Aquaculture 3: 325–354.

Lam, V.W.Y., E.H. Allison, J.D. Bell, J. Blythe, W.W. Cheung, T.L. Frölicher et al. 2020. Climate change, tropical fisheries and prospects for sustainable development. Nat. Rev. Earth Environ. 1: 440–454.

LaMonica, L.E., R.J. Fox and J.M. Donelson. 2021. Thermal sensitivity of juvenile rabbitfishes *Siganus doliatus* and *S. lineatus* (Siganidae): a key role for habitat? Coral Reefs 40: 1307–1320.

Lough, J.M. 2012. Small change, big difference: sea surface temperature distributions for tropical coral reef ecosystems, 1950–2011. J. Geophysical Res: Oceans. 117: C9.

Lugendo, B.R., A. Pronker, I. Cornelissen, A. De Groene, I. Nagelkerken, M. Dorenbosch et al. 2005. Habitat utilisation by juveniles of commercially important fish species in a marine embayment in Zanzibar, Tanzania. Aquat. Living Res. 18: 149–158.

Malcolm, H.A., W. Gladstone, S. Lindfield, J. Wraith and T.P. Lynch. 2007. Spatial and temporal variation in reef fish assemblages of marine parks in New South Wales, Australia—baited video observations. Mar. Ecol. Progr. Ser. 350: 277–290.

Malik, A., O. Mertz and R. Fensholt. 2017. Mangrove forest decline: Consequences for livelihoods and environment in South Sulawesi. Regional Environ. Change 17: 157–169.

Marras, S., A. Cucco, F. Antognarelli, E. Azzurro, M. Milazzo, M. Bariche et al. 2015. Predicting future thermal habitat suitability of competing native and invasive fish species: from metabolic scope to oceanographic modelling. Conserv. Physiol. 3: cou059.

Martin, C., E.C. McClure, K. Webber, M. Burgo and A.S. Hoey. 2024. Utilisation of macroalgal habitats by juvenile rabbitfishes on an inshore reef. Coral Reefs, doi: 10.1007/s00338-024-02581-x.

McClure, E.C., L.E. Richardson, A. Graba-Landry, Z. Loffler, G.R. Russ and A.S. Hoey. 2019. Cross-shelf differences in the response of herbivorous fish assemblages to severe environmental disturbances. Diversity 11: 23.

McCormick, M.I. and A.S. Hoey. 2004. Larval growth history determines juvenile growth and survival in a tropical marine fish. Oikos 106: 225–242.

McCormick, M.I. and L.J. Makey. 1997. Post-settlement transition in coral reef fishes: overlooked complexity in niche shifts. Mar. Ecol. Progr. Ser. 153: 247–257.

Messmer, V., M.S. Pratchett, A.S. Hoey, A.J. Tobin, D.J. Coker, S.J. Cooke et al. 2017. Global warming may disproportionately affect larger adults in a predatory coral reef fish. Glob. Change Biol. 23: 2230–2240.

Moleana, T., L. Della Patrona, T. Meziane and Y. Letourneur. 2016. First record of *Siganus randalli* (Teleost, Siganidae) in New Caledonia, and comments on its diet. Mar. Biodiv. Rec. 9: 1–5.

Motson, K. and J.M. Donelson. 2017. Limited capacity for developmental thermal acclimation in three tropical wrasses. Coral Reefs 36: 609–621.

Nakamura, Y. and M. Tsuchiya. 2008. Spatial and temporal patterns of seagrass habitat use by fishes at the Ryukyu Islands, Japan. Estuar. Coastal Shelf Sci. 76: 345–356.

Nay, T.J., J.L. Johansen, J.L. Rummer, J.F. Steffensen and A.S. Hoey. 2021. Species interactions alter the selection of thermal environment in a coral reef fish. Oecologia 196: 363–371.

Nay, T.J., J.L. Johansen, J.L. Rummer, J.F. Steffensen, M.S. Pratchett and A.S. Hoey. 2020. Habitat complexity influences selection of thermal environment in a common coral reef fish. Conserv. Physiol. 8: coaa070.

Newbold, T., L.N. Hudson, S.L.L. Hill, S. Contu, I. Lysenko, R.A. Senior et al. 2015. Global effects of land use on local terrestrial biodiversity. Nature 520: 45–50.

Olds, A.D., R.M. Connolly, K.A. Pitt and P.S. Maxwell. 2012. Habitat connectivity improves reserve performance. Conserv. Lett. 5: 56–63.

Pandolfi, J.M., R.H. Bradbury, E. Sala, T.P. Hughes, K.A. Bjorndal, R.G. Cooke et al. 2003. Global trajectories of the long-term decline of coral reef ecosystems. Science 301: 955–958.

Pankhurst, N.W. and P.L. Munday. 2011. Effects of climate change on fish reproduction and early life history stages. Mar. Freshw. Res. 62: 1015–26.

Pauly, D. 2021. The gill-oxygen limitation theory (GOLT) and its critics. Sci. Adv. 7: eabc6050.

Pecl, G.T., M.B. Araújo, J.D. Bell, J. Blanchard, T.C. Bonebrake, I.C. Chen et al. 2017. Biodiversity redistribution under climate change: Impacts on ecosystems and human well-being. Science 355: eaai9214.

Pinsky, M.L., A.M. Eikeset, D.J. McCauley, J.L. Payne and J.M. Sunday. 2019. Greater vulnerability to warming of marine versus terrestrial ectotherms. Nature 569: 108–111.

Pinto, L. and N.N. Punchihewa. 1996. Utilisation of mangroves and seagrasses by fishes in the Negombo Estuary, Sri Lanka. Mar. Biol. 126: 333–345.

Pitt, J.M. 1997. The feeding ecology of rabbitfish (Siganidae) at Green Island reef: ontogenetic and interspecific differences in diet, morphology and habitat utilisation. Doctoral dissertation, James Cook University of North Queensland.

Pombo-Ayora, L., D.J. Coker, S. Carvalho, G. Short and M.L. Berumen. 2020. Morphological and ecological trait diversity reveal sensitivity of herbivorous fish assemblages to coral reef benthic conditions. Mar. Environ. Res. 162: 105102.

Popper, D., R.C. May and T. Lichatowich. 1976. An experiment in rearing larval *Siganus vermiculatus* (Valenciennes) and some observations on its spawning cycle. Aquaculture 7: 281–290.

Pörtner, H.O. 2010. Oxygen- And capacity-limitation of thermal tolerance: a matrix for integrating climate-related stressor effects in marine ecosystems. J. Exp. Biol. 213: 881–893.

Pörtner, H.O. and A.P. Farrell. 2008. Physiology and climate change. Science 322: 690–692.

Pratchett, M.S., S.A. Blowes, D. Coker, E. Kubacki, J. Nowicki and A.S. Hoey. 2015. Indirect benefits of high coral cover for non-corallivorous butterflyfishes. Coral Reefs. 34: 665–672.

Pratchett, M.S., A.S. Hoey and S.K. Wilson. 2014. Reef degradation and the loss of critical ecosystem goods and services provided by coral reef fishes. Curr. Opinion Environ. Sust. 7: 37–43.

Pratchett, M.S., A.S. Hoey, S.K. Wilson, V. Messmer and N.A. Graham. 2011. Changes in biodiversity and functioning of reef fish assemblages following coral bleaching and coral loss. Diversity 3: 424–452.

Pratchett, M.S., P.L. Munday, S.K. Wilson, N.A. Graham, J.E. Cinner, D.R. Bellwood et al. 2008. Effects of climate-induced coral bleaching on coral-reef fishes—ecological and economic consequences. Oceanogr. Mar. Biol.: Annu. Rev. 46: 257–302.

Pratchett, M.S., C.A. Thompson, A.S. Hoey, P.F. Cowman and S.K. Wilson. 2018. Effects of coral bleaching and coral loss on the structure and function of reef fish assemblages. pp. 265–293. *In*: van Oppen, M.J.H. and J.M. Lough [eds.]. Coral bleaching: Patterns, processes, causes and consequences. Springer, Switzerland.

Priest, M.A., A.R. Halford, K.D. Clements, E. Douglas, S.L. Abellana and J.L. McIlwain. 2016. What a difference a bay makes: natural variation in dietary resources mediates growth in a recently settled herbivorous fish. Coral Reefs 35: 1187–1199.

Rajaguru, S. 2002. Critical thermal maximum of seven estuarine fishes. J. Thermal Biol. 27: 125–128.

Rasher, D.B., A.S. Hoey and M.E. Hay. 2013. Consumer diversity interacts with prey defenses to drive ecosystem function. Ecology 94: 1347–1358.

Rath, K.M., N. Fierer, D.V. Murphy and J. Rousk. 2019. Linking bacterial community composition to soil salinity along environmental gradients. ISME Journal 13: 836–846.

Reef Life Survey. 2024. Reef species of the world. https://reeflifesurvey.com/species/?families=Siganidae (accessed 07/05/2024).

Richardson, L.E., N.A. Graham and A.S. Hoey. 2020. Coral species composition drives key ecosystem function on coral reefs. Proc. R. Soc. B 287: p.20192214.

Richardson, L.E., N.A. Graham, M.S. Pratchett, J.G. Eurich and A.S. Hoey. 2018. Mass coral bleaching causes biotic homogenization of reef fish assemblages. Glob. Change Biol. 24: 3117–3129.

Rilov, G. and B. Galil. 2009. Marine bioinvasions in the Mediterranean Sea–history, distribution and ecology. pp. 549–575. *In*: Rilov, G. and J.A. Crooks [eds.]. Biological invasions in marine ecosystems: ecological, management, and geographic perspectives. Springer, Berlin.

Robinson, J.P., S.K. Wilson, S. Jennings and N.A. Graham. 2019. Thermal stress induces persistently altered coral reef fish assemblages. Glob. Change Biol. 25: 2739–2750.

Röthig, T., S.M. Trevathan-Tackett, C.R. Voolstra, C. Ross, S. Chaffron, P.J. Durack et al. 2023. Human-induced salinity changes impact marine organisms and ecosystems. Glob. Chang. Biol. 29: 4731–4749.

Sambrook, K., A.S. Hoey, S. Andréfouët, G.S. Cumming, S. Duce and M.C. Bonin. 2019. Beyond the reef: The widespread use of non-reef habitats by coral reef fishes. Fish Fisher. 20: 903–920.

Saoud, I.P., S. Kreydiyyeh, A. Chalfoun and M. Fakih. 2007. Influence of salinity on survival, growth, plasma osmolality and gill Na+–K+–ATPase activity in the rabbitfish *Siganus rivulatus*. J. Exp. Mar. Biol. Ecol. 348: 183–190.

Saoud, I.P., C. Mohanna and J. Ghanawi. 2008. Effects of temperature on survival and growth of juvenile spinefoot rabbitfish (*Siganus rivulatus*). Aquaculture Res. 39: 491–497.

Schulte, P.M. 2015. The effects of temperature on aerobic metabolism: Towards a mechanistic understanding of the responses of ectotherms to a changing environment. J. Exp. Biol. 218: 1856–1866.

Selig, E.R., D.G. Hole, E.H. Allison, K.K. Arkema, M.C. McKinnon, J. Chu et al. 2019. Mapping global human dependence on marine ecosystems. Conserv. Lett. 12: e12617.

Sing Wong, A., S. Vrontos and M.L. Taylor. 2022. An assessment of people living by coral reefs over space and time. Glob. Change Biol. 28: 7139–7153.

Siqueira, A.C., D.R. Bellwood and P.F. Cowman. 2019. The evolution of traits and functions in herbivorous coral reef fishes through space and time. Proc. R. Soc. B 286: 20182672.

Souter, D., S. Planes, J. Wicquart, M. Logan, D. Obura and F. Staub. 2021. Status of coral reefs of the world: 2020. Global Coral Reef Monitoring Network (GCRMN) and International Coral Reef Initiative (ICRI).

Stergiou, K.I. 1988. Feeding habits of the Lessepsian migrant *Siganus luridus* in the eastern Mediterranean, its new environment. J. Fish Biol. 33: 531–543.

Sugiyama, A. and A. Tomori. 1990. Variation of the catch of siganids and the seasonal run of their juveniles to the seaweed bed in Ishigaki Island. Aquaculture Sci. 2: 67–74.

Sunday, J., J.M. Bennett, P. Calosi, S. Clusella-Trullas, S. Gravel, A.L. Hargreaves et al. 2019. Thermal tolerance patterns across latitude and elevation. Phil. Trans. R. Soc. B 374: 20190036.

Sunday, J.M., A.E. Bates and N.K. Dulvy. 2011. Global analysis of thermal tolerance and latitude in ectotherms. Proc. R. Soc. B 278: 1823–1830.

Tang, S., A. Graba-Landry and A.S. Hoey. 2020. Density and height of *Sargassum* influence rabbitfish (f. Siganidae) settlement on inshore reef flats of the Great Barrier Reef. Coral Reefs 39: 467–473.

Tano, S.A., M. Eggertsen, S.A. Wikström, C. Berkström, A.S. Buriyo and C. Halling. 2017. Tropical seaweed beds as important habitats for juvenile fish. Mar. Freshw. Res. 68: 1921–1934.

Taylor, B.M., C.E. Benkwitt, J.H. Choat, K.D. Clements, N.A. Graham and M.G. Meekan. 2020. Synchronous biological feedbacks in parrotfishes associated with pantropical coral bleaching. Glob. Change Biol. 26: 1285–1294.

Tebbett, S.B., A.C. Siqueira and D.R. Bellwood. 2022. The functional roles of surgeonfishes on coral reefs: past, present and future. Rev. Fish Biol. Fisher. 32: 387–439.

Tewksbury, J.J., R.B. Huey and C.A. Deutsch. 2008. Putting the heat on tropical animals. Science 320: 1296–1297.

Tittensor, D.P., C. Mora, W. Jetz, H.K. Lotze, D. Ricard, E. Berghe et al. 2010. Global patterns and predictors of marine biodiversity across taxa. Nature 466: 1098–1101.

Tsuda, R.T. and P.G. Bryan. 1973. Food preference of juvenile *Siganus rostratus* and *S. spinus* from Guam. Copeia 1973: 604–606.

Valiela, I., J.L. Bowen and J.K. York. 2001. Mangrove forests: One of the world's threatened major tropical environments. Bioscience 51: 807–815.

Vanderklift, M.A., R.C. Babcock, F. Boschetti, M.D. Haywood, R.D. Pillans and D.P. Thomson. 2019. Declining abundance of coral reef fish in a World-Heritage-listed marine park. Sci. Rep. 9: 15524.

Vanderklift, M.A., R.D. Pillans, M. Hutton, L. De Wever, G.A. Kendrick, A. Zavala-Perez et al. 2021. High rates of herbivory in remote northwest Australian seagrass meadows by rabbitfish and green turtles. Mar. Ecol. Progr. Ser. 665: 63–73.

Vergés, A., C. Doropoulos, H.A. Malcolm, M. Skye, M. Garcia-Pizá, E.M. Marzinelli et al. 2016. Long-term empirical evidence of ocean warming leading to tropicalization of fish communities, increased herbivory, and loss of kelp. Proc. Natl Acad. Sci. USA 113: 13791–13796.

Vergés, A., P.D. Steinberg, M.E. Hay, A.G. Poore, A.H. Campbell, E. Ballesteros et al. 2014. The tropicalization of temperate marine ecosystems: climate mediated changes in herbivory and community phase shifts. Proc. R. Soc. B 281: 20140846.

Waldeland, O.R. 2017. Habitat preferences and associations of juvenile algal-associated fishes on coral reefs. MPhil dissertation, James Cook University of North Queensland.

Walther, G.R. 2010. Community and ecosystem responses to recent climate change. Phil. Trans. R. Soc. B: 365: 2019–2024.

Waycott, M., C.M. Duarte, T.J. Carruthers, R.J. Orth, W.C. Dennison, S. Olyarnik et al. 2009. Accelerating loss of seagrasses across the globe threatens coastal ecosystems. Proc. Natl Acad. Sci. USA 106: 12377–12381.

Webber, K., K. Fabricius, S.K. Wilson and A.S. Hoey. 2024. The influence of *Sargassum* biomass and thallus density on the recruitment of coral reef fishes. Coral Reefs 43: 687–699.

Wilson, S.K., M. Depczynski, R. Fisher, T.H. Holmes, R.A. O'Leary and P. Tinkler. 2010. Habitat associations of juvenile fish at Ningaloo Reef, Western Australia: the importance of coral and algae. PLoS ONE 5: e15185.

Wilson, S.K., N.A. Graham, M.S. Pratchett, G.P. Jones and N.V. Polunin. 2006. Multiple disturbances and the global degradation of coral reefs: are reef fishes at risk or resilient? Glob. Change Biol. 12: 2220–2234.

Wismer, S., A.S. Hoey and D.R. Bellwood. 2009. Cross-shelf benthic community structure on the Great Barrier Reef: relationships between macroalgal cover and herbivore biomass. Mar. Ecol. Progr. Ser. 376: 45–54.

Woodland, D.J. 1990. Revision of the fish family Siganidae with descriptions of two new species and comments on distribution and biology. Indo-Pacific Fishes 19.

Woodland, D.J. and R.C. Anderson. 2014. Description of a new species of rabbitfish (Perciformes: Siganidae) from southern India, Sri Lanka and the Maldives. Zootaxa 3811: 129–136.

Zarco-Perello, S., S. Martin and A.S. Hoey. 2024. Biogeographical diet variation within and between the rabbitfishes *Siganus corallinus*, *Siganus virgatus, Siganus doliatus* and *Siganus trispilos*. Ecology and Evolution 14: e11326.

Zarco-Perello, S., M. Pratchett and V. Liao. 2012. Temperature-growth performance curves for a coral reef fish, Acanthochromis polyacanthus. Galaxea 14: 97–103.

Zarco-Perello, S., N.E. Bosch, S. Bennett, M.A. Vanderklift and T. Wernberg. 2021. Persistence of tropical herbivores in temperate reefs constrains kelp resilience to cryptic habitats. J. Ecol. 109: 2081–2094.

8

Rabbitfishes in Temperate Ecosystems, Range-Extensions and their Socio-Ecological Impacts

Salvador Zarco-Perello,[1,*] *Adriana Verges,*[2] *Ernesto Azzurro,*[3] *Julia Santana-Garcon,*[4] *Michel Bariche,*[5] *Scott Bennett*[4] and *Yohei Nakamura*[6]

Introduction

Changes in climate are causing a general trend of poleward expansion of species. Ocean warming has accelerated in the last decades, with projections estimating temperature increases of 3°C on average by 2090 (IPCC 2021). Some temperate regions are warming faster than the global average due to the intensification of heat transport by western boundary currents from equatorial regions to higher latitudes

[1] Harry Butler Institute, Murdoch University, Perth, Western Australia, 6150, Australia.
[2] Centre for Marine Science & Innovation, School of Biological, Earth and Environmental Sciences, UNSW Sydney, NSW 2052, Australia, and Sydney Institute of Marine Sciences, Mosman NSW 2088, Australia.
Email: a.verges@unsw.edu.au
[3] National Research Council (CNR), Institute for Biological Resources and Marine Biotechnologies (IRBIM), Largo Fiera della Pesca, 2, 60125 Ancona, Italy and Integrative Marine Ecology Department, Stazione Zoologica Anton Dohrn, Naples, Italy.
Email: ernesto.azzurro@cnr.it
[4] Institute of Marine and Antarctic Studies and Centre for Marine Sociology, University of Tasmania, Hobart, Tasmania, 7001, Australia.
Emails: julia.santanagarcon@gmail.com; scott.bennett0@utas.edu.au
[5] Department of Biology, FAS, American University of Beirut, Beirut, Lebanon.
Email: michel.bariche@aub.edu.lb
[6] Graduate School of Integrated Arts and Sciences, Kochi University, Kochi, 783-8502, Japan.
Email: ynakamura@kochi-u.ac.jp
* Corresponding author: salvador.zarcoperello@gmail.com

(Hobday and Pecl 2014). In addition, marine heatwaves have increased in frequency and duration during the last century (Oliver et al. 2018). This is creating stressful environments for many species in tropical regions, whereas regions at higher latitudes are becoming more favourable for warm-affinity species. As a consequence, populations at the leading edge of distribution are extending their ranges poleward, causing declines in species diversity around the equatorial regions, while leading to increases in subtropical and temperate regions, modifying the community structure and composition in their receptive ecosystems (Chaudhary et al. 2021). As global warming proceeds, species redistributions are forecast to continue into the future (Morley et al. 2018).

The reorganization of species can modify ecological processes and the functioning of ecosystems (Whalen et al. 2020). Trophic interactions are largely driven by species diversity, composition, and temperature; thus, although range expansions will benefit species suffering thermal stress at lower latitudes (Urban 2020), they can also destabilize the functioning of receptive ecosystems (Albouy et al. 2014). Historically, primary consumption is more intense in tropical marine regions due to a combination of higher diversity and abundance of herbivores in warmer environments that increase metabolic and consumption rates (Longo et al. 2018). Herbivorous fish form assemblages of dozens of species that constantly feed on the benthos of tropical reefs, facilitating the flow of energy to higher trophic levels and regulating the competition for space between macroalgae and sessile invertebrates, such as scleractinian corals. While climate-related range contractions or reductions in the abundance of herbivorous fish in tropical latitudes have not yet been documented, such processes are likely to weaken the resilience of coral dominated ecosystem states, leading to increases in macroalgae cover, loss of habitat complexity and reef accretion (Steneck et al. 2017). In contrast, the overall density and diversity of herbivorous fishes at higher latitudes is generally much lower, and play a lesser role engineering habitat structure (Holland et al. 2020), which is typically dominated by high abundances of canopy-forming macroalgae that sustain the biological communities and ecosystem functioning (Wernberg et al. 2019). Although, significant removal of seaweed by temperate herbivorous fish has been documented in some locations (Vergés et al. 2009, Papadakis et al. 2021). Concurrently, range-extensions and increases in abundance of tropical herbivorous species at higher latitudes can lead to losses in habitat-forming macroalgae by accelerating the consumption of canopies and recruits, driving temperate habitats towards more simple configurations dominated by smaller seaweed and sessile invertebrates (Vergés et al. 2019).

Among tropical herbivorous fishes, rabbitfishes (Siganidae) are the most reported herbivorous fishes to be extending its range into multiple temperate warming hotspots of the world. Populations and individuals of different families of herbivorous fishes have been reported at higher latitudes; however, rabbitfishes have been particularly successful in establishing reproductive populations and undergone considerable range expansions. Rabbitfishes are an important group of herbivores primarily distributed in the Indo-Pacific region (Siqueira et al. 2019). They are mainly associated with coral reefs but some species are found in lagoonal habitats with seagrass meadows, mangroves and macroalgae beds (Woodland 1990). Species of the family can be deep-bodied and form pairs as adults, while others have a fusiform shape and form schools conformed from dozens to hundreds of individuals (Fox et al. 2015). Several studies have identified rabbitfishes as key consumers of macroalgae and seagrass in tropical ecosystems (Fox and Bellwood 2008, Vanderklift et al. 2021) and in many regions of Asia they are important species in fisheries markets, and aquaculture industries, having a significant economic and cultural value (Soliman and Yamaoka 2010). As such, large populations can have important ecological and socioeconomic impacts in temperate locations. This chapter provides an overview of the current state of knowledge of two major trends documented in the literature: (i) the range-expansions of rabbitfishes into novel temperate habitats and (ii) increases in abundance of existing populations at their edge of distribution in temperate ecosystems. Using case studies from the Mediterranean Sea, Western and Eastern Australia and Japan, this chapter describes population changes in space and time and the environmental and biological factors associated with their change in distribution, persistence and the current and potential socio-ecological impacts that they can have in their novel ecosystems. Finally, the last section presents a synthesis of information and future research recommendations.

Range-Expansions

Mediterranean Sea

The marine ecosystems of the Mediterranean Sea are composed of extensive and productive seagrass meadows and rocky reefs dominated by macroalgal assemblages (Sales and Ballesteros 2012, Telesca et al. 2015). Species of the brown macroalgae *Cystoseira* spp. (Fucales, Phaeophyceae) are particularly abundant and important in the region as they form prominent canopies that provide habitat and food that sustain high levels of biodiversity, akin to the kelp forests of other temperate regions (Cheminée et al. 2017, Chiarore et al. 2019). The Mediterranean Sea is a semi-enclosed basin that is naturally connected to the Atlantic Ocean through the Strait of Gibraltar; however, the construction of the Suez Canal in 1869 artificially connected the eastern region with the Red Sea, allowing the introduction and establishment of hundreds of species of subtropical origin (Zenetos et al. 2022).

The rabbitfishes *Siganus argenteus* (Quoy and Gaimard 1825), *Siganus stellatus* (Forsskål 1775), *Siganus luridus* (Rüppell 1829) and *Siganus rivulatus* (Forsskål and Niebuhr 1775) roam the coral reefs and seagrass meadows of the Red Sea (Golani and Fricke 2018). Among these, *S. luridus* and *S. rivulatus* were among the first Red Sea fishes to establish populations in Mediterranean waters. The two species have since become very common along the eastern Mediterranean coast, attaining significant relevance in regional fisheries and affecting the benthic habitats and communities. Additionally, *Siganus argentus, Siganus fuscescens* (Houttuyn 1782), *Siganus virgatus* (Valenciennes 1835) and *Siganus javus* (Linnaeus 1766) have been found in different locations in the region; however, these have been single individuals most likely introduced by shipping traffic or aquarium releases, unable to establish self-sustaining populations (Table 1).

Table 1. Species of siganids reported in the Mediterranean Sea and details of their introduction and expansion.

Species	**Origin and Year Reported**	**Distribution and Range Expansion**	**References**
Siganus rivulatus	Suez Canal and natural dispersion westwards since 1924	Expansion of 2600 km from Palestine until Tunez and Sicily in the southern coast, and Croatia in the Adriatic Sea	Azzurro et al. 2022
Siganus luridus	Suez Canal and natural dispersion westwards since 1955	Expansion of 3000 km from Israel until Tunez and Sicily in the southern coast, and Gulf of Trieste in the Adriatic Sea	Azzurro et al. 2022
Siganus argenteus	Possibly in ballast water from the Red Sea. 2015	Tobruk, Libya. Single record	Abdelghani et al. 2020 and Ahnelt 2016
Siganus virgatus	Possibly in ballast water or aquarium release. 1975	Island of Cres, Croatia. Single record	Ahnelt 2016
Siganus javus	Possibly in ballast water or aquarium release. 2009	Lattakia, Syria. Single record	Ibrahim et al. 2010
Siganus fuscescens	Possibly in ballast water. 2020	Gioia Tauro, Italy. Single record	Azzuro and Tiralongo 2020

Expansion and Persistence

Evidence suggests that *Siganus rivulatus* and *Siganus luridus* could have entered the Mediterranean Sea through a combination of ways, such as ballast water, natural drifting of their larvae and even by active swimming of adult individuals (Diamant 2010). Analyses have found no genetic differentiation between Mediterranean and Red Sea populations, suggesting that a large number of individuals have immigrated and that the gene flow through the Suez Canal could be frequent (Hassan et al. 2003). Following their successful establishment, *S. luridus* and *S. rivulatus* remained confined to the eastern region for many years. However, both species have gradually established new populations in southern and northern coasts (Fig. 1). *S. rivulatus* was first reported in the coast of Palestine in 1924 and gradually expanded westwards until Libya and Tunisia (Tortonese 1970), and northwards into the Adriatic Sea (Dulĉić and Pallaoro 2004) and more recently in Sicily (Karachle et al. 2016). *S. luridus* was first recorded in the Mediterranean coast of Israel in 1956, but a decade later it had already expanded until Tunisia (Ben-Tuvia 1964), followed by the Aegean Sea and more recently up to the Sicily Strait (Azzurro and Andaloro 2004) and the eastern Adriatic Sea, in remarkable similitude to *S. rivulatus* (Dulčić et al. 2011). Single individuals have been reported off the coast of Marseilles (Daniel et al. 2009) and the northern Adriatic Sea (Poloniato et al. 2010); however, the establishment of populations in these areas is unlikely at this point due to low seawater temperatures (Marras et al. 2015).

While the introduction of rabbitfishes was facilitated by the Suez Canal, their successful establishment and subsequent expansion is likely attributed to their eco-physiological plasticity and the ongoing warming experienced in the Mediterranean Sea (Occhipinti-Ambrogi 2007). Despite the environmental differences with the Red Sea, rabbitfishes have successfully reproduced and recruited along their Mediterranean distribution. Evidence suggests that populations have experienced significant changes in their life-histories by shifting the onset and reducing the duration of reproductive periods, slowing growth rates and increasing fecundity in comparison with Red Sea populations (Zarco-Perello et al. 2022). In their eastern range, Lebanon, *S. rivulatus* spawns in June, and *S. luridus* from May to July (Bariche et al. 2003, Bariche et al. 2009), after which hundreds of juveniles (< 30 mm total length) have been sighted in protected shallow habitats with rocky

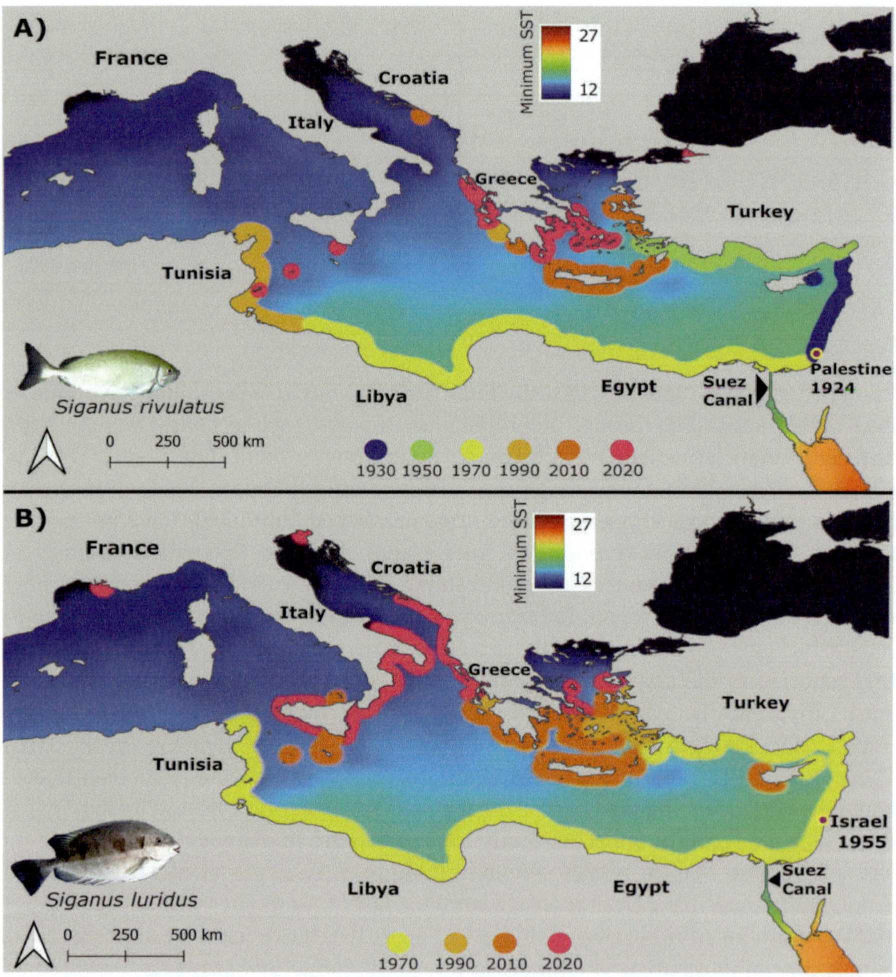

Fig. 1. Geographical expansion of the Marbled spinefoot *Siganus rivulatus* (A) and the Dusky spinefoot *Siganus luridus* (B) in the Mediterranean Sea following their introduction through the Suez Canal. Based on occurrences reported in the ORMEF database (Azzurro et al. 2022). Fish images taken by Randall J. E. and acquired from Fishbase (Froese and Pauly 2019). SST: minimum average sea surface temperature (EU's Earth Observation 2021).

substrates covered with turf and erect seaweed, muddy harbours and seagrass meadows (*Cymodocea* sp.) (Bariche et al. 2004). Further west in Rhodes Island, Greece, juvenile rabbitfishes have been seen roaming seagrass meadows, with thousands of individuals of *S. rivulatus* collected in August (Kalogirou et al. 2010). At the front of range expansion in Linosa Island, Sicily Strait, recently established populations of *S. luridus* consist of mature individuals with ripe and spent gonads, together with juveniles in rocky bottoms and seagrass meadows (Azzurro and Andaloro 2004, Azzurro et al. 2007b). The Mediterranean Sea has experienced significant warming in association with climate change, increasing in temperature by up to 0.7°C per decade in the eastern basin since 1985, and it is likely that this has benefited the physiological performance and expansion of rabbitfishes (Givan et al. 2018).

Across their Mediterranean distribution, *S. rivulatus* and *S. luridus* have adapted to local trophic resources that sustain their survival and population growth. Rabbitfishes have become common in the seagrass meadows, rocky reefs and algal meadows throughout their range-expansion, having larger and deeper home ranges (< 40 m deep) than Red Sea populations (< 6 m) (Harmelin-Vivien et al. 2005, Shakman and Kinzelbach 2007, Pickholtz et al. 2018). In these habitats, they can consume a great variety of macrophytes as revealed by stomach content analyses from Lebanon to Linosa Island, including filamentous, foliose, fleshy and canopy-forming macroalgae, as well as seagrass, bryozoans, crustaceans, polychaetes and molluscs in small quantities, depending on the availability of resources and seasonality (Bariche 2006, Azzurro et al. 2007a, Shakman et al. 2009). Analysis of C:N in seagrasses and seaweeds has shown that rabbitfishes can target macrophyte species with highest nutritional value in the local area (Santana-Garcon et al. 2023). Accordingly, rabbitfish populations have become very abundant and even dominate certain coastal communities. For instance, rabbitfishes represent 75–85 percent of total fish captures from trawlers in the coastal waters of Libya (Shakman and Kinzelbach 2007). At the range of expansion in the Sicily Strait, the abundance of *S. luridus* is still lower than native fish species; however, schools of more than 100 individuals have been recorded, suggesting population growth (Azzurro et al. 2017).

Impacts

The expansion of *S. rivulatus* and *S. luridus* in the Mediterranean Sea has been related to changes in ecosystem functioning. Their abundant populations have greatly increased the number of herbivorous individuals across their distribution, accounting for 35–100 percent of the herbivorous fish in sites of Lebanon and Israel (Diamant et al. 1986, Bariche et al. 2004, Harmelin-Vivien et al. 2005), 50–80 percent in Turkey (Sala et al. 2011, Bodilis et al. 2014) or 56–99 percent in the coasts of Greece (Kalogirou et al. 2010, Kalogirou et al. 2012, Giakoumi 2014). This has accelerated the rates of primary consumption together with native herbivorous species, as they can perform similar herbivory functions. Native herbivores only include the sparid, *Sarpa salpa* (Linnaeus 1758), an important consumer of seagrass and algal assemblages (Vergés et al. 2009, Pagès et al. 2012), and the scarid *Sparisoma cretense* (Linnaeus 1758), which can be an important component of the herbivorous communities in the eastern Mediterranean, although populations are expanding westwards

(Marletta and Lombardo 2020, Esposito et al. 2021). In addition, some native omnivorous species can also have significant impacts in macroalgal abundance (Papadakis et al. 2021). Although variations exist in diet, these species can feed on the same substrates as rabbitfishes, as confirmed by stomach contents, isotopic signatures and field observations, increasing the removal of seaweed in rocky reefs (Azzurro et al. 2007a, Vergés et al. 2014). This seems to be accelerating the energy transfer from primary producers to higher consumer levels via predation. For instance, rabbitfishes can comprise up to 68 percent of the diet of the grouper *Mycteroperca rubra* (Bloch 1793) (Aronov and Goren 2008). However, this potential benefit could come at the expense of other native species that depend on macroalgae forests.

Rabbitfish herbivory can contribute to maintaining rocky reefs denuded of macroalgae which characterize the eastern Mediterranean Sea. Across Crete and Cyprus, observations of the feeding behaviour of *S. luridus* and *S. rivulatus* indicated a strong selectivity for algal assemblages over adjacent seagrass, having feeding rates ranging between 8–18 bites individual^{-1} min^{-1}. When standardized by bite size and local herbivore density, this resulted in a combined feeding impact of up to 18 cm^2 m^2 hr^{-1} (Santana-Garcon et al. 2023). These feeding rates in combination with the oligotrophic conditions of the eastern Mediterranean can result in a high susceptibility of reef habitats to overgrazing. In addition, herbivory experiments in Turkey have shown that when reef surfaces were protected from rabbitfishes and *S. cretense*, algal biomass increased from low turfs (< 5 g m^{-2}) to abundant assemblages of macroalgae (~ 200 – 500 g m^{-2}) (Sala et al. 2011). Although this does not necessarily show that herbivory was the initial cause of the shift from macroalgae forests to barrens, it does show that rabbitfishes can suppress its recovery, maintaining lower biogenic habitat complexity and its associated biodiversity and productivity (Vergés et al. 2014). This is widely recognized now, and the management of herbivory is an important element to consider for restoration programs of macroalgal forests in the Mediterranean Sea (Cebrian et al. 2021).

Rabbitfishes also have had significant socio-economic impacts in the fishing industry, becoming among the most valuable species in the eastern Mediterranean (EastMed 2010; Fig. 2). In Lebanon, *S. rivulatus* is caught in trammel nets down to 20 m depth and thousands of juveniles can be found on the beaches, making up to 5–15 percent of the inshore fisheries (Bariche 2005), a similar proportion to Israel fisheries (Lundberg and Golani 1995). In Libya, both species are of commercial value and found regularly in catches in all coastal regions (Shakman and Kinzelbach 2007), while stocks in Egypt, Saudi Arabia and Turkey, are estimated to be overfished (Cerim et al. 2020). In Cyprus, both siganids are very popular and highly priced among recreational and commercial fishers, with overall landings estimated at ~ 300 t year^{-1}, valued at ~ €$4.5 million annually (Michailidis et al. 2020, Kleitou et al. 2022). In Turkey, *S. rivulatus* is caught in remarkable quantities by gill and trammel nets for commercial purposes, with catches likely equivalent to ~ 10 t year^{-1} (Bilecenoglu and Kaya 2002), while in Greek waters, they have been present in the local markets since the 1990s, and are of high commercial value for local fishermen in southern regions (Kalogirou et al. 2010). Despite the recognition of their importance, landings are not rigorously reported in most countries and currently their socio economic impact is likely to be largely underestimated (EastMed 2010).

Fig. 2. A: School of rabbitfishes in seagrass meadows of the eastern Mediterranean Sea. B: Adult rabbitfish individuals being caught by fisheries in the region. Photos by Adriana Verges.

Western Australia

The coast of Western Australia has extensive tropical and temperate ecosystems across a latitudinal gradient interconnected by large scale oceanographic processes. In the northern regions, important coral reef systems occur, such as the Ningaloo Reef World Heritage Area, and at mid-latitudes there is a tropical-temperate transition zone where scleractinian corals and canopy-forming macroalgae can share the domain of reef surfaces (Mulders et al. 2022). In the tropical environments, there are approximately ten species of rabbitfishes, among which some of the most abundant are *Siganus argenteus*, *Siganus virgatus*, *Siganus puellus* (Schlegel 1852), *Siganus punctatus* (Schneider and Forster 1801), *Siganus doliatus* (Guérin-Méneville 1829) and *Siganus fuscescens* (Edgar and Stuart-Smith 2014). Further south, temperatures drop and rocky reefs are dominated by the canopies of the kelp *Ecklonia radiata* (C. Agardh) J. Agardh, 1848 and fucoids such as *Sargassum* spp., which overshadow diverse communities of foliose understory macroalgae (Wernberg et al. 2005). Inshore temperate reefs are usually adjacent to seagrass meadows dominated by *Posidonia* spp. and *Amphibolis* spp. which exchange detrital material with temperate reefs and where some reef fish species migrate for foraging at different ontogenetic stages

(Ayvazian and Hyndes 1995). Ecosystems in this climate gradient are interconnected by the southward flowing Leeuwin Current, which transports tropical waters all across the coast, facilitating the dispersal of tropical species to temperate regions (Pearce et al. 2011).

Expansion and Persistence

S. fuscescens is the southernmost rabbitfish species in the region, with populations reported as south as Jurien Bay in 2007 (30.3°S; Fairclough et al. 2011). However, its distribution expanded further south when an exceptional marine heatwave occurred during the summer of 2010–2011, which transported tropical waters to high latitudes, causing sustained temperature increases of 3–5°C above the mean for several weeks (Pearce and Feng 2013). The phenomena devastated kelp forests in the tropical-temperate transition region while increasing the proportion of invertebrate and fish species of tropical affinity in temperate reefs (Wernberg et al. 2016). Among them, the rabbitfish *S. fuscescens* expanded 400 km poleward until Geographe Bay (33.6°S; Fig. 3A). High amounts of pelagic larvae, produced by hundreds of thousands per individual (Hoque et al. 1999), were likely transported by the Leeuwin Current to warmer than usual southern rocky reefs and seagrass meadows, where they managed to recruit and develop until adulthood (Zarco-Perello et al. 2017). Supporting this are genetic analyses that have found little differentiation between tropical and temperate populations of the western Australian coast (Gajdzik et al. 2021). Examinations of gonads from the newly formed populations indicated that individuals have reached sexual maturity and reproduced successfully since there have been several observations of juveniles in subsequent years (Lenanton et al. 2017). This has been further documented using otolith analyses, which have backtracked the oldest recruitment dates to the time of the heatwave in 2011, and descendant generations in subsequent years (Zarco-Perello et al. 2022). Seagrass meadows seem to be important for the recruitment of rabbitfish at higher latitudes, as surveys during summer have detected recruits and juveniles across different years and locations in abundances similar to subtropical regions (Nakamura and Tsuchiya 2008, Zarco-Perello et al. 2021). Nonetheless, juveniles have also been observed roaming in temperate reefs, and they might tend to migrate between both ecosystems once a certain body size has been reached (Zarco-Perello et al. 2021).

Similar to *S. rivulatus* and *S. luridus* in the Mediterranean Sea, the persistence of *S. fuscescens* at higher latitudes may be greatly supported by their plasticity in life-history and ecological traits. Fish surveys have shown that rabbitfish have become a common species in multiple locations in the region and as abundant as other native herbivorous fish, such as *Kyphosus cornelii* (Whitley 1944) or *Kyphosus sydneyanus* (Günther 1886) (Zarco-Perello et al. 2017). North of the Perth Metropolitan area and Rottnest Island, rabbitfish now commonly form schools from 10 to 30 individuals, with less frequent aggregations of more than 100 individuals, comparable to those reported for coral reefs (Zarco-Perello et al. 2019, Zarco-Perello et al. 2020). Populations in Cockburn Sound were so widespread and abundant, that they became a targeted species of a commercial fishery with monthly landings of up to 400 kg between 2013–2016, which were readily sold in local markets. However, this seemed to have exceeded the recruitment rates of the local population since

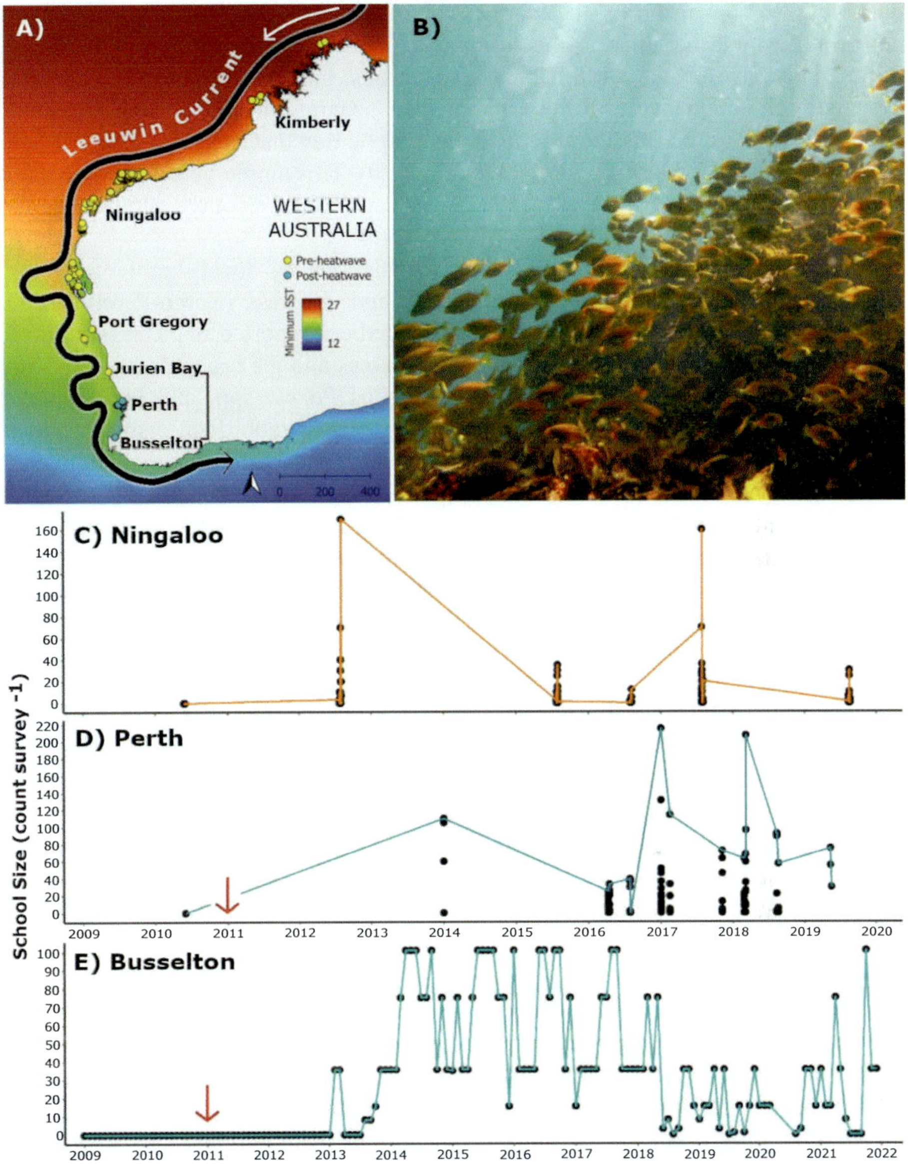

Fig. 3. A: Distribution of the rabbitfish *Siganus fuscescens* and its expansion from Jurien Bay to Busselton in western Australia after a marine heatwave (MH) in 2011. B: Aggregations of juveniles recruiting in temperate reefs of Perth in 2013. C-E: Time-series of abundance in tropical (Ningaloo: data from Reef Life Survey (Edgar and Stuart-Smith 2014)) and temperate locations in the new range of distribution: Perth (data from Zarco-Perello et al. 2017, Zarco-Perello et al. 2019, Zarco-Perello et al. 2020) and Busselton (data provided by Sophie Teede, courtesy of Busselton Jetty Inc.). SST: minimum average sea surface temperature (EU's Earth Observation 2021). Photo (B) taken by William Parkinson.

the catches have dramatically diminished in recent years (Lenanton et al. 2017). Their survival might be related to having a wider thermal tolerance, surviving winters as cold as 13°C during adulthood, and being able to adjust their reproductive phenology to spawn during the warmest months of the year, avoiding the lowest temperatures and maximizing the survival of recruits (Zarco-Perello et al. 2022). Populations have shown wide habitat associations, including temperate reefs with low and high abundance of foliose macroalgae, seagrass meadows, and man-made structures, such as the Busselton Jetty in Geographe Bay that extends 1.8 km inshore into seagrass meadows (Zarco-Perello et al. 2020). This ample versatility in habitat associations relates strongly with their broad capacity of resource acquisition and their competition with native species of herbivores.

Rabbitfish co-occur with abundant populations of local herbivorous and omnivorous fish species in temperate reefs and seagrass (Zarco-Perello et al. 2020). Currently, no competitive exclusions have been reported in the region, likely influenced by the high availability of food resources and the broad diet and feeding behaviours of rabbitfish. Temperate damselfishes *Parma* spp. and the drummer *K. cornelii* defend their territories aggressively against rabbitfish, disrupting their foraging activities momentarily, but with no significant effects on their overall consumption rates, as rabbitfish quickly regroup to feed in adjacent areas (Barrientos et al. 2021). Stomach contents and feeding observations in temperate ecosystems have found great adaptability to the local resources across the distribution range in western Australia. Different populations consume macrophytes according to resource availability, including seagrass, canopy-forming, foliose and filamentous macroalgae, and small amounts of invertebrates (e.g., snails, crustaceans, bivalves and tunicates) (Avenant 2018, Zarco-Perello et al. 2019, Barrientos et al. 2021, Gajdzik et al. 2021, Zarco-Perello et al. 2021). Likewise, analyses suggest that range-expanding populations can enhance their digestive capabilities by adapting their hindgut microbiome according to their new diets (Jones et al. 2018). Most of their trophic resources are attached in reef surfaces; however, rabbitfish can also feed on macrophytes drifting in the water column by up to 11 percent of total consumption, a behaviour documented for the first time in their new range and which increases resource partitioning by reducing resource competition in the benthic space (Zarco-Perello et al. 2019).

Impacts

Populations of *S. fuscescens* have intensified primary consumption rates in temperate ecosystems of south-western Australia. Herbivory rates in the region have been considered low historically, with only a couple of species responsible for the consumption of kelp (*K. sydneyanus, K. gladius* Knudsen and Clements 2013, and *Olisthops cyanomelas* Richardson 1850), turf consumption (*K. cornelii* and *Parma* spp.) and seagrass (*Pelates octolineatus* (Jenyns 1840)) (Zarco-Perello et al. 2020). Consequently, rabbitfish populations have increased primary consumption in many areas, inflicting thousands of bites more on macrophytes and intensifying seagrass browsing (9–15×), canopy (2–2.5×) and turf grazing (0.4–2.6×) in the Perth metropolitan area (Zarco-Perello et al. 2020), becoming the most important kelp consumer in some locations (Zarco-Perello et al. 2017; Fig. 4). Their feeding rates can be 32× higher in summer (22–23°C) than winter (15–16°C); however, even under low temperatures their consumption surpassed temperate herbivores; increasing herbivory throughout the year (Zarco-Perello et al. 2019).

The diet flexibility of rabbitfish among different macrophytes lessens their impact on kelp forests but their increases in consumption still can be ecologically relevant. High herbivory rates can weaken the resilience of kelp forests against environmental disturbances, such as eutrophication, pollution, storms and high temperatures, which can cause significant mortality rates in kelp populations, reducing their densities and increasing their vulnerability to regime-shifts (Filbee-Dexter and Wernberg 2018). Lower densities of kelp reduce productivity and weaken positive feedbacks that maintain their populations through high spore supply and the facilitation of recruitment. Thus, disturbances could push the system over the tipping point where herbivory surpasses production, leading to a canopy-free state (Layton et al. 2019). Herbivory assessments in the rocky reefs of Port Gregory which became deforested due to the marine heatwave, have shown that rabbitfish could prevent kelp recovery by exerting a strong grazing pressure on kelp recruits and juveniles (Fig. 4), consuming four times more turf seaweed than other species and constraining kelp individuals to topographical shelters that protected them from herbivores (Bennett et al. 2015, Zarco-Perello et al. 2021). Importantly, a reduction in the standing stock of seaweed not only can affect local trophic food chains, but also the subsidy of trophic resources to adjacent ecosystems, such as seagrass meadows or sandy beaches, which depend on it to for their productivity (Hyndes et al. 2014).

The acceleration of primary consumption driven by rabbitfish has correlated positively with increases in the production of herbivory-derived detritus in temperate reefs. Because herbivores ingest, shred and excrete macrophytes as particulate organic matter, they transform the function of canopy-forming macrophytes from habitat providers to sources of nutrition for other organisms, a function historically attributed to sea urchins in temperate reefs (Yorke et al. 2019). Seasonal surveys in the reefs of the Perth metropolitan area have indicated that rabbitfish have accelerated this process throughout the year by five- to eightfold, depending on fish abundances and temperature (Zarco-Perello et al. 2019). Higher availability of herbivory derived detritus could contribute to accelerate the energy flow to higher trophic levels and increase secondary productivity (Pessarrodona et al. 2021). However, more research is needed to assess the proportions and detailed pathways that tropical fish faeces could follow in temperate food webs and their impact in the productivity of the ecosystem (Zarco-Perello et al. 2019).

Increases in Abundance

Eastern Australia

The eastern coast of Australia extends across nearly 33 degrees of latitude, encompassing extensive tropical, subtropical and temperate ecosystems. In the tropics, marine ecosystems are characterized by the dominance of corals in offshore reefs, including the Great Barrier Reef, while mangroves and seagrass beds generally dominate the nearshore coastline. At higher latitudes, ecosystems are characterized by the dominance of kelp forests in shallow rocky reefs and sponge gardens in deeper reefs, with seagrass beds and saltmarshes being more dominant within estuaries. The coast is strongly influenced by the poleward-flowing East Australian Current, which

Fig. 4. *Siganus fuscescens* consuming turf seaweed and possible kelp recruits in deforested reefs in Port Gregory (A) and canopies of the kelp *Ecklonia radiata* in temperate reefs of Perth, Western Australia (B). Photos taken by Salvador Zarco-Perello.

transports warm waters from the Coral Sea to Tasmania. In recent decades, this current has been strengthening due to climate change, a phenomenon that has been linked to increases in the proportion of warm-affinity species in higher latitudes through the transport of larval propagules from tropical reefs into higher latitudes (Suthers et al. 2011). Across eastern Australia, there are about 16 species of rabbitfish restricted to tropical and subtropical latitudes; however, *Siganus fuscescens* distributes as far south as Batemans Bay, New South Wales (36.1°S; Edgar and Stuart-Smith 2014), with populations increasing in some temperate reefs (Fig. 5).

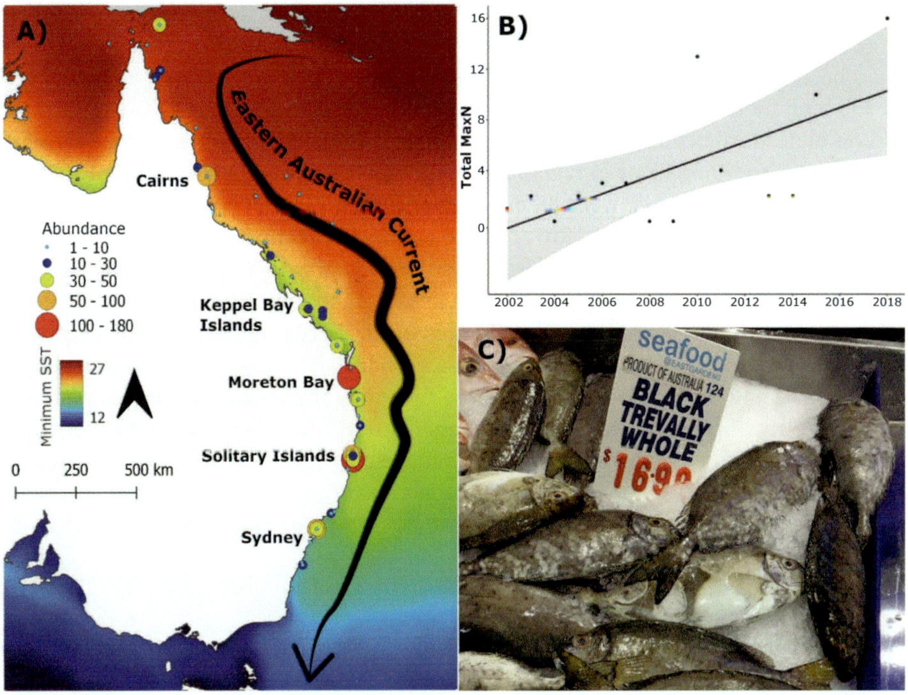

Fig. 5. Distribution and abundance of *Siganus fuscescens* in eastern Australia (as individuals per transect from Reef Life Survey (Edgar and Stuart-Smith 2014) (A), changes of its relative abundance (MaxN) in the Solitary Islands region (Smith et al., 2021) (B), and specimens caught by commercial fishers in Moreton Bay marketed as "black trevally" (photo by Adriana Vergés) (C). SST: minimum average sea surface temperature (EU's Earth Observation 2021).

Increases in Abundance

Total abundance of *S. fuscescens* is particularly high in subtropical latitudes around Moreton Bay (27°S) and northern New South Wales. Populations there can comprise 33–72 percent of the biomass of roving herbivores, and sustain a local fishery that markets rabbitfish as black trevally (Olds et al. 2012; Fig. 5). Anecdotal evidence from local net fishers in the region indicate that catches of *S. fuscescens* have become more stable in the last 10 years, suggesting that populations are increasing (Gilby et al. 2017). The best quantitative evidence of increases in abundance in high latitudes comes from the Solitary Islands region (30°S), where long-term fish monitoring using baited remote underwater video surveys (BRUVS) has showed a significant increase in the relative abundance of *S. fuscescens* (MaxN: maximum number of individuals seen in a single frame of video sampling), from 0-3 individuals in the early 2000's to schools of > 10 individuals after 2010 (Smith et al. 2021; Fig. 5).

Impacts

Siganus fuscescens is a key herbivore in subtropical eastern Australia, where it provides crucial trophic links between macroalgal forests, seagrass meadows and mangroves (Davis et al. 2014). In Moreton Bay, this species is an important consumer of *Sargassum* spp., accounting for 80% of bites on herbivory assays (Gilby et al. 2017). Further south in the Solitary Islands region, their increased abundances have been linked to the gradual decline of the kelp *Ecklonia radiata*, as evidenced by the high incidence of fish bites visible on video surveys on remaining kelps prior to complete disappearance (Vergés et al. 2016). Filmed feeding experiments with transplanted individuals of *E. radiata* in reefs where kelp used to be present showed that rabbitfish individuals were the most important consumers of kelp together with *Kyphosus bigibbus* Lacepède 1801, despite not being the most abundant herbivorous species (Vergés et al. 2016). *S. fuscescens* also displays the highest bite rates on turf algae in reefs where kelp has disappeared, which could prevent kelp recovery by consuming microscopic recruits (Vergés et al. 2016). Indeed, recent herbivore exclusion experiments on these reefs have confirmed that fish can graze on microscopic kelp recruits and limit kelp recovery in conjunction with factors such as warming and unavailability of free substrate for settlement (Smith 2021). In addition, further south in Seal Rocks (32.4°S), there is anecdotal evidence of correlations between increases in abundance of *S. fuscescens* and overgrazed kelp forests (Keith Bishop, personal communication) but these patterns are yet to be rigorously quantified.

Japan

The Japanese archipelago has diverse coastal environments that are strongly influenced by the warm Kuroshio and Tsushima currents that travel northward, and the cold Oyashio Current that moves southwards. While coral reef ecosystems develop in the subtropical environments around the Nansei Islands, rocky reefs covered in large brown seaweeds such as kelp (e.g., *Ecklonia cava Kjellman* 1885) and *Sargassum* spp., flourish in the high-latitude regions from Kyushu to Hokkaido (Fig. 6). Many tropical fish species are found during summer in rocky reefs of some southern regions (i.e., Kyushu, Shikoku and Honshu), which are thought to originate from larvae transported by warm currents from the Nansei Islands (Nakamura et al. 2013). Among these, twelve rabbitfish species have been registered in Japan, with six of those (*Siganus fuscescens*, *S. spinus* (Linnaeus 1758), *S. woodlandi* Randall & Kulbicki 2005, *S. argenteus*, *S. guttatus* (Bloch 1787), and *S. punctatus*) reported in temperate regions; however, only *S. fuscescens* has abundant populations at higher latitudes (Nakabo 2013).

Increases in Abundance

S. fuscescens is widely distributed in Japan. Reports of its presence in temperate areas along western coast exist since 1735–38, as reported in "Flora, Fauna and Crops of the Japan Islands in the Eighteenth Century" (Morinaga and Yasuda 1985). Today, its distribution is limited south of Chiba (35°N) on the Pacific side and south of Kyoto (35.7°N) on the Sea of Japan (Fujita et al. 2006; Fig. 6). Juveniles have been sighted in northern regions, with the exception of Hokkaido, but no self-sustaining populations have been established, possibly due to the extreme winter temperatures which fall below their lower critical limit (10°C–11°C), as indicated by experiments on thermal tolerance (Ueta and Tanada 2018).

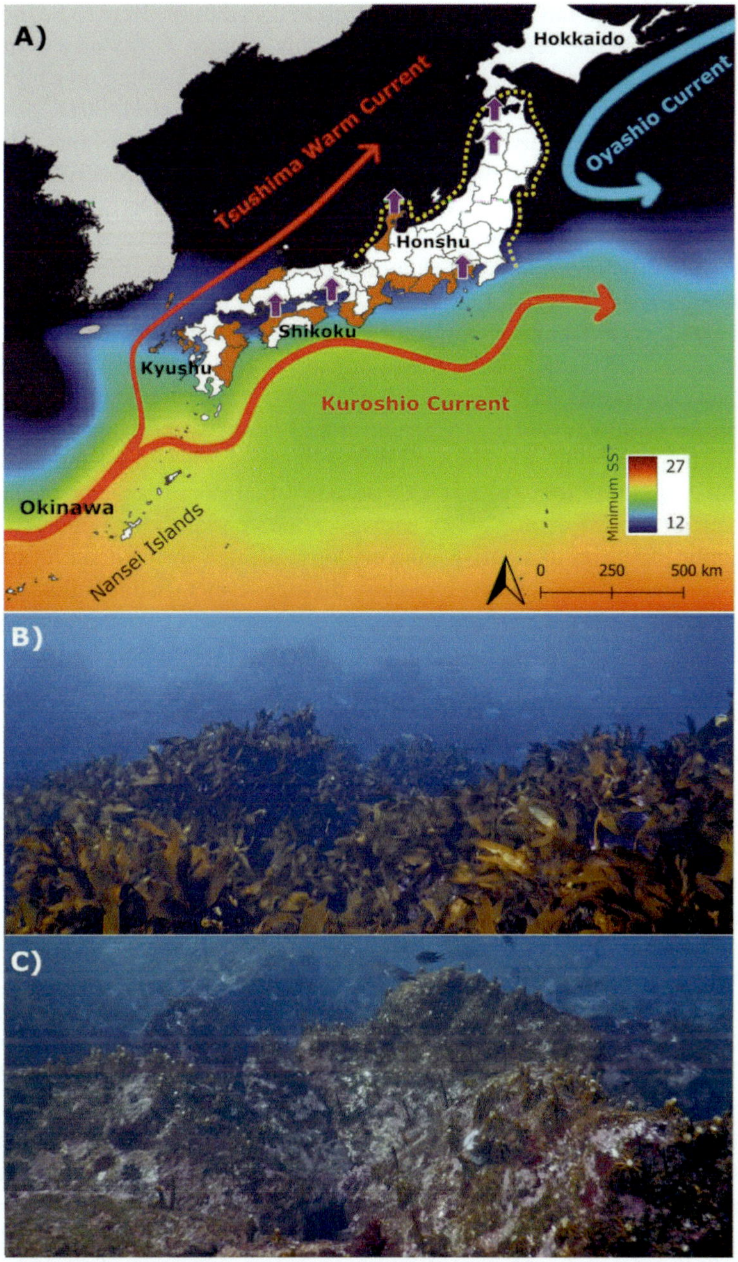

Fig. 6. (A) Map of Japan showing increases in abundance (upward arrows) and future distribution of *Siganus fuscescens* by 2091–2100 (dotted line; Sudo et al. 2022), and prefectures reporting a perceived overgrazing of kelp by rabbitfish (orange areas, Fujita et al. 2006). (B-C) Kelp forests of *Ecklonia cava* in Moroiso, Kanagawa, before and after being overgrazed by large aggregations of rabbitfish (Noda et al. 2018). SST: minimum average sea surface temperature (EU's Earth Observation, 2021).

Populations of rabbitfish within their current distribution seem to be increasing. Like in other temperate regions mentioned in this chapter, *S. fuscescens* has found plenty of trophic resources in the temperate reefs of Japan, feeding on a high variety of macroalgae, including Laminariales and Fucales (Noda et al. 2011, Noda et al. 2014). This has allowed individuals to survive by up to 13 years, the longest longevity reported for rabbitfish (Katayama et al. 2009), and reproduce successfully during July to August in Takeyama Bay (35°N), where juveniles as small as 30–50 mm FL can be seen during October (Akiyama et al. 2009). Underwater surveys in Kochi (33°N) found that rabbitfish recruits can be among the most abundant herbivorous species within macroalgal fronds (Beck et al. 2017), although they are well associated with seagrass meadows in Okinawa (Kanashiro et al. 1999). Since the 1990s, populations of rabbitfish seem to be increasing in the northeast region (Aomori and Akita) and the Seto Inland Sea (Osaka and Hiroshima), where the population had been previously considered to be low, as well as Kanagawa and Ishikawa prefectures, at the edge of their distribution (Fig. 6). These trends could be explained by the increases in temperature ranging from 0.13–0.4°C decade^{-1} that some regions of Japan are experiencing (Tanaka et al. 2012). However, perceptions of population growth are currently based on qualitative information collected through surveys with the different fisheries of the country, including diving, set-net, fishing, gill-net, market vouchers, and statistical data (Fujita et al. 2006). Quantitative ecological surveys are required to be established to evaluate future changes in rabbitfish populations.

Impacts

High abundances of *S. fuscescens* have been correlated with the overgrazing of kelp forests in some regions of Japan. The degradation and loss of macroalgae canopies ("isoyake") have been reported since the 1980s in 33 prefectures, with increases in temperature and herbivory rates by sea urchins and fish species identified as the main causes. Among herbivores, *S. fuscescens* has been highlighted as the most common consumer of kelp, seconded by *Calotomus japonicus* (Valenciennes 1840), *Prionurus scalprum* (Valenciennes 1835), and *Kyphosus* spp. (Fujita et al. 2006, Fujita 2010). Analyses of bite marks in macroalgae have indicated that kelp and sargassum canopies have been largely overgrazed by herbivorous fish in deforested reefs (Kiriyama et al. 2001). Field observations showed that rabbitfish tend to browse the kelp *E. cava* from the top of the lamina (Nakayama and Arai 1999); however, herbivory assays in aquariums revealed that they can also consume the lower zone of the lamina, fragmenting and detaching entire kelp individuals from the substrate, causing high losses of kelp biomass (Noda et al. 2002). This has been documented in the reefs of Moroiso, Kanagawa, where aggregations of rabbitfish seemed to have overgrazed 30,000 m^2 of the kelp *E. cava* within one month (Noda et al. 2018; Fig. 6), and in Hainan, Shizuoka, where populations of the kelp *Eisenia arborea* J.E. Areschoug 1876, disappeared within one month as a combination of high browsing and poor water quality (Hasegawa 2010). Large aggregations (100 to 500 individuals) of adult rabbitfish have been observed along the temperate coast of Japan, and herbivory by such aggregations may be responsible for the short-term disappearance of kelp forests (Noda et al. 2018).

High herbivory rates by rabbitfish have also been related to preventing the recovery of kelp forests and their ecosystem services. Assessments of populations of *Sargassum fusiforme* (Harvey) Setchell 1931, that have experienced declines in the coast of Nagasaki showed that a great proportion of new recruits (39–67 percent, n=87) were grazed by *S. fuscescens,* which can consume 15 to 20 times more biomass than *Kyphosus bigibbus* and *C. japonicus*, preventing the development of dense canopies (Kiriyama et al. 2005). In deforested reefs of the coasts of Hainan, Shizuoka, rabbitfish heavily grazed transplanted kelps of *E. arborea* (Nimura et al. 2007), and *Ecklonia cava*, which suffered length reductions of their lamina from 70–90 cm to 10–30 cm (Hasegawa 2010). As such, harvesting industries in the region that depend on kelp forests have suffered important economic losses. The catches of abalone have declined from 20 t year^{-1} in the 1980s to ~ 2 t year^{-1} in 1996, and the harvesting of kelp (*E. arborea)* for the food industry declined from 40 t year^{-1} in 1982 to a total collapse in 1995, inflicting important economic losses (Hasegawa 2010). For instance, healthy kelp forests can produce ~ 15,000 t year^{-1} of macroalgae biomass valued at ~ $22.7 billion yen in Hokkaido (Sudo et al. 2020), and 700 kg year^{-1} of abalone valued at ~ $11 million yen in Kochi, but all is lost if kelp forests disappear (Serisawa et al. 2004). Several efforts have been attempted to prevent high rates of herbivory on kelp by setting up protective barriers, and deterrent devices using optical, physical and auditory stimuli; however, these have been largely ineffective (Yamauchi et al. 2006). To ameliorate the impact of herbivory and compensate for the loss in fishery resources, the development of programs to incentivise the fishery and commercialization of rabbitfish, and other tropical herbivores, are being promoted (Kuwahara 2015).

Increases in temperature in Japan are expected to promote further expansions of *S. fuscescens* by 500–600 km in the Pacific coast and 200–400 km along the Sea of Japan by 2091–2100 (Sudo et al. 2022; Fig. 6). In addition, warming would lead to an acceleration of metabolic and consumption rates by tropical herbivores year-round. Herbivory assays in tank and sea-pens have shown that *S. fuscescens* reach maximum feeding rates at 26–29°C (Yamauchi et al. 2006, Kimura et al. 2007). Thus, the synergistic impacts of rising temperatures and herbivory on the structure and function of Japanese kelp forest are forecasted to accelerate in the future (Takao et al. 2015), possibly leading to novel ecosystems with higher abundances and diversity of tropical species, such as scleractinian corals (Kumagai et al. 2018).

Global Synthesis

This chapter described the history and recent developments in shifts in distribution and increases in abundance of rabbitfishes in diverse temperate regions of the world. Studies across the world highlight that *Siganus rivulatus*, *Siganus luridus* and *Siganus fuscescens* share the characteristics of having broad and plastic biological and ecological traits that likely favoured their dispersal, survival and persistence in their new environments. Although their thermal optima (aerobic scope) range between 27–29°C, adult individuals have overwintered in temperatures as low as 13°C at higher latitudes. Phenological and phenotypic changes, such as restricting their reproduction to spring, increasing batch fecundity and living for longer than tropical populations, can lead to high reproductive output and offspring survival. Moreover, adult and juvenile rabbitfishes were associated with multiple habitats, such as reefs with low and high cover of macroalgae, sandy substrates, jetties, harbours and even estuaries with low salinity. This is concomitant to their broad feeding behaviours and diets, ranging from epilithic algal matrix, filamentous, foliose and canopy-forming macroalgae, to seagrass and small amounts of invertebrates. All traits combined seem to make these rabbitfish species highly competitive under a diverse set of circumstances in high latitudes.

Accumulated evidence shows that high rabbitfish abundances can lead to significant increases in consumption rates on habitat-forming macrophytes (e.g., *Ecklonia* spp. and *Cystoseira* spp.). This accelerates the transformation of macrophytes from habitat providers to energy sources, which can contribute to the loss of kelp forests and prevention of their recovery if consumption rates surpass primary productivity. As expected, the balance between both processes varies in space and time, with some locations showing no evident effects on macroalgae cover, and others having significant negative correlations between abundances of rabbitfish and foundation species of macroalgae. Currently, the strongest impacts reported are the kelp deforestations in Japan in correlation with extraordinarily high aggregations of rabbitfish (> 500 individuals), which would exert an extraordinary consumption pressure *per capita* on kelp, considering that a school of rabbitfish commonly consists of 30 or less individuals. Several transplantation experiments have shown that rabbitfishes can contribute to the maintenance of canopy-free states after the collapse of kelp forests by intensive feeding on turf seaweed and potentially kelp recruits, and by consuming solitary kelp individuals.

Rabbitfishes have had diverse effects on the availability of natural resources for human societies in temperate regions. On one hand, rabbitfishes have become part of local fish markets in all temperate regions they have expanded into, similarly to some tropical countries. The best example of this is the eastern Mediterranean Sea, where rabbitfishes have been well integrated into recreational and commercial fisheries, with catches of hundreds of tonnes valued at millions of euros per year. On the other hand, by contributing to kelp deforestations and preventing its recovery, rabbitfishes can reduce the services and resources provided by these ecosystems. The abalone and kelp harvesting industries in Japan have suffered the most because of this, causing great losses in economic and social benefits. However, despite the well understood pressing need for population control, rabbitfishes have not yet become a widely targeted species in the commercial fisheries of the country.

Range-expansions of tropical herbivores are changing the paradigm of conservation planning of temperate marine ecosystems, where historically herbivorous fish did not pose a threat in most of the world. The consideration of herbivory in management plans for the prevention of deforestations and successful restoration of certain kelp forests will become more important as climate change progresses. Similitudes exist with the phenomena of sea urchin barrens, where the solution for kelp deforestation is population control through increased natural predation or fishing industry. However, solutions likely would not be straightforward, and the ecological and social impacts of range-expansions would need to be assessed with cost-benefit analyses, particularly considering the mid- and long-term evolution of kelp forests under future thermal stress. Thus, it is important to conduct further investigations on the range-expanding capabilities (thermal scope, plasticity and adaptability of life-histories) and possible impacts of more species to help predict possible outcomes in future scenarios (ecosystem tipping points and hysteresis) and make pre-emptive and adaptive management plans. Particularly considering synergies of biological and environmental factors, such as the interaction with competitors and predators, and increases in temperature, which simultaneously affects primary productivity and the physiology of herbivores.

Acknowledgements

Thanks to Sophie Teede and Busselton Jetty Inc. for providing abundance data of rabbitfish in Busselton, western Australia, and Shannen Smith and Hamish Malcolm (New South Wales Department of Primary Industries) for collection and processing of rabbitfish data from eastern Australia. Thanks also to the volunteers and staff from the Reef Life Survey initiative for their global fish survey dataset. Data used for the distribution of rabbitfish in the Mediterranean were provided by the project @ Cnr UseIT – Utilizzo di Sinergie operative per la gestione integrata di specie aliene Invasive in Italia, in collaboration with the ORMEF geoportal www.ormef.eu.

References

Abdelghani, A., S.A.A. Al Mabruk, F. Crocetta and D. Golani. 2020. The Streamlined Rabbitfish *Siganus argenteus* (Quoy and Gaimard, 1825) in the Mediterranean Sea. Thalassas. 37: 287–290.

Ahnelt, H. 2016. Translocations of tropical and subtropical marine fish species into the Mediterranean. A case study based on *Siganus virgatus* (Teleostei: Siganidae). Biologia. 71: 952–959.

Akiyama, S., M. Naganuma and S. Katayama. 2009. Annual life cycle of rabbitfish *Siganus fuscescens* in Tateyama Bay, Chiba prefecture. Fish. Eng. 46: 107–115.

Albouy, C., L. Velez, M. Coll, F. Colloca, F. Le Loc'h, D. Mouillot et. al. 2014. From projected species distribution to food-web structure under climate change. Glob. Chang. Biol. 20: 730–741.

Aronov, A. and M. Goren. 2008. Ecology of the mottled grouper (*Mycteroperca rubra*) in the eastern Mediterranean. Electronic Journal of Ichthyology 2: 43–55.

Avenant, C. 2018. Diet of the Tropicalised Herbivore *Siganus fuscescens* Across a Broad Latitudinal Gradient and Comparisons with Temperate Seagrass-associated Omnivorous Fishes. Ms. Thesis. Edith Cowan University, Perth, Australia.

Ayvazian, S.G. and G.A. Hyndes. 1995. Surf-zone fish assemblages in south-western Australia: do adjacent nearshore habitats and the warm Leeuwin Current influence the characteristics of the fish fauna? Mar. Biol. 122: 527–536.

Azzuro, E. and F. Tiralongo. 2020. First record of the mottled spinefoot *Siganus fuscescens* (Houttuyn, 1782) in Mediterranean waters: a Facebook based detection. Medit. Mar. Sci. 21: 448–451.

Azzurro, E. and F. Andaloro. 2004. A new settled population of the lessepsian migrant *Siganus luridus* (Pisces: Siganidae) in Linosa Island—Sicily Strait. J. Mar. Biol. Ass. 84: 819–821.

Azzurro, E., E. Fanelli, E. Mostarda, M. Catra and F. Andaloro. 2007a. Resource partitioning among early colonizing *Siganus luridus* and native herbivorous fish in the Mediterranean: an integrated study based on gut-content analysis and stable isotope signatures. J. Mar. Biol. Ass. 87: 991–998.

Azzurro, E., O. Carnevali, M. Bariche and F. Andaloro. 2007b. Reproductive features of the non-native *Siganus luridus* (Teleostei, Siganidae) during early colonization at Linosa Island (Sicily Strait, Mediterranean Sea). J. Appl. Ichthyol. 23: 640–645.

Azzurro, E., G. Franzitta, M. Milazzo, M. Bariche and E. Fanelli. 2017. Abundance patterns at the invasion front: the case of *Siganus luridus* in Linosa (Strait of Sicily, Central Mediterranean Sea). Mar. Freshwater Res. 68: 697–702.

Azzurro, E., S. Smeraldo, A. Minelli and M. D'Amen. 2022. ORMEF: a Mediterranean database of exotic fish records. Sci. Data 9: 363.

Bariche, M. 2005. Age and growth of Lessepsian rabbitfish from the eastern Mediterranean. J. Appl. Ichthyol. 21: 141–145.

Bariche, M. 2006. Diet of the Lessepsian fishes, *Siganus rivulatus* and *S. luridus* (Siganidae) in the eastern Mediterranean: A bibliographic analysis. Cybium 30: 41–49.

Bariche, M., M. Harmelin-Vivien and J.P. Quignard. 2003. Reproductive cycles and spawning periods of two Lessepsian siganid fishes on the Lebanese coast. J. Fish Biol. 62: 129–142.

Bariche, M., Y. Letourneur and M. Harmelin-Vivien. 2004. Temporal fluctuations and settlement patterns of native and lessepsian herbivorous fishes on the lebanese coast (eastern mediterranean). Environ. Biol. Fishes 70: 81–90.

Bariche, M., R. Sadek and E. Azzurro. 2009. Fecundity and condition of successful invaders: *Siganus rivulatus* and *S. luridus* (Actinopterygii: Perciformes: Siganidae) in the eastern Mediterranean Sea. Acta Icth et Piscat. 39: 11–18.

Barrientos, S., S. Zarco-Perello, C. Piñeiro-Corbeira, R. Barreiro and T. Wernberg. 2021. Feeding preferences of range-shifting and native herbivorous fishes in temperate ecosystems. Mar. Environ. Res. 172: 105508.

Beck, H.J., D.A. Feary, Y. Nakamura and D.J. Booth. 2017. Temperate macroalgae impacts tropical fish recruitment at forefronts of range expansion. Coral Reefs 36: 639–651.

Ben-Tuvia, A. 1964. Two siganid fishes of Red Sea origin in the eastern Mediterranean. Bulletin of the Ministry of Agriculture, Department of Fisheries, Sea Fisheries Research Station 37: 29–30.

Bennett, S., T. Wernberg, E.S. Harvey, J. Santana-Garcon and B.J. Saunders. 2015. Tropical herbivores provide resilience to a climate-mediated phase shift on temperate reefs. Ecol. Lett. 18: 714–723.

Bilecenoglu, M. and M. Kaya. 2002. Growth of marbled spinefoot *Siganus rivulatus* Forsskål, 1775 (Teleostei: Siganidae) introduced to Antalya Bay, eastern Mediterranean Sea (Turkey). Fish. Res. 54: 279–285.

Bodilis, P., P. Louisy, M. Draman, H.O. Arceo and P. Francour. 2014. Can citizen science survey non-indigenous fish species in the eastern Mediterranean Sea? Environ. Manage. 53: 172–180.

Cebrian, E., L. Tamburello, J. Verdura, G. Guarnieri, A. Medrano, C. Linares et al. 2021. A roadmap for the restoration of mediterranean macroalgal forests. Front. Mar. Sci. 8: 709219.

Cerim, H., O. Soykan and A. Gülşahin. 2020. Mortality and exploitation of marbled spinefoot, *Siganus rivulatus* (Actinopterygii: Perciformes: Siganidae), from southern Aegean Sea small-scale fishery. Acta Icth et Piscat. 50: 183–190.

Chaudhary, C., A.J. Richardson, D.S. Schoeman and M.J. Costello. 2021. Global warming is causing a more pronounced dip in marine species richness around the equator. Proc. Natl. Acad. Sci. USA. 118: e2015094118.

Cheminée, A., J. Pastor, O. Bianchimani, P. Thiriet, E. Sala, J.M. Cottalorda et al. 2017. Juvenile fish assemblages in temperate rocky reefs are shaped by the presence of macro-algae canopy and its three-dimensional structure. Sci. Rep. 7: 14638.

Chiarore, A., I. Bertocci, S. Fioretti, A. Meccariello, G. Saccone, F. Crocetta and F.P. Patti. 2019. *Syntopic Cystoseira* taxa support different molluscan assemblages in the Gulf of Naples (southern Tyrrhenian Sea). Mar. Freshwater Res. 70: 1561–1575.

Daniel, B., S. Piro, E. Charbonnel, P. Francour and Y. Letourneur. 2009. Lessepsian rabbitfish *Siganus luridus* reached the French Mediterranean coasts. Cybium. 33: 163–164.

Davis, J.P., K.A. Pitt, B. Fry, A.D. Olds and R.M. Connolly. 2014. Seascape-scale trophic links for fish on inshore coral reefs. Coral Reefs. 33: 897–907.

Diamant, A. 2010. Red-Med immigration: a fish parasitology perspective, with special reference to the Myxosporea. pp. 85–97. *In:* Golani, D. and B. Appelbaum-Golani [eds.]. Fish Invasions of the Mediterranean Sea: Change and Renewal. Pensoft Publishers. Moscow, Russia.

Diamant, A., A.B. Tuvia, A. Baranes and D. Golani. 1986. An analysis of rocky coastal eastern Mediterranean fish assemblages and a comparison with an adjacent small artificial reef. J. Exp. Mar. Biol. Ecol. 97: 269–285.

Dulĉić, J. and A. Pallaoro. 2004. First record of the marbled spinefoot *Siganus rivulatus* (Pisces: Siganidae) in the Adriatic Sea. J. Mar. Biol. Ass. 84: 1087–1088.

Dulčić, J., B. Dragičević, R. Grgičević and L. Lipej. 2011. First substantiated record of a Lessepsian migrant—the dusky spinefoot, *Siganus luridus* (Actinopterygii: Perciformes: Siganidae), in the Adriatic Sea. Acta Icth et Piscat. 41: 141–143.

EastMed. 2010. Report of the Sub-Regional Technical meeting on the Lessepsian migration and its impact on Eastern Mediterranean fishery. Nicosia, Cyprus: FAO–EastMed Project.

Edgar, G.J. and R.D. Stuart-Smith. 2014. Systematic global assessment of reef fish communities by the Reef Life Survey program. Sci. Data. 1: 140007.

Esposito, G., M. Prearo, V. Menconi, D. Mugetti, D. Meloni, M. Tomasoni et al. 2021. Northward spread of the Parrotfish *Sparisoma cretense* (Teleostei: Scaridae) in the Mediterranean Sea: An Update on Current Distribution with Two New Records from Sardinia. JMSE. 9: 536.

EU's Earth Observation. 2021. Copernicus Marine Service. DOI: https://doi.org/10.48670/moi-00016.

Fairclough, D.V., I.C. Potter, E. Lek, A.K. Bivoltsis and R.C. Babcock. 2011. The fish community and main fish populations of the Jurien Bay Marine Park. Strategic Research Fund for the Marine Environment (SRFME) Collaborative Research Project. Murdoch University, Western Australia.

Filbee-Dexter, K. and T. Wernberg. 2018. Rise of turfs: A new battlefront for globally declining kelp forests. Bioscience 68: 64–76.

Fox, R.J. and D.R. Bellwood. 2008. Remote video bioassays reveal the potential feeding impact of the rabbitfish *Siganus canaliculatus* (f: Siganidae) on an inner-shelf reef of the Great Barrier Reef. Coral Reefs 27: 605–615.

Fox, R.J., D.R. Bellwood and M.D. Jennions. 2015. Why pair? Evidence of aggregative mating in a socially monogamous marine fish (*Siganus doliatus*, Siganidae). R. Soc. Open Sci. 2: 150252.

Froese, R. and E. Pauly [Eds.]. 2024. FishBase. World Wide Web electronic publication. www.fishbase.org. (06/2024).

Fujita, D. 2010. Current status and problems of isoyake in Japan. Bull. Fish. Res. Agen. 32: 33–42.

Fujita, D., M. Noda and H. Kuwahara. 2006. Marine Herbivorous Fish: Ecology Fishery and Utilization. Seizando-Shoten Publishing, Tokyo (in Japanese).

Gajdzik, L., T.M. DeCarlo, A. Koziol, M. Mousavi-Derazmahalleh, M. Coghlan, M.W. Power et al. 2021. Climate-assisted persistence of tropical fish vagrants in temperate marine ecosystems. Commun. Biol. 4: 1231.

Giakoumi, S. 2014. Distribution patterns of the invasive herbivore *Siganus luridus* (Rüppell, 1829) and its relation to native benthic communities in the central Aegean Sea, Northeastern Mediterranean. Mar. Ecol. 35: 96–105.

Gilby, B.L., I.R. Tibbetts and T. Stevens. 2017. Low functional redundancy and high variability in Sargassum browsing fish populations in a subtropical reef system. Mar. Freshwater Res. 68: 331–341.

Givan, O., D. Edelist, O. Sonin and J. Belmaker. 2018. Thermal affinity as the dominant factor changing Mediterranean fish abundances. Glob. Chang. Biol. 24: e80–e89.

Golani, D. and R. Fricke. 2018. Checklist of the Red Sea Fishes with delineation of the Gulf of Suez, Gulf of Aqaba, endemism and Lessepsian migrants. Zootaxa. 4509: 1–215.

Harmelin-Vivien, M.L., G. Bitar, J.G. Harmelin and P. Monestiez. 2005. The littoral fish community of the Lebanese rocky coast (eastern Mediterranean Sea) with emphasis on Red Sea immigrants. Biol. Inv. 7: 625–637.

Hasegawa, M. 2010. Isoyake studies in Shizuoka Prefecture, Japan. Bull. Fish Res. Agen. 32: 109–114.

Hassan, M., M. Harmelin-Vivien and F. Bonhomme. 2003. Lessepsian invasion without bottleneck: example of two rabbitfish species (*Siganus rivulatus* and *Siganus luridus*). J. Exp. Mar. Biol. Ecol. 291: 219–232.

Hobday, A.J. and G.T. Pecl. 2014. Identification of global marine hotspots: sentinels for change and vanguards for adaptation action. Rev. Fish Biol. Fish. 24: 415–425.

Holland, M.M., J.A. Smith, J.D. Everett, A. Vergés and I.M. Suthers. 2020. Latitudinal patterns in trophic structure of temperate reef-associated fishes and predicted consequences of climate change. Fish Fish. 21: 1092–1108.

Hoque, M.M., A. Takemura, M. Matsuyama, S. Matsuura and K. Takano. 1999. Lunar spawning in *Siganus canaliculatus*. J. Fish Biol. 55: 1213–1222.

Hyndes, G.A., I. Nagelkerken, R.J. McLeod, R.M. Connolly, P.S. Lavery and M.A. Vanderklift. 2014. Mechanisms and ecological role of carbon transfer within coastal seascapes. Biol. Rev. Camb. Philos. Soc. 89: 232–254.

Ibrahim, A., M. Lahlah, M.Y. Kassab, W. Ghanem and S. Ogaily. 2010. *Signatus javus*, a new record from the Syrian waters, with reference to growth and feeding of two Lessepsian fish. Rapp. Comm. int. Mer Médit. 39: 554.

IPCC. 2021. Climate Change 2021: The Physical Science Basis. Contribution of Working Group I to the Sixth Assessment Report of the Intergovernmental Panel on Climate Change. pp 2391. (Masson-Delmotte, V., P. Zhai, A. Pirani, S.L. Connors, C. Péan, S. Berger et al. [eds.]). Cambridge University Press, Cambridge, UK.

Jones, J., J.D. DiBattista, M. Stat, M. Bunce, M.C. Boyce, D.V. Fairclough et. al. 2018. The microbiome of the gastrointestinal tract of a range-shifting marine herbivorous fish. Front. Microbiol. 9: 2000.

Kalogirou, S., M. Corsini-Foka, A. Sioulas, H. Wennhage and L. Pihl. 2010. Diversity, structure and function of fish assemblages associated with *Posidonia oceanica* beds in an area of the eastern Mediterranean Sea and the role of non-indigenous species. J. Fish Biol. 77: 2338–2357.

Kalogirou, S., H. Wennhage and L. Pihl. 2012. Non-indigenous species in Mediterranean fish assemblages: Contrasting feeding guilds of *Posidonia oceanica* meadows and sandy habitats. Estuar. Coast. Shelf Sci. 96: 209–218.

Kanashiro, K., F. Motonaga and M. Kimura. 1999. Settlement of white-spotted spinefoot, *Siganus canaliculatus* (Pisces: Siganidae), in the coastal waters off Okinawa Island, Japan. Nippon Suisan Gakkaishi. 65: 19–25.

Karachle, P.K., A. Angelidis, G. Apostolopoulos, D. Ayas, M. Ballesteros, C. Bonnici et al. 2016. New mediterranean biodiversity records (march 2016). Medit. Mar. Sci. 17: 230–252.

Katayama, S., S. Akiyama, M. Naganuma and R. Shibata. 2009. Age and growth of rabbitfish *Siganus fuscescens* in Tateyama Bay, Japan. Aquaculture Sci. 57: 417–422.

Kimura, H., M. Yamauchi and M. Notoya. 2007. Early harvests of *Undaria undarioides* using net cages prevention of fish consumption. Aquaculture Sci. 55: 467–473.

Kiriyama, T., M. Noda and A. Fuji. 2001. Grazing and bite marks on *Ecklonia kurome* caused by several herbivorous fishes. Suisanzoshioku. 49: 431–438.

Kiriyama, T., A. Fujii and Y. Fujita. 2005. Feeding and characteristic bite marks on *Sargassum fusiforme* by several herbivorous fishes. Aquaculture Sci. 53: 355–365.

Kleitou, P., D.K. Moutopoulos, I. Giovos, D. Kletou, I. Savva, L.L. Cai et al. 2022. Conflicting interests and growing importance of non-indigenous species in commercial and recreational fisheries of the Mediterranean Sea. Fish. Manag. Ecol. 29: 169–182.

Kumagai, N.H., J.G. Molinos, H. Yamano, S. Takao, M. Fujii and Y. Yamanaka. 2018. Ocean currents and herbivory drive macroalgae-to-coral community shift under climate warming. Proc. Natl. Acad. Sci. USA. 115: 8990–8995.

Kuwahara, H. 2015. The evaluation of current status for grazing control techniques of herbivorous fish. Fisheries Eng. 51: 253–257.

Layton, C., V. Shelamoff, M.J. Cameron, M. Tatsumi, J.T. Wright and C.R. Johnson. 2019. Resilience and stability of kelp forests: The importance of patch dynamics and environment-engineer feedbacks. PLoS ONE. 14: e0210220.

Lenanton, R.C.J., C.E. Dowling, K.A. Smith, D.V. Fairclough and G. Jackson. 2017. Potential influence of a marine heatwave on range extensions of tropical fishes in the eastern Indian Ocean—Invaluable contributions from amateur observers. Reg. Stud. Mar. Sci. 13: 19–31.

Longo, G.O., M.E. Hay, C.E.L. Ferreira and S.R. Floeter. 2018. Trophic interactions across 61 degrees of latitude in the Western Atlantic. Global Ecol. Biogeogr. 28: 107–117.

Lundberg, B. and D. Golani. 1995. Diet Adaptations of Lessepsian Migrant Rabbitfishes, *Siganus luridus* and *Siganus rivulatus*, to the Algal Resources of the Mediterranean Coast of Israel. Mar. Ecol. 16: 73–89.

Marletta, G. and A. Lombardo. 2020. Assessment of grazing impact on deep canopy-forming species in the western Ionian Sea, Central Mediterranean. Int. J. Aquat. Biol. 8: 365–376.

Marras, S., A. Cucco, F. Antognarelli, E. Azzurro, M. Milazzo, M. Bariche et al. 2015. Predicting future thermal habitat suitability of competing native and invasive fish species: from metabolic scope to oceanographic modelling. Conserv. Physiol. 3: cou059.

Michailidis, N., S. Katsanevakis and N. Chartosia. 2020. Recreational fisheries can be of the same magnitude as commercial fisheries: The case of Cyprus. Fish. Res. 231: 105711.

Morinaga, T. and K. Yasuda. 1985. Flora, fauna and crops of the Japan islands in the eighteenth century. Kagaku Shoin, Inc. Tokyo, Japan.

Morley, J.W., R.L. Selden, R.J. Latour, T.L. Frölicher, R.J. Seagraves and M.L. Pinsky. 2018. Projecting shifts in thermal habitat for 686 species on the North American continental shelf. PLoS ONE. 13: e0196127.

Mulders, Y., K. Filbee-Dexter, S. Bell, N.E. Bosch, A. Pessarrodona, D. Sahin et. al. 2022. Intergrading reef communities across discrete seaweed habitats in a temperate-tropical transition zone: Lessons for species reshuffling in a warming ocean. Ecol. Evol. 12: e8538.

Nakabo, T. 2013. Fishes of Japan with pictorial keys to the species. 3rd ed. Tokai University Press. Tokyo, Japan.

Nakamura, Y. and M. Tsuchiya. 2008. Spatial and temporal patterns of seagrass habitat use by fishes at the Ryukyu Islands, Japan. Estuar. Coast. Shelf. Sci. 76: 345–356.

Nakamura, Y., D.A. Feary, M. Kanda and K. Yamaoka. 2013. Tropical fishes dominate temperate reef fish communities within western Japan. PLoS ONE. 8: e81107.

Nakayama, Y. and S. Arai. 1999. Grazing of the brown alga *Ecklonia cava* by three herbivorous fishes on the coast of Nakagi, South Izu, central Japan. Jpn. J. Phycol. 47: 105–112.

Nimura, K., H. Takatsuji, S. Masuda and J. Shimamoto. 2007. Growth and maturation of *Ecklonia cava* and Eisenia arborea seedlings transplanted along the coast of Hainan, Shizuoka prefecture and the grazing caused by herbivorous fish *Siganus fuscescens*. Aquaculture Sci. 55: 541–546.

Noda, M., C. Hasegawa and T. Kyuno. 2002. Unique feeding behavior of thallus of the kelp *Eisenia bicyclis* by adult *Siganus fuscescens* captive in an experimental tank. Journal of National Fisheries University 50: 151–159.

Noda, M., H. Ohara, K. Urakawa, N. Murase and K. Yamamoto. 2011. Diet and prey availability of *Siganus fuscescens* occurring in a Sargassum bed at Futaoi Island in the Sea of Hibiki with respect to feeding on large brown macroalgae. Nippon Suisan Gakkaishi. 77: 1008–1019.

Noda, M., H. Ohara, N. Murase, I. Ikeda and K. Yamamoto. 2014. The grazing of *Eisenia bicyclis* and several species of Sargassaceous and Cystoseiraceous seaweeds by *Siganus fuscescens* in relation to the differences of species composition of their seaweed beds. Nippon Suisan Gakkaishi. 80: 201–213.

Noda, M., J. Kinoshita, N. Tanada and N. Murase. 2018. Characteristics of bite scars observed in kelp forests of Lessoniaceae denuded by short-term foraging damages of the herbivorous fish *Siganus fuscecens*. Journal of National Fisheries University 66: 111–122.

Occhipinti-Ambrogi, A. 2007. Global change and marine communities: alien species and climate change. Mar. Pollut. Bull. 55: 342–352.

Olds, A.D., K.A. Pitt, P.S. Maxwell and R.M. Connolly. 2012. Synergistic effects of reserves and connectivity on ecological resilience. J. Appl. Ecol. 49: 1195–1203.

Oliver, E.C.J., M.G. Donat, M.T. Burrows, P.J. Moore, D.A. Smale, L.V. Alexander et al. 2018. Longer and more frequent marine heatwaves over the past century. Nat. Commun. 9: 1324.

Pagès, J.F., S. Farina, A. Gera, R. Arthur, J. Romero and T. Alcoverro. 2012. Indirect interactions in seagrasses: fish herbivores increase predation risk to sea urchins by modifying plant traits. Funct. Ecol. 26: 1015–1023.

Papadakis, O., K. Tsirintanis, V. Lioupa and S. Katsanevakis. 2021. The neglected role of omnivore fish in the overgrazing of Mediterranean rocky reefs. Mar. Ecol. Prog. Ser. 673: 107–116.

Pearce, A.F. and M. Feng. 2013. The rise and fall of the "marine heat wave" off Western Australia during the summer of 2010/2011. J. Mar. Syst. 111: 139–156.

Pearce, A., D. Slawinski, M. Feng, B. Hutchins and P. Fearns. 2011. Modelling the potential transport of tropical fish larvae in the Leeuwin Current. Cont. Shelf Res. 31: 2018–2040.

Pessarrodona, A., A. Vergés, N.E. Bosch, S. Bell, S. Smith and M.P. Sgarlatta. 2021. Tropicalization unlocks novel trophic pathways and enhances secondary productivity in temperate reefs. Funct. Ecol. 36: 659–673.

Pickholtz, R.S.M., M. Kiflawi, A.M. Friedlander and J. Belmaker. 2018. Habitat utilization by an invasive herbivorous fish (*Siganus rivulatus*) in its native and invaded range. Biol. Inv. 20: 3499–3512.

Poloniato, D., S. Ciriaco, R. Odorico and J. Dulčić. 2010. First record of the dusky spinefoot *Siganus luridus* (Ruppell, 1828) in the Adriatic Sea. Annales Ser. Hist. Nat. 20: 161–166.

Sala, E., Z. Kizilkaya, D. Yildirim and E. Ballesteros. 2011. Alien marine fishes deplete algal biomass in the Eastern Mediterranean. PLoS ONE. 6: e17356.

Sales, M. and E. Ballesteros. 2012. Seasonal dynamics and annual production of *Cystoseira crinita* (Fucales: Ochrophyta)-dominated assemblages from the northwestern Mediterranean. Sci. Mar. 76: 391–401.

Santana-Garcon, J.B.S., N. Marbà, A. Vergés, R. Arthur and T. Alcoverro. 2023. Tropicalization shifts herbivore pressure from seagrass to rocky reef communities. Proc. R. Soc. B. 290: 20221744.

Serisawa, Y., Z. Imoto, T. Ishikawa and M. Ohno. 2004. Decline of the *Ecklonia cava* population associated with increased seawater temperatures in Tosa Bay, southern Japan. Fish Sci. 70: 189–191.

Shakman, E.A. and R. Kinzelbach. 2007. Distribution and characterization of Lessepsian migrant fishes along the coast of Libya. Acta Icth et Piscat. 37: 7–15.

Shakman, E., C. Boedeker, M. Bariche and R. Kinzelbach. 2009. Food and feeding habits of the Lessepsian migrants *Siganus luridus* Rüppell, 1828 and *Siganus rivulatus* Forsskal, 1775 (Teleostei: Siganidae) in the southern Mediterranean (Libyan coast). J. Biol. Res.-Thessalon. 12: 115 – 124.

Siqueira, A.C., D.R. Bellwood and P.F. Cowman. 2019. Historical biogeography of herbivorous coral reef fishes: The formation of an Atlantic fauna. J. Biogeogr. 46: 1611–1624.

Smith, S.M., H.A. Malcolm, E.M. Marzinelli, A.L. Schultz, P.D. Steinberg and A. Vergés. 2021. Tropicalisation and kelp loss shift trophic composition and lead to more winners than losers in fish communities. Glob. Chang. Biol. 27: 2537–2548.

Smith, S.M. 2021. Novel interactions in tropicalised ecosystems. Ph.D. Thesis. University of New South Wales, Sydney, Australia.

Soliman, V.S. and K. Yamaoka. 2010. Assessment of the fishery of siganid juveniles caught by bagnet in Lagonoy Gulf, Southeastern Luzon, Philippines. J. Appl. Ichthyol. 26: 561–567.

Steneck, R.S., D.R. Bellwood and M.E. Hay. 2017. Herbivory in the marine realm. Curr. Biol. 27: R484–R489.

Sudo, K., K. Watanabe, N. Yotsukura and M. Nakaoka. 2020. Predictions of kelp distribution shifts along the northern coast of Japan. Ecol. Res. 35: 47–60.

Sudo, K., S. Maehara, M. Nakaoka and M. Fujii. 2022. Predicting future shifts in the distribution of tropicalization indicator fish that affect coastal ecosystem services of Japan. Front. Built Environ. 7: 788700.

Suthers, I.M., J.W. Young, M.E. Baird, M. Roughan, J.D. Everett, Brassington et al. 2011. The strengthening East Australian Current, its eddies and biological effects—an introduction and overview. Deep Sea Res. Part II Top. Stud. Oceanogr. 58: 538–546.

Takao, S., N.H. Kumagai, H. Yamano, M. Fujii and Y. Yamanaka. 2015. Projecting the impacts of rising seawater temperatures on the distribution of seaweeds around Japan under multiple climate change scenarios. Ecol. Evol. 5: 213–223.

Tanaka, K., S. Taino, H. Haraguchi, G. Prendergast and M. Hiraoka. 2012. Warming off southwestern Japan linked to distributional shifts of subtidal canopy-forming seaweeds. Ecol. Evol. 2: 2854–2865.

Telesca, L., A. Belluscio, A. Criscoli, G. Ardizzone, E.T. Apostolaki, S. Fraschetti et al. 2015. Seagrass meadows (*Posidonia oceanica*) distribution and trajectories of change. Sci. Rep. 5: 12505.

Tortonese, E. 1970. On the occurrence of Siganus (Pisces) along the coast of North-Africa. Doriana. 4: 1–2.

Ueta, Y. and N. Tanada. 2018. Influences of low water temperature and food availability in winter on the survival and feeding of the rabbitfish *Siganus fuscecens* in captivity. Bull. Tokushima Pref. Fish. Res. Ins. 12: 11–19.

Urban, M.C. 2020. Climate-tracking species are not invasive. Nat. Clim. Chang. 10: 382–384.

Vanderklift, M.A., R.D. Pillans, M. Hutton, L. De Wever, G.A. Kendrick, A. Zavala-Perez et al. 2021. High rates of herbivory in remote northwest Australian seagrass meadows by rabbitfish and green turtles. Mar. Ecol. Prog. Ser. 665: 63–73.

Vergés, A., T. Alcoverro and E. Ballesteros. 2009. Role of fish herbivory in structuring the vertical distribution of canopy algae Cystoseira spp. in the Mediterranean Sea. Mar. Ecol. Prog. Ser. 375: 1–11.

Vergés, A., F. Tomas, E. Cebrian, E. Ballesteros, Z. Kizilkaya, P. Dendrinos et. al. 2014. Tropical rabbitfish and the deforestation of a warming temperate sea. J. Ecol. 102: 1518–1527.

Vergés, A., C. Doropoulos, H.A. Malcolm, M. Skye, M. Garcia-Pizá, E.M. Marzinelli et al. 2016. Long-term empirical evidence of ocean warming leading to tropicalization of fish communities, increased herbivory, and loss of kelp. Proc. Natl. Acad. Sci. USA. 113: 13791–13796.

Vergés, A., E. McCosker, M. Mayer-Pinto, M.A. Coleman, T. Wernberg, T. Ainsworth et al. 2019. Tropicalisation of temperate reefs: Implications for ecosystem functions and management actions. Funct. Ecol. 33: 1000–1013.

Wernberg, T., G.A. Kendrick and B.D. Toohey. 2005. Modification of the physical environment by an *Ecklonia radiata* (Laminariales) canopy and implications for associated foliose algae. Aquat. Ecol. 39: 419–430.

Wernberg, T., S. Bennett, R.C. Babcock, T. de Bettignies, K. Cure, M. Depczynski et al. 2016. Climate-driven regime shift of a temperate marine ecosystem. Science 353: 169–172.

Wernberg, T., K. Krumhansl, K. Filbee-Dexter and M.F. Pedersen. 2019. Status and trends for the world's kelp forests. pp. 57–78. *In:* Ch. Sheppard [ed.]. World seas: an environmental evaluation (second edition). Academic Press.

Whalen, M.A., R.D.B. Whippo, J.J. Stachowicz, P.H. York, E. Aiello, T. Alcoverro et al. 2020. Climate drives the geography of marine consumption by changing predator communities. Proc. Natl. Acad. Sci. USA. 117: 28160–28166.

Woodland, D.J. 1990. Revision of the fish family Siganidae with descriptions of two new species and comments on distribution and biology. Indo-Pacific Fishes. Vol. 19. Bishop Museum. Honolulu, Hawaii.

Yamauchi, M., H. Kimura and D. Fujita. 2006. Seasonal and diurnal feeding patterns of the herbivorous fish *Siganus fuscescens* and scaring by optic and auditory stimuli. Fish. Eng. 43: 65–68.

Yorke, C.E., H.M. Page and R.J. Miller. 2019. Sea urchins mediate the availability of kelp detritus to benthic consumers. Proc. Biol. Sci. 286: 20190846.

Zarco-Perello, S., T. Wernberg, T.J. Langlois and M.A. Vanderklift. 2017. Tropicalization strengthens consumer pressure on habitat-forming seaweeds. Sci. Rep. 7: 820.

Zarco-Perello, S., T.J. Langlois, T. Holmes, M.A. Vanderklift and T. Wernberg. 2019. Overwintering tropical herbivores accelerate detritus production on temperate reefs. Proc. R. Soc. B. 286: 20192046.

Zarco-Perello, S., D. Fairclough, C. Dowling, J. DiBattista, R. Austin, T. Wernberg and B. Taylor. 2022. Maximization of fitness by phenological and phenotypic plasticity in range expanding rabbitfish (Siganidae). J. Anim. Ecol. 91: 1666–1678.

Zarco-Perello, S., G. Carroll, M. Vanderklift, T. Holmes, T.J. Langlois and T. Wernberg. 2020. Range-extending tropical herbivores increase diversity, intensity and extent of herbivory functions in temperate marine ecosystems. Funct. Ecol. 34: 2411–2421.

Zarco-Perello, S., N.E. Bosch, S. Bennett, M.A. Vanderklift and T. Wernberg. 2021. Persistence of tropical herbivores in temperate reefs constrain kelp resilience to cryptic habitats. J. Ecol. 109: 2081–2094.

Zenetos, A., P.G. Albano, E. López Garcia, N.I.R. Stern, K. Tsiamis and M. Galanidi. 2022. Established non-indigenous species increased by 40% in 11 years in the Mediterranean Sea. Medit. Mar. Sci. 23.

9

Rabbitfish Aquaculture

A Review of Knowledge and Prospects for Future Sustainable Farming Practices

Thomas Camus,[1,*] *Brett M. Taylor*[2] and *Andrew S. Hoey*[3]

Introduction

As the world's population continues to grow, the harvest and production of aquatic animals through fisheries and aquaculture is an increasingly important component of global food security, source of protein, and essential micronutrients (Hicks et al. 2019, FAO 2020). Indeed, estimated global consumption of aquatic foods has increased at almost double the rate of global human population growth (3.0% and 1.6% per annum, respectively) over the past 60 years (FAO 2022). This increase was initially driven by wild capture fisheries, with global harvest increasing from *ca.* 30 million tonnes in the 1960's to a peak of 130 million tonnes in 1996, after which it has remained relatively stable (Pauly and Zeller 2016). In contrast, aquaculture production has increased 10-fold over the past three decades, from *ca.* 12 million tonnes in 1990 to 122.6 million tonnes in 2020, and now accounts for almost half (49.2%) of the global production of aquatic seafood products (FAO 2022).

Initially, there were some suggestions, or hope, that increasing aquaculture production would reduce demand and pressure on wild stocks (e.g., Pomeroy et al. 2006), however there is little evidence to support this. The vast majority (up to 90%) of wild fish stocks are considered to be fully exploited, overexploited or collapsed

[1] UMR 241 SECOPOL, IFREMER/UPF/ILM/IRD, Centre Ifremer du Pacifique Vairao, Taiarapu-Ouest 98719, French Polynesia.
[2] University of Guam Marine Laboratory, and University of Guam Sea Grant, Mangilao, Guam, 96923, USA.
Email: brettmtaylor@gmail.com
[3] College of Science and Engineering, James Cook University, Townsville, Queensland 4810, Australia.
Email: Andrew.hoey1@jcu.edu.au
* Corresponding author: Thomas.Camus@ifremer.fr

(Pitcher and Cheung 2013, Longo et al. 2019). While aquaculture is often seen as an alternative to wild capture fisheries, the focus on culturing carnivorous species with high market value and growth rates has placed considerable pressure on the harvest of some wild stocks (e.g., sardines and anchovy) for feed (Essington et al. 2015, Froehlich et al. 2018). Culturing species from lower trophic levels (i.e., primary consumers or herbivores) is increasingly viewed as being a viable and sustainable alternative that alleviates pressure on wild stocks as a source of food (Anastasiades 2010, Li et al. 2018).

Rabbitfishes (f. Siganidae) are a group of primarily herbivorous marine fishes (Hoey et al. 2013, Clements – Chapter 4) that have long been recognized as a potential candidate for aquaculture (Ablan and Rosario 1961, Lam 1974, von Westernhagen and Rosenthal 1976). Rabbitfishes are a common component of nearshore coral reef and non-reef habitats throughout much of the Indo-Pacific region, and are absent from the east Pacific and Atlantic Oceans (Woodland 1990, Chapter 1). There are 29 extant species of rabbitfish, with all species belonging to a single genus, *Siganus*, and are often separated into two clades: the first of brightly-coloured reef-associated pairing species (e.g., *Siganus doliatus*, *Siganus stellatus*) and the second of drab species that typically occur in schools and associate with non-reef habitats (e.g., *Siganus canaliculatus, Siganus sutor*; Siquiera and Cowman – Chapter 3). Members of the schooling clade are targeted by commercial, subsistence and/or recreational fishers throughout much of their range (Lam 1974). This demand, coupled with their rapid initial growth, fast maturation, high fecundity (Taylor et al. – Chapter 5), and tolerance to varied environmental conditions (Hoey et al. – Chapter 7) make rabbitfishes an ideal candidate for aquaculture.

Despite over sixty years of interest in developing the aquaculture of rabbitfishes; the first trials of culturing rabbitfishes being conducted in the 1960's (Ablan and Rosario 1961), and a coordinated effort to promote and implement rabbitfish aquaculture in the early- to mid-1970's (Lam 1974, von Westernhagen and Rosenthal 1976), the commercial culture of rabbitfishes is still in its infancy. Reported aquaculture production of rabbitfishes has increased steadily from the mid-1990's reaching a peak of 397 tonnes in 2020, however this represents less than 0.3% of total harvest of wild capture fisheries for rabbitfishes (2020: 132,775 tonnes; Fig. 1). In this chapter we synthesize the current status and future potential of rabbitfish aquaculture, and in doing so identify key knowledge gaps and challenges to the implementation of commercial scale aquaculture.

Rabbitfish Aquaculture around the World

Rabbitfishes are recognized and used as food fishes in many parts of their native range throughout the Indo-Pacific region, and also the Mediterranean Sea following their invasion of this region (Bariche 2005, Grandcourt et al. 2007, Maina et al. 2013, Mehanna et al. 2018, Zarco-Perello et al. – Chapter 8). However, their demand and market potential vary considerably across regions, and this variability has direct implications for the spatial distribution of commercial aquaculture production of rabbitfishes among and within regions. Rabbitfish aquaculture has been reported from 11 countries (Philippines, Indonesia, Egypt, India, Saudi Arabia, Bahrain, Fiji,

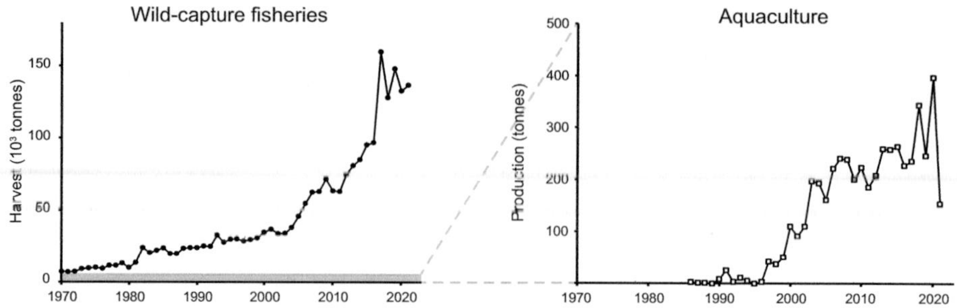

Fig. 1. Comparison of the reported global production of rabbitfishes (f. Siganidae) from wild-capture fisheries versus aquaculture over the past five decades (1970–2022). Data sourced from Food and Agriculture Organization of the United Nations – Fisheries and Aquaculture (https://www.fao.org/fishery/statistics-query/en/global_production/global_production_quantity).

Mauritius, Qatar, Singapore, and United Arab Emirates), with the major countries in terms of total aquaculture production being the Philippines and the Kingdom of Saudi Arabia (FAO 2023; Fig. 2). For example, in 2017, the Philippines accounted for the vast majority (82%) of global rabbitfish aquaculture production (193.5 t of 236t; Anuraj et al. 2021). However, rabbitfish aquaculture has been at least trialled in all major areas of the Indo-Pacific region, including Asia (Philippines, Indonesia, China, Thailand, India, Singapore), the Pacific (Cook Islands, Fiji, Guam, Micronesia, New Caledonia, Palau, Solomon Islands), the Mediterranean (Cyprus, Middle East (Bahrain, Qatar, Saudi Arabia, United Arab Emirates), and Africa (Egypt, Mauritius, Tanzania). In each of these regions, the cultured species varies based on their availability and market demand, with the most commonly cultured species being *Siganus guttatus*, *Siganus rivulatus*, *Siganus argenteus* and *S. canaliculatus* (Table 1). Below, we summarise rabbitfish aquaculture activities from the major areas of the Indo-Pacific and Mediterranean regions, noting that detail of the scale and extent of the activities is often extremely limited.

Asia

Asia dominates global aquaculture production, with over 91% of global production stemming from the region and nearly two-thirds of production from China alone. While rabbitfishes are a favoured fish in China that command high prices because of their delicacy and nutritional value (Xu et al. 2011, Li et al. 2018), Chinese aquaculture production of rabbitfishes is low (Fig. 2). Rabbitfish aquaculture in Asia first began in Hong Kong in 1972 with the cage culture of *S. canaliculatus* (Lai and Yu 2002). Today *S. canaliculatus* remains the most cultured rabbitfish species, and one of the few marine herbivorous fishes farmed in China (Li et al. 2018, Xie et al. 2019) although its production remains relativerly low (Fig. 2).

In contrast, the Philippines is the top global producer of both wild-caught and cultured rabbitfish, with reported aquaculture production of 197 tonnes (2020) and marine landings of 115 937 tonnes (2017), representing the bulk of both aquaculture and wild-capture fisheries production in those years (OECD 2021). *Siganus fuscescens* (the sister taxon to *S. canaliculatus*) is the dominant species caught

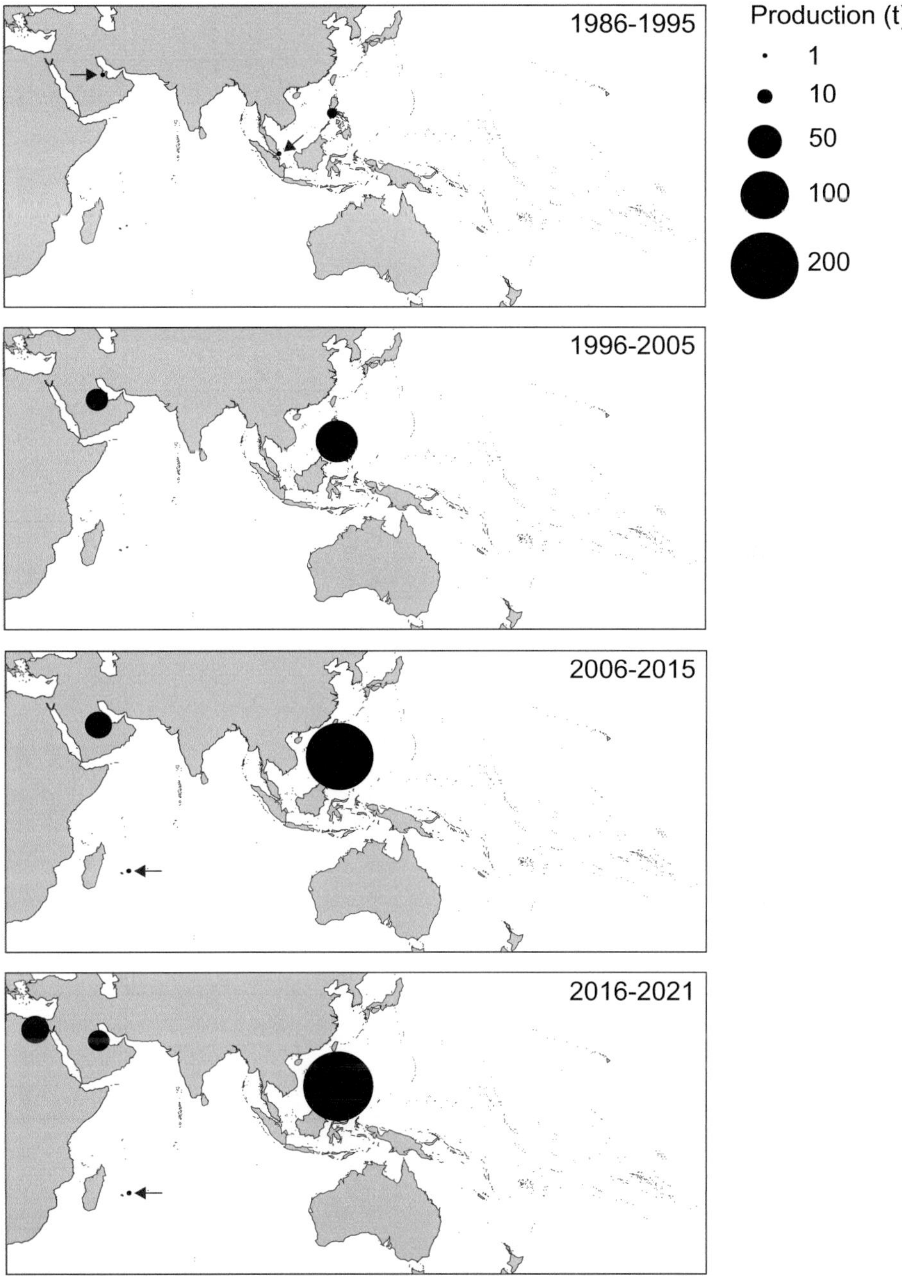

Fig. 2. Temporal changes in the distribution of aquaculture production of rabbitfishes (f. Siganidae) over the past 35 years (1986-2021). Data sourced from Food and Agriculture Organization of the United Nations – Fisheries and Aquaculture (https://www.fao.org/fishery/statistics-query/en/global_production/global_production_quantity)

Table 1. Summary of reported aquaculture trials and production of rabbitfishes (f. Siganidae) across five regions (Asia, Pacific, Middle East, Africa, Mediterannean) and 27 countries.

Region	Country	Species	Commercial aquaculture	Larval rearing	Juvenile grow-out	FAO reported production	References
Asia	Philippines	*S. argenteus* *S. canaliculatus* *S. guttatus* *S. spinus* *S. vermiculatus*	capture based	Successful for *S. guttatus* and *S. vermiculatus* Trialled for *S. argenteus*	Sea cages and ponds used for grow out of wild caught juveniles	Yes	Abalos 1999, Carumbana and Luchavez, 1979, Gonzales et al. 2018, van Beijnen et al. 2022
	China	*S. canaliculatus* *S. guttatus*	capture based	Successful production of *S. canaliculatus* and *S. guttatus* larvae, however low survival	Sea cages and ponds used for grow out of wild caught juveniles	No	Hong and Zhang 2003, Li et al. 2018, Xie et al. 2018
	India	*S. canaliculatus* *S. vermiculatus*		Trialled for *S. vermiculatus*, however low survival	Sea cages	No	Jaikumar 2012, Anuraj et al. 2021
	Indonesia	*S. canaliculatus* *S. doliatus* *S. guttatus* *S. lineatus*	capture based	Trialled with very low survival rates	Sea cages and brackish ponds	No	Rachmansyah et al 2007, Syah et al. 2020, Darsiani et al. 2022
	Hong Kong	*S. canaliculatus*			Sea cages	No	Tseng and Chan 1982, Lai and Yu 2002
	Japan	*S. canaliculatus* *S. fuscescens* *S. guttatus*			Trials only	No	Waycott 2023
	Malaysia	*S. canaliculatus* *S. guttatus* *S. javus* *S. vermiculatus*		Successful rearing of *S. canaliculatus, S. guttatus, S. javus, S. vermiculatus*		No	Tan 2021

	Singapore	*S. canaliculatus* *S. guttatus* *S. javus*		Experimental trials		Yes – 1990 only (7t)	Shen et al. 2021
	Vietnam	*S. guttatus*		Trials with 100% mortality		No	Binh et al 2014
Pacific	Australia	*S. fuscescens*		Experimental		No	Thepot 2021
	Cook Islands	*S. argenteus*		Scoping only		No	Malpct et al. 2008, Teitelbaum et al. 2008
	Fiji	*S. vermiculatus*			Polyculture trials with milkfish, mullet and prawns in Earthen ponds	No	May et al. 1974, Lichatowich and Popper 1975
	Guam	*S. argenteus* *S. randalli* *S. spinus*		Trials with *S. randalli*, however production limited	Experimental in outdoor tanks, floating cages, ponds and raceways.	No	Tobias 1976, Nelson et al. 1992
	New Caledonia	*S. fuscescens* *S. lineatus*		Successful rearing of *S. fuscescens* and *S. lineatus*	Trails using hatchery-sourced juveniles	No	Teitelbaum et al. 2008, Luong et al. 2017
	Palau	*S. canaliculatus* *S. lineatus*		Successful rearing of *S. canaliculatus*	Floating cages	No	Bryan and Madraisau 1977, Camperio et al. 2020
	Pohnpei	*S. argenteus* *S. randalli*		Trial with wild-caught juveniles	Floating cages	No	Tobey et al 2004, Seal and Ellis 2019
	Solomon Isl	*S. argenteus*		Trial with wild-caught juveniles	Sea cages	No	Teitelbaum and Hilly 2009

Table 1 contd. ...

...Table 1 contd.

Region	Country	Species	Commercial aquaculture	Larval rearing	Juvenile grow-out	FAO reported production	References
Middle East	Saudi Arabia	*S. canaliculatus* *S. rivulatus*		Commercial hatchery production of *S. rivulatus*	Sea cages and ponds	Yes	Lichatowich et al. 1984, Bukhari 2005
	Bahrain	*S. canaliculatus*		Trials only	Trials only	No	Uwate and Shams 1999
	Iran	*S. canaliculatus*		Trials only	Trials only	No	Regunathan and Kitto 2005
	Israel	*S. argenteus* *S. luridus* *S. rivulatus*			Tanks and submerged cages used for grow out of wild caught juveniles	No	Popper et al. 1979, Pham and Van Phan 2016
	Saudi Arabia	*S. canaliculatus* *S. rivulatus*	capture based	Successful for *S. canaliculatus*	Sea cages and ponds used for grow out of wild caught juveniles	Yes	Lichatowich et al. 1984, Bukhari 2005
	United Arab Emirates	*S. canaliculatus* *S. javus*		Successful rearing of *S. canaliculatus*		No	Dickson 2022
Africa	Egypt	*S. rivulatus*			Trials using wild caught juveniles	No	El-Dakar et al. 2011
	Tanzania	*S. sutor*			Trials using wild caught juveniles	No	Bwathondi 1982
	Mauritius	Not reported, likely *S. sutor*	capture based		Embanked lagoons used for grow out of wild caught juveniles	Yes	Jaikumar 2012
Mediterannean	Cyprus	*S. rivulatus*		Commercial production of fingerlings	Limited production in offshore sea cages	No	Stephanou and Georgiou 2000
	Lebanon	*S. rivulatus*			Trials using wild caught juveniles in floating cages	No	Nasser et al 2018, Saoud et al. 2014

by fishers, contributing up to 40% of the total catch of wild harvest fisheries in some regions (Hernandez et al. 1990), while *S. fuscescens*, *S. guttatus* and *Siganus vermiculatus* are the main aquaculture species. Filipinos have a rich history of culturing marine herbivores, particularly milkfish, using extensive culture methods. Traditional farming of rabbitfishes has occurred for many decades (Pillai 1962, Ponce 1983, Urmaza 1983), however many of these early reports were related to incidental introduction and subsequent grow-out of rabbitfish. Indeed, fishpond owners rarely acquired rabbitfish and only anecdotal culture occurred when fry were accidentally introduced into milkfish ponds (von Westernhagen and Rosenthal 1975). Commercial aquaculture of rabbitfish has occurred for several decades in the Philippines using either wild-caught or hatchery reared juveniles. Some fish farmers raise *S. guttatus* and *S. vermiculatus* from captured juveniles at commercial scale, with higher production reported in sea-cages than in brackish water ponds (Abalos 1999, Gonzales et al. 2018, Caballero et al. 2022). The Southeast Asian Fisheries Development Center (SEAFDEC) began culturing rabbitfish in 1983 and now produces *S. guttatus* fry throughout the year, although the precise quantity produced is unknown (Caballero et al. 2022). While, *S. argenteus*, *S. canaliculatus* and *S. guttatus* are all considered good candidates for culture within the Philippines, *S. guttatus* is considered the most suitable since it has the fastest growth of the candidate species (when cultured in floating cages) and can attain a weight of more than 1kg (Juario et al. 1985).

Indonesia is the second highest producer of aquaculture globally, and over 125 marine finfish hatcheries are now operating in Indonesia, principally located in Bali, Lampung, and East Java (Rimmmer et al. 2013, FAO 2022). These hatcheries have the capability to switch cultured species depending on market demand, however the aquaculture of rabbitfish is mostly dependent on wild seed stocks collected in Sumatra, Java, and Sulawesi (Kongkeo et al. 2010). The primary species cultured in Indonesia are S. *canaliculatus, S. guttatus, S. javus*, and *S. vermiculatus* (Laining et al. 2016). Domestication of *S. guttatus* was reported at the Barru Tiger Shrimp Farm (South Sulawesi) with larval rearing conducted in outdoor ponds using mesocosm culture techniques (Darsiani et al. 2022).

In other regions throughout south-east Asia the predominant rabbitfish species used for aquaculture are *S. canaliculatus* (or its sister taxon *S. fuscescens*), *S. javus*, *S. guttatus*, and *S. vermiculatus*. In Malaysia, successful larval rearing of *S. javus* to the juvenile stage was reported in 2021 using enriched rotifers and copepod nauplii (Tan 2021). A year later it was reported in the press that three species of rabbitfish (*S. guttatus*, *S. canaliculatus* and *S. vermiculatus*) were successfully hatchery-reared using eco-friendly and sustainable production methods in Tuaran (Department of Fisheries Sabah 2022). In nearby Singapore, *S. canaliculatus*, *S. javus* and *S. guttatus* have been minor species cultured in floating farms (Chou and Lee 2008), although little information is available as to whether they were sourced from larval rearing or wild capture of juveniles. Interestingly, a recently review of the status of aquaculture in Singapore made no mention of rabbitfishes (Shen et al. 2021). In Vietnam, *S. guttatus* was successfully reared from fertilized eggs to 31-day old larvae (Van Phan and Pham 2016), with the authors stating this had been a major impediment to commercial scale aquaculture in the country. A polyculture of two rabbitfish, *S. guttatus* and *S. canaliculatus* in combination with the giant tiger prawn *Penaeus monodon* was developed as a sustainable aquaculture model (Binh et al. 2014). In India, hatchery production of *S. vermiculatus* using enriched rotifers and copepods (Anuraj et al. 2021) and grow-out trials of *S. canaliculatus* in floating cages were reported as promising (Jaikumar et al. 2011), yet we could not find any detail of commercial aquaculture in India. Rabbitfish were not popular with Japanese consumers in the past, due to poisonous spines and what was perceived as an unpleasant smell emanating from the intestines. This has recently evolved and some investigation is now being conducted on *S. fuscescens* at the Aquaculture Research Institute of Kindai University, with the goal of culturing them using low cost, sustainable feeds, such as vegetable wastes (Waycott 2023).

The Pacific

Rabbitfish aquaculture has a long-standing tradition in the Pacific Islands, driven by their high market demand and the unique characteristics of their life cycle. With periodic appearances of large schools of juvenile rabbitfishes on shallow reef flats, these fish have been exploited through the grow-out of wild-captured fry for decades. Despite the traditional reliance on the capture of fry from the wild, the successful larval rearing of *S. canaliculatus* using cultured rotifers (*Brachionus plicatilis*) and wild copepods was reported in the early 1970's in Palau (May et al. 1974). Guam also has a history of rabbitfish aquaculture research with many studies conducted between the 1970s and the 1990s, including hatchery production of *Siganus randalli* (Nelson et al. 1992), however, there is currently no rabbitfish aquaculture production reported in Guam.

New Caledonia is currently developing hatchery-based aquaculture of *S. fuscescens* and *Siganus lineatus* (sister taxon to *S. guttatus*) at the Caledonian Center for Development and Transfer of Marine Aquaculture, with subsequent commercialization trials. Notably, co-culture trials have been explored within shrimp ponds (Luong et al. 2017). In Fiji juvenile grow-out in earthen ponds were reported for *S. vermiculatus*, including polyculture trials in combination with other finfish species (Lichatowich and Popper 1975). There has been no commercial production reported in Australia but hatchery rearing of *S. fuscescens* has been used for experimental research purposes (Thepot 2021). Development of rabbitfish aquaculture has also been explored and reported from Solomon Islands, Cook Islands (Teitelbaum et al. 2008), and Pohnpei (Seale and Ellis 2019). Ultimately, these operations appear to have been experimental or established for small-scale farming, underlining the potential of rabbitfish aquaculture for subsistence practices and emerging small-scale commercial ventures.

Middle East

Demand for rabbitfish is high in all countries bordering the Arabian/Persian Gulf (Dickson 2022) with a total of 400 tonnes of high-value tropical marine finfish, including rabbitfish, being farmed in the United Arab Emirates and Saudi Arabia in 2018 (Dickson 2022). The interest in culturing rabbitfishes on the Arabian Peninsula has been directed mainly towards *S. rivulatus* and *S. canaliculatus* (Wassef and Adbul Hady 1997, Uwate and Shams 1999, Saoud and Ghanawi 2007). In the United Arab Emirates, the Aquaculture and Marine Studies Centre (AMSC) was tasked in 2015 to support the growing aquaculture industry in the country and is now capable of producing 3 million fingerlings per year, including *S. canaliculatus* (Dickson 2022). Similarly, the National Mariculture Center (NMC) in Bahrain successfully completed the mass propagation of *S. canaliculatus*, among other species, and became a regional finfish seed supply hub for the Gulf countries, with consumer responses indicating a preference for cultured over wild rabbitfish (Uwate and Shams 1999). In Saudi Arabia, the life cycle of *S. rivulatus* was successfully closed in the early 2000's and seed supply became available without the need for wild juvenile collection (Bukhari 2005). Production of *S. rivulatus* in floating cages is reported in Saudi Arabia although with little details (Saoud and Ghanawi 2007).

Africa

In Africa, *S. sutor* is one of the most economically important marine fishes captured along the coasts of Tanzania, Kenya, Mozambique, Seychelles, Mauritius and Rodrigues (Grandcourt and Cesar 2003, Bijoux et al. 2013). The sustained catches despite relatively heavy fishing effort may have hampered attempts to culture rabbitfishes in this region, although there was some early investigation into the culture of *S. sutor* (previously mis-identified as *S. canaliculatus*) in floating cages in Tanzania revealed high culture potential for the species in the region (Bwathondi 1982). Egypt has one of the oldest and most developed aquaculture industries, and the highest aquaculture production of all African countries (Dadzie 1992). *Siganus rivulatus* is the main species of rabbitfish cultured in Egypt, with considerable investigation into the influence of dietary substitutes and supplements on the growth, survival and cost of rearing juveniles (e.g., El-Dakar et al. 2007, Dawood et al. 2014).

Mediterranean

Rabbitfish, although only relatively recent invaders of the Mediterranean Sea, are important components of commercial wild harvest fisheries in Lebanon (Nasser et al. 2019), Cyprus (Stephanou and Georgiou 2000) and Israel (Ben-Tuvia et al. 1972). While demand in these countries is high, catches are declining, making rabbitfish economically valuable and interesting to aquaculturists (Stephanou and Georgiou 2000, Hamze et al. 2014). On a commercial level, hatchery production of rabbitfish in the Mediterranean only exists in Cyprus, where *S. rivulatus* fingerlings are supplied to a private company by the Meneou Marine Aquaculture Research Station (MeMARS) for grow-out in off-shore sea cages (Anastasiades 2010). Research on the aquaculture of *S. rivulatus* was conducted for more than 15 years with focus on the reproduction and larval rearing (Stephanou and Georgiou 2000). *Siganus rivulatus* aquaculture in the Mediterranean has the potential to become an alternative to high trophic level species such as seabream and seabass, allowing to reduce the environmental costs as well as the dependence on small pelagic fish products for feeding.

Advantages and Limitations of Rabbitfish Aquaculture

Rabbitfish as Commercial Aquaculture Candidates

Rabbitfishes possess many characteristics and traits that make them amenable to culture practices, namely their rapid initial growth, high fecundity, tolerance to a range of environmental characteristics (temperature and salinity, Hoey et al. – Chapter 7), and a predominantly herbivorous diet. Additionally, they are esteemed food fishes and yield a high market value in many regions (Ignatus 2009, Anastasiades 2010, Roumieh et al. 2013). They also contain high concentrations of some micronutrients, such as arachidonic acid (ARA), docosahexaenoic acid (DHA) and long-chain polyunsaturated fatty acids (LC-PUFA), which provide essential benefits for human health, particularly in child development (Ogata et al. 2004, Jiarpinijnun et al 2017, Li et al. 2018).

Rabbitfishes are generally considered to be primary consumers (Hoey et al. 2013, Zarco-Perello et al. 2024, Clements – Chapter 4), but show omnivorous feeding habits in captivity and readily accept many food items of vegetable and animal origin (Lam 1974, Darsono 1990, Yousif et al. 2004, Ignatus 2009). This is a significant advantage for aquaculture as they likely require less protein when compared to carnivorous candidate species (Tacon et al. 1990), thus reducing their feed costs and alleviating pressure on stocks of wild forage fish (Stephanou and Georgiou 2000, Rachmansyah et al. 2007, El-Dakar et al. 2011). Most rabbitfishes can tolerate a wide range of temperature and salinity in their natural habitat (Ignatus 2009, Jaikumar 2012, Seale and Ellis 2019), which is advantageous under aquaculture conditions. Several rabbitfish species have been successfully reared in either brackish or in marine waters (Gundermann et al. 1983, Saoud et al. 2007, Anuraj et al. 2021), can survive extended periods of low oxygen concentrations (Gundermann et al. 1983), and tolerate relatively high concentrations of nitrite (Saoud et al. 2014) and ammonia (Roumieh et al. 2012). Ultimately, they are hardy fishes known to adapt well to captivity and to be tolerant of crowded conditions, with stocking densities of up to 100–250 fish per cubic metre being recommended for some species (Abalos 1999, Bukhari 2005, Saoud et al. 2007, Jaikumar et al. 2011, Syah et al. 2020). Their tolerance of high-stocking densities could be due to their gregarious nature, especially among naturally schooling species.

Their non-aggressive feeding behavior and lack of territoriality also make them ideal candidates for co-culture and polyculture (Seale and Ellis 2019). Being herbivorous they also tend to reduce fouling on underwater infrastructure, and juvenile rabbitfish are often co-cultured in grouper cages to reduce algal fouling on the nets (van Beijnen et al. 2022). Co-culture trials conducted in shrimp ponds showed that the inclusion of rabbitfish led to more stable water quality, thereby increasing the growth performance, productivity and profitability of shrimp culture, and improved utilization of supplied feed while minimizing pollution of coastal waters caused by pond effluents (Luong et al. 2017). In the Philippines, rabbitfishes are co-cultured with milkfish and shrimp (Tacon et al. 1990). In China, cultures of rabbitfish have developed from a monoculture mode to a polyculture mode and they are now commonly cultured along prawns (*Penaeus monodon*), crabs

(*Scylla paramamosain*) or finfish (*Chanos chanos*; *Sparus latus*; *Epinephelus* spp.; Li et al. 2018). Finally, large numbers of juvenile rabbitfish can be collected from coastal waters during seasonal recruitment events, allowing for vast capture-based aquaculture operations to be developed.

Many of the inherent biological traits of rabbitfishes facilitate their successful aquaculture. For example, rabbitfish are extremely fecund, with estimates ranging from 245,000 to 1.2 million eggs in large female individuals (Popper and Gundermann 1976, Rachmansyah et al. 2007, Malugao 2019). Collectively, their relative investment in gonads during spawning (i.e., gonado-somatic index) exceeds most other tropical reef fishes (Taylor et al. – Chapter 5). Rabbitfishes are also lunar-synchronized spawners (Park et al. 2006), which makes possible for aquaculture farmers to anticipate the timing of spawning, a significant advantage to prepare larval rearing. In addition, rabbitfishes readily reproduce in captivity, whether spontaneously or when stimulated by hormonal treatments. The first successful reported hormone injections were conducted on *S. rivulatus* and *S. argenteus* (Popper et al. 1979), while natural spawning in captivity has been achieved in other species (Lam 1974). Year-round spawning in captivity has been achieved for various species (Hara et al. 1986) and off-season spawning has been achieved and successfully repeated using environmental manipulation (Stephanou and Georgiou 2000). Successful spawning in *S. argenteus* was also achieved by lowering the water level of the broodstock tank (Burgan and Zseleczky 1979).

Rabbitfish are gonochores, whereby sex differentiation happens early and remains fixed through life. Having fixed sex ratios is advantageous for broodstock as hermaphrodites (common in tropical fisheries) require maintenance of spawning ratios. While generally they have no visual markers differentiating sexes, sex of rabbitfish can be easily distinguished during spawning seasons as females are typically larger than males and have enlarged abdominal regions on account of the ripening ovaries (Tseng and Chan 1982, Ignatus 2009). Finally, despite their limited size for most species, the growth rates of rabbitfish compare favorably with those of other commercially farmed fishes (Seale and Ellis 2019) and commercial size can be achieved within a year.

Drawbacks of Rabbitfishes for Commercial Aquaculture

Given the long history of rabbitfish aquaculture and the list of favourable characteristics above, one may wonder why this practice is still considered as emerging, with a low worldwide production insufficient to warrant their own category in annual FAO reports. The main limitation hampering the development of rabbitfish aquaculture is the difficulty in producing hatchery-reared fry using reliable and repeatable methods (Lam 1974, Anastasiades 2010). Since the *Siganid Mariculture Implementation Conference* was held in Hawaii in 1973, many attempts have been made to rear larvae to metamorphosis with only few successes reported (Bryan et al. 1974, Bryan and Madraisau 1977, Juario et al. 1985, Stephanou and Georgiou 2000, Ignatius 2009, Anuraj et al. 2021, Tan 2021, Darsiani et al. 2022). Hence, successful rabbitfish aquaculture practices have predominantly occurred through the grow-out of wild caught juveniles (i.e., capture-based aquaculture).

Difficulties in larval rearing of rabbitfish are mostly linked to their relatively small mouth gape (100–125 μm) at first feeding (Ignatius 2009). This mechanical limitation significantly reduces the types of live prey suitable for their first feeding and renders their larval rearing much more complicated than most of the other established aquaculture marine finfish species. A greater understanding of the nutritional requirements of rabbitfish larvae is also warranted, as rotifers, even when enriched, are known to be sub-optimal prey items to various larvae of marine finfish species (Camus 2012). The final issue with closing the life cycle is that ripe eggs of rabbitfishes are naturally adhesive (the only exception being *S. argenteus*), a property that renders their collection for counting and incubation very difficult (Ignatius 2009). Egg adhesiveness is found to increase with decreasing salinity regimes, such that at 12 ppt they can barely be removed without damage (von Westernhagen and Rosenthal 1975). Several workarounds have been developed, such as introducing a removable structure (e.g., pieces of shade cloth) in the broodstock tank during spawning periods and to remove them after spawning to collect the eggs. Nevertheless, this feature makes the process more difficult and costly.

There are other biological drawbacks to successful rabbitfish aquaculture. First, food conversion efficiency of rabbitfishes is not particularly good, and generally falls short of that of other cultured fishes. This is not surprising as feed conversion in herbivores is generally low, whether feeding on naturally occurring or artificial diets (von Westernhagen and Rosenthal 1976). Secondly, rabbitfish undergo early sexual maturation and can mature faster in captivity than in nature, with some species able to spawn by three months of age and at length as little as 110–120 mm (Lam 1974, Tharwat 2004). Growth is often reduced upon reaching sexual maturity, as dietary energy is diverted from somatic growth to sexual development (von Westernhagen and Rosenthal 1975). This may not be an important drawback for rabbitfish aquaculture as commercial size is relatively small (100–150 g) and growth rate of sexually mature fish are still considered compatible with commercial aquaculture. Third, most existing rabbitfish aquaculture activities rely on commercial diets formulated for other species (e.g., carp, sea bass, shrimps). This can be problematic as commercial farms using sea bass feeds tend to produce very fatty rabbitfish that are not readily accepted by consumers (Abou-Daoud et al. 2014). To the authors' knowledge, only one commercial diet formulated specifically for rabbitfish exists (Tateh Aquafeed®), available in the Philippines.

In addition, few studies exist regarding disease management of rabbitfishes in aquaculture settings. Captive rabbitfishes develop various pathologies including exophthalmia, cataracts, nodule infections, abnormal coloration such as whitening of the roof of the skull, bloated abdomens, parasitic infection by trematodes on gills, body lesions and fin rot (Lam 1974, Carumbana and Luchavez 1979). In cage culture, mass mortality of *S. canaliculatus* in Singapore was attributed to a gram-positive bacterium similar to *Streptococcus faecium*, and an outbreak of *Pseudomonas purrefaciens* was reported in *S. rivulatus* cultured in the Red Sea (Carumbana and Luchavez 1979). The importance of rigorous studies of diseases of rabbitfishes and their control under standardized conditions was urged fifty years ago by Lam (1974) but are still lacking. Finally, rabbitfish are difficult to handle due to their venomous spines, and although painful the damage inflicted is not considered dangerous to human adults (Seale and Ellis 2019). Rabbitfishes usually spread their dorsal and ventral spines when they are frightened, and can become entangled in the nets during catching operations (von Westernhagen and Rosenthal 1976).

Rabbitfish are considered invasive species in some region of the world like the Mediterranean and precautionary measures need to be put into place to mitigate the potential impact of escapees from aquaculture facilities on local fauna and flora. Escaped rabbitfish may have deleterious effects on natural habitat, for example by increasing the herbivorous pressure on algae beds or affecting fragile habitats such as seagrass meadows (Sanchez-Jerez 2013). Fertilized eggs spawned by broodstock can also be considered as a potential vector for introduction of farmed fish and disease into marine ecosystems (Sanchez-Jerez 2013).

Ultimately, it is clear that rabbitfish possess a unique aquaculture potential conferred by many of their biological characteristics, but the inability to successfully close the life-cycle may limit application to areas where wild-capture based practices can take advantage of seasonal recruitment runs of commercial species. As a result, commercial aquaculture of rabbitfish is still largely based on the capture and subsequent grow out of wild juveniles (*capture-based aquaculture*) with relatively few operations using hatchery-reared juveniles (*hatchery-based aquaculture*) (Fig. 3).

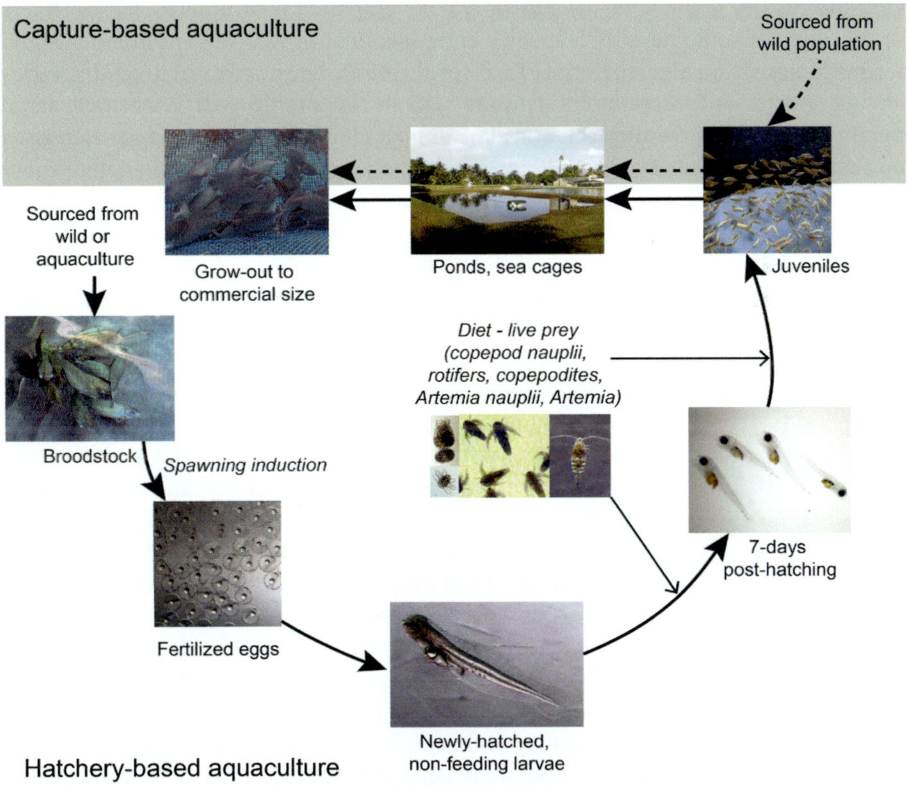

Fig. 3. Production cycle for the aquaculture of rabbitfishes (f. Siganidae) highlighting the major differences between *capture-based* (grey shading and dashed lines) and *hatchery-based* (solid lines) methods. Images: T. Camus

Meanwhile, the increasing demand for rabbitfish is promoting their overfishing in many regions (Wahyuningtyas et al. 2017, Tresnati et al. 2020), and some populations are being impacted by habitat degradation and/or loss (Hoey et al. – Chapter 7). For these reasons, developing hatchery-based aquaculture of these species has become more urgent than ever before.

Capture-Based Aquaculture of Rabbitfishes

Although rabbitfish have long been identified as appropriate aquaculture candidates, hatchery-based production is still problematic and thus the majority of their aquaculture is capture-based (Villanueva et al. 2021). This method has the advantage of not requiring facilities, technology, or economical investment, so rural or peri-urban areas are well-suited to the grow-out of wild caught juveniles, whether in floating cages, pens or ponds (Teitelbaum et al. 2008, Seale and Ellis 2019; Fig. 3). Capture-based aquaculture is also a fast technique as rabbitfish can reach a market size of 100 g in 6–7 months from wild caught juveniles (Tacon et al. 1990, Saoud and Ghanawi 2007), and sometimes less than three months in pond culture (Nelson et al. 1992).

Capture-based aquaculture of rabbitfishes is facilitated by an important feature of their life-history in the wild: after the larval development is completed, large schools of juveniles of some rabbitfish species enter lagoonal or reef flat environments and feed voraciously on algal material (Tacon et al. 1990). The number of juveniles varies among species and recruitment seasons, and in favourable years extremely large numbers of young juveniles can consume almost all of the seaweeds and seagrasses on reef flats (Thacker et al. 1997). It is relatively easy for fishers to capture these juvenile rabbitfishes using cast-net or large landing nets, as the rabbitfishes tend to spend several weeks in dense patches in shallow coastal environments (Tobias 1976). Large numbers of rabbitfish juveniles can be collected during these periods before being transferred to a grow-out facility such as ponds, tanks, or floating cages and grown with relatively cheap labour cost (Tharwat 2004).

Despite the relative simplicity of capture-based aquaculture of rabbitfishes, many aspects of this culture method are problematic. While the timing of recruitment (i.e., period of the year and moon phase) is predictable, the quantity (magnitude) of recruiting juveniles is stochastic, varying considerably from one year to the next (Soliman et al. 2010). Natural recruitment can be very low some years, whereby only very small quantities of juveniles—and sometimes none—are available for capture. In the absence of hatchery-based alternatives, no aquaculture development can take place, as was the case in Guam in 1973 (Tsuda et al. 1974). Predicting trends of recruitment has proven difficult as yearly fluctuations are observed but data sets spanning many years show unclear patterns to these fluctuations (Tobias 1976).

The capture of wild rabbitfish juveniles can also be problematic as it will negatively impact future recruitment, leaving fewer individuals to establish themselves in the local environment and spawn in subsequent years. This has been observed in the Philippines as once abundant rabbitfish fry became scarce in some regions because of overfishing and a lack of management (Caballero et al. 2022). Similarly in Palau, anecdotal reports indicate a significant reduction in rabbitfish schools between 1930 and the 1970s that were attributed to destructive fishing practices (Hasse et al. 1977).

The collection of juvenile rabbitfishes *en masse* from the natural environment may also disrupt the flow of energy and balance of natural ecosystems. Reducing the number of rabbitfish juveniles may increase the standing biomass of algal assemblages, while also reducing the availability of potential prey for piscivorous fishes. Capture-based aquaculture of juvenile rabbitfish can also lead to conflict and competition among local residents and fishers in regions like the Philippines or Guam, where juveniles are considered delicacies and sold at high prices as fresh, paste, or pickled in jars (May et al. 1974, Tsuda et al. 1974).

The classical method used to culture capture-based rabbitfish juveniles is described at length in a free manual recently published by the SEAFDEC (Caballero et al. 2022). The manual details the major steps required for nursery and grow-out culture of *S. guttatus* although it could likely be applied to other rabbitfish species. It was published with the objective of sharing knowledge with stakeholders from the private sector, academia, and government agencies. Overall, the only definitive solution to the many problems linked to the capture-based approach is to develop hatchery-based culture and rearing of rabbitfish larvae. Only through this approach would it then be possible to offer a continuous and sufficient supply of hatchery-bred rabbitfish larvae to support grow-out facilities, and hence market demand, while reducing dependence on wild-caught juveniles and promoting sustainable aquaculture practices (May et al. 1974, Kongkeo et al. 2010, Tan 2021). Yet, until hatchery breeding is fully developed, fish farmers will continue to favour capture-based aquaculture of rabbitfishes despite the disadvantages.

Hatchery-Based Aquaculture of Rabbitfishes

Hatchery-based aquaculture offers more control over production and alleviates uncertainties around the magnitude of recruitment events when compared to the capture-based approach, however it necessitates significant financial and time investment into the development and refinement of rearing techniques as well as the establishment of appropriate rearing facilities (Fig. 3). This approach would likely be difficult to implement in remote regions such as some Pacific islands or in rural communities far removed from industrialized areas. Hatchery development of rabbitfish shows high promise for servicing niche export markets as well as local markets while sustaining market demand and decreasing fishing pressure (Carumbana and Luchavez 1979, Teitelbaum et al. 2008).

History

Attempts to breed rabbitfish in captivity were first reported by Manacop (1937) and by Fujita and Ueno (1954) but it was not until the early 1970's that a cooperative effort was initiated between the Hawaii Institute of Marine Biology and the Micronesian Mariculture Demonstration Center (MMDC) in Palau to advance hatchery-based aquaculture of rabbitfishes (May et al. 1974). The first successful rabbitfish larval rearing reported in the literature was conducted at the MMDC using a green water technique (May et al. 1974). Following this pioneering work, several species of rabbitfish have been reared in hatcheries throughout several regions of the world, albeit with varying success. For example, the first larval rearing trials of *S. guttatus* were conducted in the Philippines in the early 1980s with highly variable survival rates reported at metamorphosis (Juario et al. 1985). Since these early attempts, there has been continued efforts to develop rabbitfish hatchery protocols (Teitelbaum et al. 2008) and today, mass production of rabbitfish fry has been possible. However, low survival rates during the early stages of development are common (Cabanilla-Legaspi et al. 2021). Despite the progress, it is still in its nascent stages of development for some targeted species such as *S. sutor* (Fourooghifard et al. 2020).

Broodstock Conditioning and Spawning

For successful commercial-scale hatchery-based production of rabbitfishes, broodstock should produce an adequate volume of good quality eggs year-round to support hatchery operations and offset the significant investments required (Anastasiades 2010). Rabbitfish are known to spawn naturally in captivity and hormonal treatment may not be necessary if spawning is attempted on the right days of the lunar month and optimal or adequate nutrition is provided (Darsono 1990).

Several studies have established that lunar periodicity is the major factor in stimulating rabbitfish reproductive activity, although it is still not well understood what proximal cues are used by the fish (Rahman et al. 2003, Park et al. 2006). It is advantageous for rabbitfish broodstock to be maintained outdoors with natural photoperiods and no nocturnal light pollution as this will favour year-round spontaneous spawning, providing adapted feeding is provided. By contrast, broodstock maintained indoors will tend to lose their lunar synchronicity, resulting in haphazard timing of spawning (T. Camus pers. obs.).

Rabbitfish can be easily induced to spawn using spawning agents such as human chorionic gonadotropin (HCG; Cabanilla-Legaspi et al. 2021) and to a lesser extent luteinizing hormone-releasing hormone (LHRHa; Duray 1998) or a pituitary like hormone (Bryan et al. 1974). Induced spawning has been achieved in multiple rabbitfish species, including *S. randalli* (HCG at a dose of 1 IU/g; Nelson et al. 1992), *S. rivulatus* (HCG at a dose of 250 and 500 IU/Kg; Popper et al. 1979), *S. argenteus* (HCG at a dose of 250 IU/kg; Popper et al. 1979), and *S. guttatus* (HCG at a dose of 2 IU/g; Georgiou and Stephanou 1991).

Larval Rearing and Limitations

After spawning, fertilized eggs develop rapidly and larvae generally hatch within 24 hours. Newly-hatched larvae of rabbitfishes are generally small (1–2 mm), possess a small yolk sac containing one to several lipid droplets and lack a functional mouth or pigmented eyes (Popper et al. 1973). The larvae tend to orient themselves head down in the water column, frequently wriggling downward in short bursts, with sinking velocity strongly dependent on salinity (von Westernhagen and Rosenthal 1975). Yolk resorption occurs within 1–2 days while lipid droplets can last up to 6 days depending on egg quality and temperature.

A critical period in which most of the larval mortality occurs is the transition from the yolk-sac stage to first feeding, and mortality during this stage must be reduced in order to achieve the production levels required to support commercial aquaculture (Nelson et al. 1992). Appropriate timing to initiate feeding of rabbitfish larvae is crucial (Darsiani et al. 2022), and failure to provide adequate live food of a suitable size to larvae by 3 days post-hatching will result in high mortality as endogenous energy reserves become depleted (T. Camus pers. obs.).

Initial larvae feeding usually consist of rotifers, with *B. plicatilis* being the most commonly used. However, the average size of *B. plicatilis* is generally too large to be properly ingested by first feeding larvae of most rabbitfish species. A workaround is to sieve *B. plicatilis* cultures so that only the smallest size-classes (80–90 µm: Nelson and Fitzgerald 1990, and < 125 µm: Juario et al. 1985) are offered to the larvae. This is a considerable issue as only a small proportion of the *B. plicatilis* population constitutes this size group, mainly consisting of newly hatched individuals. Alternatives consist in feeding the larvae using a super small (SS) strain of rotifer, such as *B. rotundiformis*, although this strain is not available in all regions of the world and is more complicated to culture than the standard strain. Still, even SS rotifers were found to be too large for *S. rivulatus* larvae to ingest (Stephanou and Georgiou 2000). Copepod, and more specifically their naupliar stages, have an excellent track record for being suitable prey for first feeding of marine fish larvae with small mouth gape (Camus 2012), including rabbitfish (Anuraj et al. 2021). Despite early suggestions that copepods could be easily cultured (Palau: May et al. 1974), subsequent attempts failed (Bryan and Madraisau 1977). The ability to culture copepods in quantities compatible with commercial hatchery practices remains problematic. Clearly, future research is needed to develop reliable large-scale copepod cultures (Nguyen et al. 2015), and/or the production of alternate prey of a suitable size for the initial feeding of rabbitfish.

Weaning and Metamorphosis

Cultured larvae metamorphose into fry at 18 to 26 mm total length (Gundermann 1983). The process of metamorphosis includes changes in coloration, behavior, and feeding habits: as the body colour turns from an opaque grey to brown, the larvae spend more time feeding on algae and less time swimming in search of live food, hence changing from a carnivorous to a herbivorous diet (Bryan and Madraisau 1977). These changes occurring at and around metamorphosis may be separated into three stages: (i) the dark-head carnivorous stage (pre-metamorphosis), (ii) the brown-head omnivorous stage (concurrent with metamorphosis), and (iii) the juvenile herbivorous stage (post-metamorphosis) (Bryan and Madraisau 1977).

Captive rabbitfishes are easy to wean from larval to juvenile diets as they readily accept artificial feeds with little or no training (von Westernhagen and Rosenthal 1975, Tacon et al. 1990, Seale and Ellis 2019). Cannibalism is not considered an issue during larval rearing, but post-metamorphosis rabbitfishes tend to displayed distinct territorial behavior, and larger individuals can chase and bite smaller individuals to the extent that there have been reports of caudal fins of smaller individuals being bitten down to the peduncle (von Westernhagen and Rosenthal 1975).

Diet and Formulated Feeds

Optimal Dietary Content for Rabbitfishes

The natural dietary composition for rabbitfishes at different developmental stages are still largely unknown which hinders the development of appropriate artificial feeds. Gonzales et al. (2018) state that although some studies were conducted on the nutritional requirements, digestive systems, and feeding habits of rabbitfishes in the Philippines, there is still no commercial diet specifically formulated for rabbitfishes available. That said, there is a commercial feed available in the Philippines (Tateh® aquafeed – first produced in 2010) but with very little detail regarding its composition.

Crude Protein Requirements

The first step in formulating an artificial diet is knowledge of the nutritional requirements of the cultured species, and first and foremost its crude protein requirements, which are predominantly used for growth and maintaining good health (Monzer et al. 2017). Protein requirements are typically species- and age-specific, with larvae and juveniles having higher requirements than adult fish (Anastasiades 2010). Other factors include water temperature, feeding rate, dietary energy level, genetic inheritance, digestibility, and amino acid composition of the dietary protein (Abou-Daoud et al. 2014). Although herbivorous fish, such as rabbitfish, are assumed to have a lower protein requirement than carnivorous fish, studies to date suggest that the protein requirements of rabbitfishes is broadly comparable to that of carnivorous fishes (Abou-Daoud et al. 2014).

The protein requirements of *S. rivulatus* have been the focus of considerable research, with most studies indicating optimal dietary protein of *ca.* 40% for juveniles across a range of body sizes. For example, a diet of 40% protein (and 6% lipid) led to optimal growth and digestive functions in very small (< 1 g) *S. rivulatus* juveniles (Salem et al. 2021). Similarly, growth in small (1.49 ± 0.2 g) and medium-sized (7.4 ± 0.1 g) juveniles was optimized when fed diets with 40% protein (digestible energy 16–18 MJ/Kg; El-Dakar et al. 2011) and 40% crude protein (14.2 MJ/kg; Abou-Daoud et al. 2014), respectively. Finally, Anastasiades (2010) reported no significant difference in growth rates when *S. rivulatus* was fed with feed of 35% or 44% protein content, although the experimental feeds were not isoenergetic. Studies have recommended a broadly similar protein content for optimizing growth of juvenile *S. guttatus* (weaning of post-larvae: microparticles with 40% crude protein, Juario et al. 1985; young juveniles [< 1 g]: 35% protein [3882 kcal/Kg energy], Parazo 1990; fry: 33% protein, 10% fat and 20% ash [3700 kcal/Kg energy], Soletchnik 1984 in Boonyaratpalin 1997).

Information of the dietary protein content for later life-stages of cultured rabbitfishes is limited. Growth and feed efficiency of sub-adult (initial weight 48–68 g) *S. canaliculatus* was optimal when fed a dry crumble containing 31% dietary crude protein (Tacon et al. 1990). However, it is unclear whether this lower protein content was related to life-stage or was specific to this species. Similar and slightly lower levels of dietary protein (25–30%) were found to limit growth of *S. argenteus* over 174 days (Teitelbaum et al. 2008).

Lipid Requirements

Dietary lipids are major energy providers and often used to supplement dietary protein for growth (Monzer et al. 2017), with optimal lipid requirements for rabbitfish being suggested to be *ca.* 6–14% (Li et al. 2018). Optimal dietary lipid requirement for growth of wild-caught juvenile (2.9 g ± 0.01) *S. rivulatus* was 98 g/Kg, considerably lower than the requirements of cultured carnivorous species (Ghanawi et al. 2011). Zhu et al. (2013) reported an optimal dietary lipid level of 13.94% for larger (28.09 ± 0.60g) *S. guttatus*.

Plant-Based Substitutes for Rabbitfish Feeds

Globally, aquaculture uses an estimated 75 million tonnes of artificial feed each year, amounting to approximately 250–300 million tons of fresh ingredients, principally in the form of fish meal (Nasser et al. 2019). In 2015, an estimated 332 tonnes of rabbitfish were produced via aquaculture, necessitating approximately 530 tonnes of commercial feeds to be manufactured (Nasser et al. 2019), although this has since declined to approximately 154 tonnes of rabbitfish in 2021 (Fig. 1).

Proteins in aquafeeds are predominantly sourced from fish meal, itself sourced from pelagic fish by-catch, forage fish (e.g., herring, sardines, anchovy), as well as herring and salmon processing (Monzer et al. 2017). Fish meal is rich in highly digestible proteins, rich in omega-3 poly unsaturated fatty acids, and is an important source of essential vitamins and minerals. These properties have made fish meal the benchmark ingredient in artificial diets for the aquaculture industry (Assaf 2018). However, fish meal is not a sustainable resource due to the relatively static supply of wild forage fishmeal coupled with its increase in demand (Ciani et al. 2020) and now it is estimated that about a quarter of all fish landed globally is destined for the fish meal industry, causing a significant pressure on wild fish stocks (van Beijnen et al. 2022).

In addition to the use of unsustainable ingredients (i.e., fish meal), artificial feeds are very expensive, generally representing upwards of 50–80% of operating costs, with fish meal being the major contributor to feed cost (El-Dakar et al. 2011, Ciani et al. 2020). Increasing costs coupled with uncertainties regarding the availability and unsustainable nature of fishmeal and fish oil, have been identified as limiting the development of aquaculture (Anastasiades 2010, Xu et al. 2012). The replacement of fish meal with alternative plant-based (or algal-based) ingredients is seen as a viable alternative for the culture of herbivorous species (Assaf 2018), and is of paramount importance for the future of aquaculture.

Plant-based Protein Substitution

The formulation of artificial feeds that are nutritive, palatable, inexpensive, and environmentally sustainable, while yielding good performance is a major challenge facing marine aquaculture today (Li et al. 2018). Rabbitfishes, being herbivorous, can potentially be reared using low protein diets under captivity (Abou-Daoud et al. 2014), reducing their environmental impact and nitrogen footprint (Roumich et al. 2012, Syah et al. 2020). The natural diet of rabbitfish, based largely on phototrophs (i.e., algae, cyanobacteria, and seagrasses; Hoey et al. 2013, Clements – Chapter 4), and the fact that they readily feed on leaves of terrestrial plants and vegetables in culture conditions, opens up the potential to successfully replace fish meal and fish oil with proteins and lipids of vegetable origin (Carumbana and Luchavez 1979, Anastasiades 2010). Moreover, replacing fish meal with plant-based materials and low-cost by-products may promote the development of small-scale aquaculture operations in remote areas (Teitelbaum et al. 2008).

Plant-based Materials

Various plant-based meals such as soybean, canola, cottonseed, sunflower, lupin and linseed as well as various seaweeds (or macroalgae) have been successfully used to completely replace or to partially substitute fish meal in aquaculture feeds. Rabbitfishes are an ideal candidate to test the efficacy of using plant-based substitutes.

Macroalgae are a natural dietary component of rabbitfishes. It is perhaps not surprising that macroalgae have been identified as a potential alternative source of proteins and lipids and in doing so developing low-cost but effective formulated feeds for rabbitfish (Li et al. 2018, Sulaeman and Laining 2022). The benefits of incorporating macroalgae into aquaculture feeds likely extends beyond the cost savings, and may facilitate the recycling of low-value macroalgae and/or promote the development of macroalgae farming (Li et al. 2018). However, the nutritional composition of macroalgae can vary markedly, such that the same species may have different effects, such as targeting different innate immune responses, both among and within fish species (Thepot 2021).

The partial or complete replacement of fish meal with plant-sourced proteins is a challenging task. The inclusion of macroalgae in artificial diets might negatively impact fish performance due to deficiencies in their amino acid profile or the presence of potentially high levels of carbohydrates and anti-nutritional factors, thereby decreasing digestibility and assimilation by the fish (El-Dakar et al. 2007, Anastasiades 2010, Dawood et al. 2014, Laining et al. 2016, Assaf 2018). For example, a diet based on a mixture of *Enteromorpha* and the plant Leucaena (*Leucoce phala*) in pellet form reduced the grow of *S. argenteus* and *S. spinus* by 25 to 30% when compared to fish fed traditional diet consisting of trout chow on (Tsuda et al. 1974).

Yousif et al. (2004) compared four isonitrogenous (39% crude protein) and isoenergetic (19 kJ/g) diets with various rates of dehydrated *Enteromorpha* incorporated (0, 10, 20 and 30%) and a fifth diet of fresh *Enteromorpha* on the growth and composition of *S. canaliculatus* fry (0.25 ± 0.04 g) for 12 weeks. The highest survival, growth performance and feed utilization efficiency were realized in *S. canaliculatus* fed the control diet (0% inclusion) and fresh *Enteromorpha*, and declined with increasing inclusion levels of dehydrated *Enteromorpha*. The body moisture and protein content of the cultured fry were not influenced by diet, however, the lipid and energy content were higher in the fish fed the control diet and fresh *Enteromorpha* (Yousif et al. 2004).

Several studies in China have also investigated the efficacy of incorporating plant- and algal-based materials into feeds on the growth performance of *S. canaliculatus*. For example, the growth of *S. canaliculatus* was 3- to 4-fold greater when fed a formulated diet compared to the green alga *Enteromorpha prolifra*, or the red alga *Gracilaria lemaneiformis* (Xu et al. 2014 in Li et al. 2018). Similarly, the inclusion of 5–15% of *E. prolifra* in feeds led to relatively small reductions in growth of *S. canaliculatus*, however these reductions were partly offset with the addition of non-starch polysaccharide enzymes (Zhou et al. 2013). Perhaps the most promising plant-based material in feeds for rabbitfish is the use of soybean oil as a partial substitute for fish oil. Growth, condition and protein efficiency ratio of *S. canaliculatus* was greatest when fed a diet in which 45% of the fish oil was replaced with soybean oil, compared to diets with 0%, 23% or 67% of fish oil replaced with soybean oil (Xu et al. 2012 in Li et al. 2018).

More targeted studies to understand natural diets (e.g., Hoey et al. 2013) and feeding preferences when offered a range of algae (e.g., Pillans et al. 2004) are needed to advance our understanding of algal preferences of rabbitfish species, allowing for a more efficient use of macroalgae in aquafeed and provide a foundation for the development of low-cost and effective formulated feeds for rabbitfish (Xie et al. 2018).

Perspectives for the Sustainable Aquaculture of Rabbitfishes

As the world faces an environmental crisis and the demand for fish production rises with increasing human population, improving the sustainability of aquaculture is of major importance. Aside from implementing more efficient means of farming fish, consideration needs to be given to shifting emphasis from tertiary consumers (i.e., high order carnivores) to primary consumers (i.e., herbivores) (Abou-Daoud et al. 2014, van Beijnen et al. 2022). Aquaculture of marine herbivores has enormous potential to feed the growing population while lowering the environmental footprint when compared to traditional animal production sectors.

Rabbitfish exhibit significant potential for promoting sustainable growth in marine aquaculture. This potential stems from their herbivorous/omnivorous feeding behaviour, together with their tolerance of a wide range of environmental conditions (Tacon et al. 1990). Yet, the development of rabbitfish aquaculture has been hindered primarily by the limited and/or unpredictable supply of wild-caught fry, the challenges of larval rearing (especially the supply of live prey of a suitable size), and a dearth of comprehensive information regarding commercial production protocols (Ghanawi et al. 2010). Furthermore, inadequate investments in hatcheries are anticipated to sustain low production outputs for marine species, including rabbitfishes (Ortega 2021). This scarcity of research and development, particularly in the realm of larval rearing methodologies, has resulted in rabbitfish aquaculture around the world being largely reliant on capture-based methods.

Capture-based aquaculture of rabbitfishes is relatively easy to perform as it does not require specialised skills or equipment, however the quantity and quality of production are unpredictable due to the stochastic nature of rabbitfish recruitment. Perhaps more importantly, capture-based aquaculture of rabbitfishes is not a sustainable practice (see discussion above). Future research aimed at closing the life cycle of rabbitfish species is urgently required to achieve a continuous and reliable supply of quality hatchery-reared juveniles, which would encourage fish farmers to develop rabbitfish culture (Rachmansyah et al. 2007).

While notable progress has been achieved in broodstock conditioning and spawning induction, research on larval rearing methods remains comparatively limited, impeding hatchery production of most rabbitfish species due to persistently low larval survival rates (Darsiani et al. 2022). A pivotal juncture during the larval rearing of rabbitfish is the initial feeding of the larvae, owing to their small mouth gape (Rachmansyah et al. 2007, Malugao 2019, Anuraj et al. 2021). Numerous studies have demonstrated that traditional live prey organisms such as rotifers and Artemia, when used in isolation, yield markedly low larval survival rates (Stephanou and Georgiou 2000, Tan 2021). Therefore, the advancement of capabilities in the production of alternate live feeds (e.g., copepods) is likely to be a linchpin for improving survival rates in forthcoming larval rearing experiments.

Beyond the realm of larval rearing, future endeavours to quantify the nutritional requisites of rabbitfish at various life stages is needed. This entails the testing of macroalgae and other phototroph species (e.g., cyanobacteria) as potential sources of dietary protein. However, concurrent research must address the potential adverse effects of anti-nutrient factors frequently associated with macroalgae consumption. These efforts collectively aim to formulate species-specific and stage-specific feeds tailored to rabbitfish aquaculture, enhancing sustainability and curbing feeding expenses.

Incorporating rabbitfish into polyculture is also practicable, and likely to yield high economic returns (Li et al. 2018). Polyculture can enhance water quality stability, increase productivity and profitability, improve use of supplied feed, and minimize pollution of coastal waters by reducing pond effluents (Luong et al. 2017, Verdian et al. 2020). Rabbitfish are promising candidates for polyculture, although trials are needed to assess their potential in this culture method.

In sum, rabbitfishes are an ideal candidate for aquaculture and their potential has been long recognized (e.g., May et al. 1974). Despite their potential, the aquaculture of rabbitfishes has remained relatively low (Fig. 1), and likely reflects the challenges of developing successful larval rearing and hence reliable supply of juveniles to grow-out facilities. Developing reliable larval culture protocols and low-cost and sustainable feeds will be critical to the expansion of the eco-friendly aquaculture of a high valued marine species.

References

Abalos, R.S. 1999. Growth and yield of rabbitfish (*Siganus guttatus*) reared in river floating net cages at various stocking densities and feeding schemes. PCAMRD Book Series (Philippines).

Ablan, G.L. and W.M. Rosario. 1961. Teuthid fish for marine culture in the Philippines. Fish. Gaz. Philipp 5: 23–24.

Abou-Daoud, Y., J. Ghanawi, M. Farran, D.A. Davis and I.P. Saoud. 2014. Effect of dietary protein level on growth performance and blood parameters of marbled spinefoot *Siganus rivulatus*. J. Appl. Aquaculture 26: 103–118.

Anastasiades, G. 2010. Biodiversity in eastern Mediterranean marine aquaculture: an approach to new species. Ph.D. Thesis, University of Insubria, Varese, Italy.

Anuraj, A., B.P.P. Suresh, J. Loka, B. Ignatius, B. Santhosh, K.R. Ramudu et al. 2021. Induced breeding and larval rearing of vermiculated spinefoot, *Siganus vermiculatus* (Valenciennes, 1835) in indoor conditions. Aquaculture 539: 736600.

Anuraj, A., P.P. Suresh Babu and K.R. Ramudu. 2022. Training Manual on Hatchery production and farming of marine finfishes. Karwar Regional Station of ICAR-CMFS (eprints.cmfri.org.in/16452/1/Training%20Manual%20Hatchery%20production%20and%20farming%20of%20marine%20finfishes_2022_Karwar%20RC.pdf).

Assaf, N.A. 2018. Replacing fishmeal by safflower meal in the diet of the marine algaevorous teleost, *Siganus rivulatus*. Ph.D. thesis, American University of Beirut, Beirut, Lebanon.

Barakat, A., R. Roumieh, N.E. Abdel Meguid, J. Ghanawi and I.P. Saoud. 2011. Feed regimen affects growth, condition index, proximate analysis and myocyte ultrastructure of juvenile spinefoot rabbitfish *Siganus rivulatus*. Aquaculture Nutrition 17: e773–e780.

Bariche, M. 2005. Age and growth of Lessepsian rabbitfish from the eastern Mediterranean. J. Appl. Ichthyology 21: 141–145.

Ben-Tuvia, A., G. Wm. Kissil and D. Popper. 1972. Experiment in rearing rabbitfish (*Siganus rivulatus*) in sea water. Aquaculture 1: 359–364.

Bijoux, J.P., L. Dagorn, G. Berke, P.D. Cowley, M. Soria, J.C. Gaertner et al. 2013. Temporal dynamics, residency and site fidelity of spawning aggregations of a herbivorous tropical reef fish *Siganus sutor*. Mar. Ecol. Progr. Ser. 475: 233–247.

Binh, N., C.M. Rebancos, R.D. Guerrero and V.R. Cardenas. 2014. Sustainability of Polyculture model for small aquaculture farmers in Phu Vang district, Thua Thien Hue Province, Vietnam. Int. J. Agricult. Innov. Res. 2: 798–806.

Boonyaratpalin, M. 1997. Nutrient requirements of marine food fish cultured in Southeast Asia. Aquaculture 151: 283–313.

Bryan, P.G., R.C. May, B. Madraisau and J.P. McVey. 1974. Induced spawning and larval rearing of the rabbitfish *Siganus canaliculatus*, South Pacific Commission. Seventh Technical Meeting on Fisheries (Nuku'alofa, Tonga, 15–19 July 1974): 1–10.

Bryan, P.G. and B.B. Madraisau. 1977. Larval rearing and development of *Siganus lineatus* (Pisces: Siganidae) from hatching through metamorphosis. Aquaculture 10: 243–252.

Bukhari, F.A. 2005. Trials of rabbitfish *Siganus rivulatus* production in floating cages in the Red Sea. Emirates J. Food Agricult. 17: 23–29.

Burgan, B.G. and K.A. Zseleczky. 1979. Induced spawning and early development of the rabbitfish, *Siganus argenteus* (Quoy and Gaimard), in the Philippines. Silliman J. 26: 163–171.

Bwathondi, P.O. 1982. Preliminary investigations on rabbitfish, *Siganus canalicullatus*, cultivation in Tanzania. Aquaculture 27: 205–210.

Caballero, P.A., E.B. Coniza and R. Dayrit. 2022. Nursery and grow-out culture of rabbitfish *Siganus guttatus* in brackishwater ponds. Aquaculture Department, Southeast Asian Fisheries Development Center. Aquaculture extension manual no. 70: 1–30

Cabanilla-Legaspi, M.I.C., R.F.M. Traifalgar, E.G.T. de Jesus-Ayson, K.G.S. Andrino-Felarca and R.E.P. Mamauag. 2021. Growth, metamorphosis and survival of orange-spotted rabbitfish (*Siganus guttatus*) larvae fed sodium iodide-supplemented brine shrimp (*Artemia* sp.). Aquaculture 536: 736443.

Camus, T. 2012. The improvement of copepods intensive culture protocols as live feeds for aquaculture hatcheries. Ph.D. Thesis, James Cook University, Townsville, Australia.

Carumbana, E. and J. Luchavez. 1994. Siganid culture in floating cages. Aqua Farm News 12: 16–17.

Chou, R. and H.B. Lee. 1997. Commercial marine fish farming in Singapore. Aquaculture Res. 28: 767–776.

Ciani, J.F.C., R. Jones and F.T. Barrows. 2020. Performance of fishmeal-free rabbitfish (*Siganus lineatus*) feed in Palau1. Micronesica 3: 1–11.

Dadzie, S. 1992. An overview of aquaculture in eastern Africa. Hydrobiologia 232: 99–110.

Darsiani, D., M. Setiawati, D. Jusadi, M.A. Suprayudi and A. Laining. 2022. Development of siganid (*Siganus guttatus*) larvae during the transition period. Depik Jurnal Ilmu-Ilmu Perairan, Pesisir dan Perikanan 11: 23–28.

Darsono, P. 1993. Culture potential of rabbitfishes, *Siganus* (siganidae). Oseana 18: 1–24.

Dawood, M.A., A. El-Dakar, M. Mohsen, E. Abdelraouf, S. Koshio, M. Ishikawa et al. 2014. Effects of using exogenous digestive enzymes or natural enhancer mixture on growth, feed utilization, and body composition of rabbitfish, *Siganus rivulatus*. J. Agricult. Sci. Technol. B 4: 180–187.

Department of Fisheries Sabah. 2022. Company receives three records for producing fish species. .https://fishdept.sabah.gov.my/2022/05/company-receives-three-records-for-producing-fish-species/.

Dickson, M. 2022. Regional review on status and trends in aquaculture development in the Near East and North Africa – 2020. FAO Fisheries and Aquaculture Circular No. 1232/5. Food and Agriculture Organization of the United Nations, Rome, Italy. https://doi.org/10.4060/cb7818en.

Duray, M.N. 1998. Biology and culture of siganids. Aquaculture Department, Southeast Asian Fisheries Development Center (SEAFDEC). Tigbauan, Philippines. 54 pp.

El-Dakar, A.Y., S.M. Shalaby and I.P. Saoud. 2007. Assessing the use of a dietary probiotic/prebiotic as an enhancer of spinefoot rabbitfish *Siganus rivulatus* survival and growth. Aquacult. Nutrition 13: 407–412.

El-Dakar, A.Y., S.M. Shalaby and I.P. Saoud. 2011. Dietary protein requirement of juvenile marbled spinefoot rabbitfish *Siganus rivulatus*. Aquacult. Res. 42: 1050–1055.

Essington, T.E., P.E. Moriarty, H.E. Froehlich, E.E. Hodgson, L.E. Koehn, K.L. Oken et al. 2015. Fishing amplifies forage fish population collapses. Proc. Natl Acad. Sci, USA. 112: 6648–6652.

FAO. 2020. The State of World Fisheries and Aquaculture 2020. Sustainability in action. Food and Agriculture Organization of the United Nations, Rome, Italy. https://doi.org/10.4060/ca9229en.

FAO. 2022. The State of World Fisheries and Aquaculture 2022. Towards Blue Transformation. Food and Agriculture Organization of the United Nations, Rome, Italy. https://doi.org/10.4060/cc0461en.

FAO. 2023. Cultured Aquatic Species Information Programme – Siganus spp Food and Agriculture Organization of the United Nations, Rome, Italy. https://www.fao.org/fishery/en/culturedspecies/siganus_spp/en.

Fourooghifard, H., A. Matinfar, K. Roohani Ghadikolaee, M.M. Moezzi, E. Abdolalian, M.R. Zahedi et al. 2020. Effect of salinity, light intensity and tank size on larval survival rate of shoemaker rabbitfish *Siganus sutor* (Valenciennes, 1835). Iranian J. Fisheries Sci. 19: 340–351.

Froehlich, H.E., N.S. Jacobsen, T.E. Essington, T. Clavelle and B.S. Halpern. 2018. Avoiding the ecological limits of forage fish for fed aquaculture. Nature Sustainability 1: 298–303.

Fujita, S. and M. Ueno. 1954. On the development of the egg and prelarval stages of *Siganus fuscescens* (Houttuyn) by artificial insemination. Japanese J. Ichthyol. 3: 129–132.

Ghanawi, J., I.P. Saoud and S.M. Shalaby. 2010. Effect of size sorting on growth performance of juvenile spinefoot rabbitfish, *Siganus rivulatus*. J. World Aquacult. Soc. 41: 565–573.

Ghanawi, J., L. Roy, D.A. Davis and I.P. Saoud. 2011. Effects of dietary lipid levels on growth performance of marbled spinefoot rabbitfish *Siganus rivulatus*. Aquaculture 310: 395–400.

Gonzales, R.D., S.S. Parreño, R.S. Abalos, L.A. Santos, C.C. Salayog, P.J.B. Ramirez et al. 2018. Comparative analysis of siganid (*Siganus guttatus*) value chains from aquaculture in regions 1 and 2, Philippines. Int. J. Sci. Technol. Res. 7: 145–150.

Grandcourt, E.M. and H.S. Cesar. 2003. The bio-economic impact of mass coral mortality on the coastal reef fisheries of the Seychelles. Fisheries Res. 60: 539–550.

Grandcourt, E., T. Al Abdessalaam, F. Francis and A. Al Shamsi. 2007. Population biology and assessment of the white-spotted spinefoot, *Siganus canaliculatus* (Park, 1797), in the southern Arabian Gulf. J. Appl. Ichthyol. 23: 53–59.

Gundermann, N., D.M. Popper and T. Lichatowich. 1983. Biology and life cycle of *Siganus vermiculatus* (Siganidae, Pisces). Pacific Sci. 37: 165–180.

Hamze, R., N. Nasser, M. Osta, Z. Kassaify and I.P. Saoud. 2014. Investigations on susceptibility of marbled rabbitfish *Siganus rivulatus* to various infectious marine bacteria. Animal Veterinary Sci. 2: 161–168.

Hara, S., M.N. Duray, M. Parazo and Y. Taki 1986. Year-round spawning and seed production of the rabbitfish, *Siganus guttatus*. Aquaculture 59: 259–272.

Hasse, J.J., B.B. Madraisau and J.P. McVey. 1977. Some aspects of the life history of *Siganus canaliculatus* (Park) (Pisces: Siganidae) in Palau. Micronesica 13: 297–312.

Hernandez, H.B., P.M. Aliño and A. Jarre. 1990. The daily food consumption of *Siganus fuscescens*, an important herbivore in seagrass communities at Bolinao, Pangasinan Province, Philippines. ICES CM 16: 1–10.

Hicks, C.C., P.J. Cohen, N.A.J. Graham, K.L. Nash, E.H. Allison, C. D'Lima et al. 2019. Harnessing global fisheries to tackle micronutrient deficiencies. Nature 574: 95–98.

Hoey, A.S., S.J. Brandl and D.R. Bellwood. 2013. Diet and cross-shelf distribution of rabbitfishes (f. Siganidae) on the northern Great Barrier Reef: implications for ecosystem function. Coral Reefs 32: 973–984.

Ignatius, B. 2009. Broodstock development, breeding and larval rearing of *Siganus canaliculatus*. pp. 54–55. *In:* Madhu, K. and R. Madhu [eds.]. Recent Advances in Breeding and Larviculture of Marine Finfish and Shellfish. Central Marine Fisheries Institute, Kerala, India.

Jaikumar, M., L. Kanagu, C. Stella and B. Gunalan. 2011. Culturing a rabbit fish (*Siganus canalicullatus*) in cages: A study from Palk Bay, South East Coast of India. Int. J. Water Resources Environ. Engineer. 3: 251–257.

Jaikumar, M. 2012. A review on biology and aquaculture potential of rabbit fish in Tamilnadu (*Siganus canaliculatus*). Int. J. Plant Animal Environ. Sci. 2: 57–64.

Jiarpinijnun, A., J.T. Geng, N. Thanathornvarakul, S. Keratimanoch, N. Üçyol, E. Okazaki et al. 2022. Identifying volatile compounds in rabbit fish (*Siganus fuscescens*) tissues and their enzymatic generation. European Food Res. Technol. 248: 1485–1497.

Juario, J.V., M.N. Duray, V.M. Duray, J.F. Nacario and J.M.E. Almendras. 1985. Breeding and larval rearing of the rabbitfish, *Siganus guttatus* (Bloch). Aquaculture 44: 91–101.

Kongkeo, H., C. Wayne, M. Murdjani, P. Bunliptanon and T. Chien. 2010. Current practices of marine finfish cage culture in China, Indonesia, Thailand and Vietnam. Aquacult. Asia Mag. 15: 32–39.

Komatsu, T., S. Nakamura and M. Nakamura. 2006. A sex cord-like structure and some remarkable features in early gonadal sex differentiation in the marine teleost *Siganus guttatus* (Bloch). J. Fish Biol. 68: 236–250.

Lai, L.W. and B.T. Yu. 2002. The evolution of the fry market in the marine fish culture industry of Hong Kong: an economic perspective. Aquacult. Econ. Mgt. 6: 191–214.

Laining, A., U. Usman and R. Syah. 2016. Aquatic weed *Ceratophyllum* sp. as a dietary protein source: its effects on growth and fillet amino acid profile of rabbitfish, *Siganus guttatus*. Aquacult. Aquar. Conserv. Legislat. 9: 352–359.

Lam, T.J. 1974. Siganids: their biology and mariculture potential. Aquaculture 3: 325–354.

Li, Y., Q. Zhang and Y. Liu. 2018. Rabbitfish–an emerging herbivorous marine aquaculture species. pp. 329–334. *In:* Gui, J.-F., Q. Tang, Z. Li, J. Liu, S.S. De Silva [eds.]. Aquaculture in China: Success Stories and Modern Trends. John Wiley and Sons, Oxford, UK.

Lichatowich, T. and D. Popper. 1975. Report on the growth of rabbit fish in fish ponds in Fiji. Aquaculture 5: 211–212.

Longo, S.B., B. Clark, R. York and A.K. Jorgenson. 2019. Aquaculture and the displacement of fisheries captures. Conserv. Biol. 33: 832–841.

Luong, T.C., H. Lemonnier, S. Hochard, F. Royer and Y. Letourneur. 2017. Effects of blue shrimp *Litopenaeus stylirostris* and goldlined rabbitfish *Siganus lineatus* in mono-and polyculture on production and environmental conditions. Aquaculture Res. 48: 1368–1379.

Maina, G.W., M. Samoilys, H. Alidina and K. Osuka. 2013. Targeted fishing of the shoemaker spinefoot rabbitfish, Siganus sutor, on potential spawning aggregations in southern Kenya. pp. 13–26. *In*: Robinson, J. and M. Samoilys [eds.] Reef fish spawning aggregations in the Western Indian Ocean: Research for management. WIOMSA, CORDIO, Kenya.

Malugao, M.E., E.B. Metillo and J.G. Gorospe. 2019. Functional response of goldlined spinefoot rabbitfish *Siganus guttatus* (Siganidae, Perciformes) early juveniles to live zooplankton prey. Internat. J. Biosci. 14: 238–250.

Manacop, P.R. 1937. The artificial fertilization of dangit, *Amphacanthus oramin* (Bloch and Schneider). Philipp. J. Sci 62: 229–237.

May, R.C., D. Popper and J.P. McVey. 1974. Rearing and larval development of *Siganus canaliculatus* (Park) (Pisces: Siganidae). Micronesica 10: 285–298.

Mehanna, S., S.S. Mohammad, M.S. El-Mahdy and A.A.Y. Osman. 2018. Stock assessment and management of the rabbitfish *Siganus rivulatus* from the Southern Red Sea, Egypt. Egyptian J. Aquatic Biol. Fisheries 22: 323–329.

Monzer, S., N. Nasser, J. Babikian and I.P. Saoud. 2017. Substitution of fish meal by soybean meal in diets for juvenile marbled spinefoot, *Siganus rivulatus*. J. Applied Aquacult. 29: 101–116.

Nasser, N., J. Babikian, M.G. Hatem, I.P. Saoud and M.G. Abiad. 2019. Evaluation of post-consumer food waste as partial replacement of commercial feed in marbled rabbitfish, *Siganus rivulatus* aquaculture. Int. J. Environ. Sci. Technol. 16: 4059–4068.

Nelson, S.G. and W.J. FitzGerald. 1990. Development of Aquaculture in the Commonwealth of the Northern Mariana Islands: A Feasibility Study. Technical Report 96, University of Guam Marine Laboratory. 177pp.

Nelson, S.G., S.A. Lock and L.A. Collins. 1992. Growth of the rabbitfish *Siganus randalli* (Woodland) in relation to the feasibility of its culture on Guam. Technical Report 97, University of Guam Marine Laboratory. 30 pp..

OECD (Organisation for Economic Co-operation and Development). 2021. Aquaculture production. In ECD. Stat. Paris. Cited 26 November 2021. https://stats.oecd.org/Index.aspx?DataSetCode=FISH_AQUA#.

Ogata, H.Y., A.C. Emata, E.C. Garibay and H. Furuita. 2004. Fatty acid composition of five candidate aquaculture species in Central Philippines. Aquaculture 236: 361–375.

Ortega, R.C. 2021. Trends in the major aquaculture food fish production in the Philippines. pp. 46–62. *In*: Proceedings of the International Workshop on the Promotion of Sustainable Aquaculture, Aquatic Animal Health, and Resource Enhancement in Southeast Asia. Aquaculture Department, Southeast Asian Fisheries Development Center, Philippines.

Parazo, M.M. 1990. Effect of dietary protein and energy level on growth, protein utilization and carcass composition of rabbitfish, *Siganus guttatus*. Aquaculture 86: 41–49.

Park, Y.J., A. Takemura and Y.D. Lee. 2006. Annual and lunar-synchronized ovarian activity in two rabbitfish species in the Chuuk lagoon, Micronesia. Fisheries Sci. 72: 166–172.

Pauly, D. and D. Zeller. 2016. Catch reconstructions reveal that global marine fisheries catches are higher than reported and declining. Nature Comm. 7: 10244.

Pillai, T.G. 1962. Siganid fish farming. pp. 51–52. *In*: Fish Farming Method in the Philippines, Indonesia and Hongkong. Food and Agriculture Organization of the United Nations, Rome, Italy (Technical paper/FAO Fisheries Biology: no. 18).

Pitcher, T.J. and W.W. Cheung. 2013. Fisheries: hope or despair? Mar. Poll. Bull.74: 506–516.

Pillans, R.D., C.E. Franklin and I.R. Tibbetts. 2004. Food choice in *Siganus fuscescens*: influence of macrophyte nutrient content and availability. J. Fish Biol. 64: 297–309.

Popper, D., H. Gordin and G.W. Kissil. 1973. Fertilization and hatching of rabbitfish *Siganus rivulatus*. Aquaculture 2: 37–44.

Pomeroy, R.S., J.E. Parks and C.M. Balboa. 2006. Farming the reef: Is aquaculture a solution for reducing fishing pressure on coral reefs? Mar. Policy 30: 111–130.

Ponce, G.S. 1983. Growth performance of the species of rabbitfish in nursery ponds. Ilocos Fish. J. 1: 191–196.

Popper, D., R. Pitt and Y. Zohar. 1979. Experiments on the propagation of Red Sea siganids and some notes on their reproduction in nature. Aquaculture 16: 177–181.

Popper, D. and N. Gundermann. 1975. Some ecological and behavioural aspects of siganid populations in the Red Sea and Mediterranean coasts of Israel in relation to their suitability for aquaculture. Aquaculture, 6: 127–141.

Rachmansyah, U., S. Lante and T. Ahmad. 2007. Rabbitfish *Siganus guttatus* breeding and larval rearing trial. Aquaculture Asia 12: 39–40.

Rahman, M.S., M. Morita, A. Takemura and K. Takano. 2003. Hormonal changes in relation to lunar periodicity in the testis of the forktail rabbitfish, *Siganus argenteus*. Gen. Compar. Endocrinol. 131: 302–309.

Rimmer, M.A., K. Sugama, D. Rakhmawati, R. Rofiq and R.H. Habgood. 2013. A review and SWOT analysis of aquaculture development in Indonesia. Rev. Aquacult. 5: 255–279.

Roumieh, R., A. Barakat, N.E. Abdelmeguid, J. Ghanawi and P.I. Saoud. 2013. Acute and chronic effects of aqueous ammonia on marbled spinefoot rabbitfish, *Siganus rivulatus* (Forsskål 1775). Aquaculture Res. 44: 1777–1790.

Saade, E., H.A. Sulaeman and N.M. Jannah. 2022. The effect of corn starch substitution with *Sargassum* sp. Starch in Diet on Grow-Out of Cultivated Rabbitfish, (*Siganus guttatus*) in Floating Net Cages. J. Aquacult. Fish Health 10: 11–20.

Salem, M.E.S., H.M. Abdel-Ghany, A.E. Sallam and M.M. El-Feky. 2021. Effects of different dietary protein and lipids levels on growth performance and digestive enzymes of the rabbitfish (*Siganus rivulatus*), reared in well water. Egyptian J. Aquatic Biol. Fisheries Zool. Dept. 25: 645–660.

Sanchez-Jerez, P. 2013. Fish Escapes from Cage Aquaculture in the Mediterranean. World Aquaculture December: 30–35.

Santos, S., J. Ghanawi and I.P. Saoud. 2015. Effects of water temperature and body weight on anaesthetic efficiency in marbled rabbitfish (*Siganus rivulatus*). Aquaculture Res. 46: 928–936.

Saoud, I.P., S. Kreydiyyeh, A. Chalfoun and M. Fakih. 2007. Influence of salinity on survival, growth, plasma osmolality and gill Na+–K+–ATPase activity in the rabbitfish *Siganus rivulatus*. J. Exp. Mar. Biol. Ecol. 348: 183–190.

Saoud, P., S. Naamani, J. Ghanawi and N. Nasser. 2014. Effects of acute and chronic nitrite exposure on rabbitfish *Siganus rivulatus* growth, hematological parameters, and gill histology. J. Aquaculture Res. Development 5: 1000263.

Saoud, I.P., J. Ghanawi and N. Lebbos. 2007. Effects of stocking density on the survival, growth, size variation and condition index of juvenile rabbitfish *Siganus rivulatus*. Aquacult. Int. 16: 109–116.

Seale, A.P. and S. Ellis. 2019. Sustainable capture-based aquaculture of rabbitfish in Pacific island lagoons. Aquaculture and Aquaponics 1: 1–9.

Shen, Y., K. Ma and G.H. Yue. 2021. Status, challenges and trends of aquaculture in Singapore. Aquaculture 533: 736210.

Soliman, V.S., H. Yamada and K. Yamaoka. 2010. Early life-history of the spiny siganid *Siganus spinus* (Linnaeus 1758) inferred from otolith microstructure. J. Appl. Ichthyol. 26: 540–545.

Stephanou, D. and G. Georgiou. 2000. Recent experiences on the culture of rabbitfish *Siganus rivulatus* in Cyprus. Cahiers Options Mediterraneennes 47: 295–301.

Sulaeman, H.A. and A. Laining. 2022, December. Evaluation of seaweed, *Ulva lactuca* as a fresh diet for nursery stage of golden rabbitfish, *Siganus guttatus*. In IOP Conference Series: Earth and Environmental Science (Vol. 1119, No. 1, p. 012057). IOP Publishing.

Syah, R., B.R. Tampangallo, M.C. Undu, A.I.J. Asaad and A. Laining. 2020. Rabbitfish (*Siganus guttatus*) culture in floating net cage with different stocking densities. In IOP Conference Series: Earth and Environmental Science (Vol. 564, No. 1, p. 012022). IOP Publishing.

Tacon, A.G.J., N. Rausin, M. Kadari and P. Cornelis. 1990. The food and feeding of marine finfish in floating net cages at the National Seafarming Development Centre, Lampung, Indonesia: rabbitfish, *Siganus canaliculatus* (Park). Aquaculture Res. 21: 375–390.

Tan, J. 2021. First hatchery bred streaked spinefoot rabbitfish (*Siganus javus*) in Sabah, Malaysia. Borneo J. Mar. Sci. Aquacult. 5: 1–7.

Taylor, B.M., J. Gourley and M.S. Trianni. 2016. Age, growth, reproductive biology and spawning periodicity of the forktail rabbitfish (*Siganus argenteus*) from the Mariana Islands. Marine Freshw. Res. 68: 1088–1097.

Teitelbaum, A. and Z. Hilly. 2009. Cage culture of rabbitfish (Siganids) in Solomon Islands, with emphasis on locally available diets. Mini-project MS0802. Secretariat of the Pacific Community, Noumea, New Caledonia.

Teitelbaum, A., T. Prior, F. Legarrec, C. Oengpepa and P. Mesia. 2008. Rabbitfish: a candidate for aquaculture in the Pacific. Secretariat of the Pacific Community Fisheries Newsletter 127: 40–44.

Thacker, R.W., D.G. Nagle and V.J. Paul. 1997. Effects of repeated exposures to marine cyanobacterial secondary metabolites on feeding by juvenile rabbitfish and parrotfish. Mar. Ecol. Progr. Ser. 147: 21–29.

Tharwat, A. 2004. Reproductive cycle and mariculture potential of the rabbitfish *Siganus canaliculatus* in Saudi Arabia. Egyptian J. Aquatic Biol. Fisheries 8: 123–143.

Thépot, V. 2021. Seaweed as a functional ingredient in the diet of farmed fish. Ph.D. thesis, University of the Sunshine Coast, Sippy Downs, Queensland.

Tobias, W.J. 1976. Ecology of *Siganus argenteus* (Pisces: Siganidae) in relation to its mariculture potential in Guam. Ph.D. Thesis, University of Guam, Guam.

Tresnati, J., S.B.A. Omar and A. Tuwo. 2020. Population dynamics of the white spotted rabbitfish (*Siganus canaliculatus* Park, 1797) in Makassar Strait and Gulf of Bone, Indonesia. IOP Conference Series: Earth and Environmental Science 492: 012093.

Tseng, W.Y. and K.L. Chan. 1982. The reproductive biology of the rabbitfish in Hong Kong. J. World Mariculture Soc. 13: 313–321.

Tsuda, R.T., P.G. Bryan, W.J. FitzGerald and W.J. Tobias. 1974. Juvenile-adult rearing of *Siganus* (Pisces: Siganidae) in Guam. South Pacific Commission Seventh Technical Meeting of Fisheries, Nuku'alofa, Tonga. July 15–19.

Urmaza, M. Sr. 1983. Growing siganid on a commercial scale. Greenfields 13: 29–31.

Uwate, K.R. and A.J. Shams. 1999. A long-term approach to aquaculture development: Bahrain. Aquacult. Econom. Manag. 3: 193–203.

van Beijnen, J., K. Hoevenaars and G. Yan 2022. Rabbitfish farming in Asia, Bangkok. Food and Agriculture Organization of the United Nations, Rome, Italy. https://www.fao.org/apfic/publications/detail/ar/c/1600912/.

Van Phan, U. and H.Q. Pham. 2016. Embryonic and larval development and effect of salinity levels on egg and ovary performance in rabbit fish (*Siganus guttatus*). Israeli J. Aquaculture-Bamidgeh 68.

Verdian, A.H., I. Effendi, T. Budiardi and I. Diatin. 2020. Production performance improvement of white shrimp (*Litopenaeus vannamei*) culture with integrated multi trophic aquaculture system in Seribu Islands, Jakarta, Indonesia. Iranian J. Fish. Sci. 19: 1415–1427.

Villanueva, E.G., K. Hoevenaars, J. van Beijnen, A.P. Gonzales, L.A. Creencia and R.G. Dolorosa. 2021. Simplified hatchery protocols for culture of orange-spotted spinefoot *Siganus guttatus* (Bloch, 1787) in Palawan, Philippines. Asian Fish. Sci. 34: 217–224.

Von Westernhagen, H. and H. Rosenthal. 1975. Rearing and spawning siganids (Pisces: Teleostei) in a closed seawater system. Helgoland Mar. Res. 27: 1–18.

Von Westernhagen, H. and H. Rosenthal. 1976. Some aspects of the suitability of various Philippine siganid species (Siganidae) for mariculture. Aquaculture 9: 297–311.

Wahyuningtyas, L.A., M. Nurilmala, M.F.A. Sondita, A.A. Taurusman and A.O. Sudrajat. 2017. Nutritional profile of rabbitfish (*Siganus* spp.) from the Kepulauan Seribu (Thousand Islands), Jakarta, Indonesia. Int. Food Res. J. 24: 685–690.

Wassef, E.A. and H.A.A. Hady. 1997. Breeding biology of rabbitfish *Siganus canaliculatus* (Siganidae) in mid Arabian Gulf. Fish. Res. 33: 159–166.

Waycott, B. 2023. Kindai University researchers say rabbitfish can thrive in a form of 'circular aquaculture' *Global Seafood Alliance.* https://www.globalseafood.org/advocate/does-rabbitfish-farming-offer-a-solution-to-food-waste-in-japan/.

Woodland, D.J. 1990. Revision of the fish family Siganidae with descriptions of two new species and comments on distribution and biology. Indo-Pacific Fishes no. 19.

Xie, D., X. Li, C. You, S. Wang and Y. Li. 2019. Supplementation of macroalgae together with non-starch polysaccharide-degrading enzymes in diets enhanced growth performance, innate immune indexes, and disease resistance against *Vibrio parahaemolyticus* in rabbitfish *Siganus canaliculatus.* J. Appl. Phycol. 31: 2073–2083.

Xu, S., L. Zhang, Q. Wu, X. Liu, S. Wang, C. You et al. 2011. Evaluation of dried seaweed *Gracilaria lemaneiformis* as an ingredient in diets for teleost fish *Siganus canaliculatus*. Aquacult. Int. 19: 1007–1018.

Xu, S., S. Wang, L. Zhang, C. You and Y. Li. 2012. Effects of replacement of dietary fish oil with soybean oil on growth performance and tissue fatty acid composition in marine herbivorous teleost *Siganus canaliculatus*. Aquacult. Res. 43: 1276–1286.

Yousif, O.M., M.F. Osman, A.R. Anwahi and M.A. Zarouni. 2004. Growth response and carcass composition of rabbitfish, *Siganus canaliculatus* (Park) fed diets supplemented with dehydrated seaweed, *Enteromorpha* sp. Emirates J. Food Agricult. 16: 18–26.

Zhou, S., C. You, S. Wang and Y. Li. 2013. Effects of dietary seaweed *Enteromorpha prolifera* on growth performance, physiological and biochemical characteristics of rabbitfish *Siganus canaliculatus*. J. Fishery Sci. China 20: 1257–1265.

Zhu, W., J.Y. Liu, P. Zhuang, F. Zhao, C. Song, X. Zou et al. 2013. Effects of dietary lipid level on growth and body composition of *Siganus guttatus*. Mar. Fisheries 35 : 65–71.

10

Rabbitfishes, Analogues of their Terrestrial Namesakes or Key Components of Future Food Security? Emerging Topics and Future Research Directions

Andrew S. Hoey[1,*] and *Brett M. Taylor*[2]

Introduction

Rabbitfishes (Siganidae) are ubiquitous members of coastal and estuarine fish assemblages throughout the tropical and subtropical Indian and Pacific Oceans. Despite the relatively limited taxonomic diversity of rabbitfish (31 spp all belonging to a single genus *Siganus*) compared to other families of piscine herbivores (i.e., parrotfishes: 100 spp, Parenti and Randall 2011; surgeonfishes: 86 spp, Froese and Pauly 2024), and the morphological similarities among species, rabbitfishes have received considerable research attention over the past 60 years. Indeed, the number of scientific publications since 1960 that have focused on or considered rabbitfishes (722 publications), although lower than that of parrotfishes (1,288 publications), is directly comparable to that of the surgeonfishes (722 publications; Fig. 1). Much of the early research on rabbitfishes focused on their potential for aquaculture production, while more recent research has focused on their contribution to the consumption of macroalgal biomass (largely using tethered pieces of macroalgae; e.g., Loffler et al. 2015, Bauman et al. 2017) and the movement of rabbitfishes into novel environments

[1] College of Science and Engineering, James Cook University, Townsville, Queensland 4811, Australia.
[2] University of Guam Sea Grant and Marine Laboratory, University of Guam, Mangilao, Guam, 96923, USA.
Email: brettmtaylor@gmail.com
* Corresponding author: andrew.hoey1@jcu.edu.au

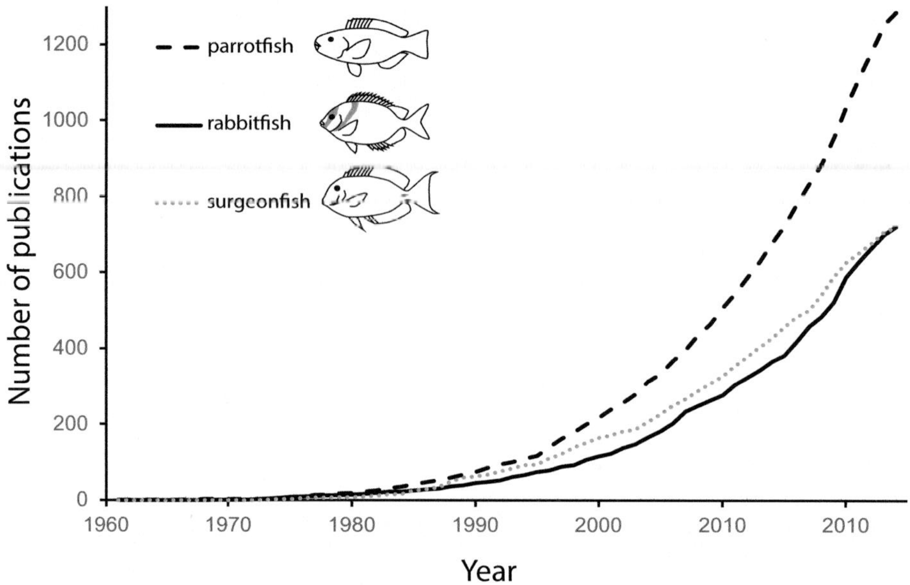

Fig. 1. Comparison of the cumulative number of scientific papers published on each of the three major herbivorous fish taxa: parrotfishes (f. Labridae, t. Scarini), surgeonfishes (f. Acanthuridae), and rabbitfishes (f. Siganidae) from 1960 – 2023. Data are based on a Web of Science search (June 2024).

(e.g., Verges et al. 2016, Zarco-Perello et al. 2021). Collectively, this body of work has provided a firm grounding to understand the evolution, ecology, and aquaculture and fisheries potential of this group, however our understanding of several other topics is limited.

Despite a relatively rich fossil fauna (ten species from eight genera) compared to other coral reef fish families (Friedman and Carnevale 2018), there are only 31 species of extant rabbitfishes from a single genus (*Siganus*) currently recognised (Woodland – Chapter 1). Rabbitfishes, although present throughout the Cenozoic (~ 57 Ma to present) and sharing similar early habitats to that of other groups (e.g., surgeonfishes), appear to have not undergone episodes of diversification that are evident in other piscine herbivore lineages (namely parrotfishes, surgeonfishes and damselfishes; Choat et al. 2012, Sorenson et al. 2013, McCord et al. 2021). It is not only their interesting evolutionary history, but also their ecology that sets rabbitfishes apart from other piscine herbivores. With few exceptions (namely *S. argenteus* and *S. woodlandi*) rabbitfishes are most abundant and speciose on sheltered coastal or inshore reef and macrophyte habitats (e.g., Hoey et al. 2013), whereas parrotfishes and surgeonfishes tend to be most abundant and speciose on wave exposed outer reef habitats (Hoey and Bellwood 2008, Cheal et al. 2012, Tebbett et al. 2022). Moreover, rabbitfishes display a high degree of flexibility in habitat and diet, and appear to have a broad tolerance range of environmental conditions (Hoey et al. – Chapter 7). It is this flexibility, and in particular their ability to exploit soft benthic habitats and tolerate a broad range of temperatures, that has likely contributed to the ability of rabbitfishes to invade novel habitats (i.e., sub-tropical and temperate reefs, the Meditarannean Sea) where they have contributed to the decline in native habitat-forming macrophytes. Despite the negative effects on macrophytes in some regions, there has been a suggestion that a rabbitfish invasion may be beneficial, or even critical, to the future of Caribbean coral reefs (Bellwood and Goatley 2017). While we do not subscribe to this view, the ability of rabbitfishes to tolerate a broad range of environmental conditions, coupled with their largely herbivorous diet and rapid growth rates, may make rabbitfishes an increasingly important component of food security and a source of protein and micronutrients under ongoing climate change and habitat degradation.

Taxonomy and Evolution

Currently there are 29 described species of extant rabbitfish, with Woodland (Chapter 1) recognising an additional two species, namely recognising *S. stellatus* and *S. laqueus* as full species rather than subspecies of *S. stellatus*, and two incipient species of *S. corallinus* (although these are yet to be described). There are also some uncertainties regarding the taxonomic status of the sister taxa *S. canaliculatus*—*S. fuscescens*, and disjunct populations of other species. Despite the relatively low species richness of the rabbitfishes there are no complete morphological or molecular phylogenies. For example, the most complete molecular phylogeny includes 24 extant species of rabbitfish (Siqueira and Cowman – Chapter 3), with the mapping of some ecological traits inferred from closely related species rather than from the individual species themselves (e.g., Siqueira et al. 2019). Moreover, a detailed investigation of the internal and external ecomorphology of rabbitfishes is lacking and is presumably related to the view that they exhibit limited morphological variation (Woodland 1990, Tyler and Bannikov 1997). However, comparisons of external morphology, and colour and patterns show considerable variation among rabbitfishes that broadly aligns with behaviour and habitat associations (Huertas and Hemmingson – Chapter 2). Sourcing genetic samples for the remaining species and incorporating these into a molecular phylogeny, a comprehensive examination of the internal and external morphology of all rabbitfish species, and quantifying species-specific ecological traits should be a focus of future research.

Food Production

Rabbitfishes have long been recognized as a potential candidate for aquaculture (reviewed in Camus et al. – Chapter 9), and some species (primarily schooling species) have a long history of being targeted by fishers (e.g., Lam 1974, Maina et al. 2013). Despite this long interest and considerable research effort, the commercial culture of rabbitfishes is still in its infancy; at its peak accounting for less than 0.5 percent of the production from wild capture fisheries of rabbitfishes (2020 aquaculture production: 397 tonnes, wild-capture fisheries: 133,000 tonnes, Camus et al. – Chapter 9). While many of the inherent biological traits of rabbitfishes facilitate their successful aquaculture (namely high fecundity, fast growth, tolerance of environmental conditions and high stocking densities, herbivorous/omnivorous diet) difficulties in producing hatchery-reared juveniles has hampered production. These difficulties are primarily attributed to their small oral gape at first feeding (Ignatius 2009) and the limited availability, or lack of, suitable sized prey, and have led to most aquaculture of rabbitfishes relying on the grow-out of wild caught juveniles. Indeed, numerous studies have shown that widely used live prey organisms (i.e., rotifers and *Artemia*) are too large for first-feeding larvae and yield low survival rates, and have pointed to the need to develop the culture of smaller live feeds. Future research into the larval rearing of rabbitfishes and, in particular, the initial feeding of larvae is required if rabbitfish aquaculture is to reach its potential and lower the environmental footprint when compared to the culture of higher-order consumers.

Meanwhile, wild-caught rabbitfishes continue to represent an important fishery-resource throughout the Indo-Pacific and even in the Mediterranean for invasive Lessepsian migrant species. Their relative importance to economies and food security is greatest in the Western Indian Ocean where a single highly-productive species, *Siganus sutor*, greatly contributes to coastal fisheries catches in several countries, largely through the systematic targeting of spawning aggregations (Robinson et al. 2011). The range of target species in rabbitfish fisheries becomes more diverse in the Coral Triangle region and Western Pacific, reflecting the higher species richness and habitat diversity in this region. Adults and sometimes juveniles of primarily 'schooling' species are most heavily targeted, whereby pair-forming species are generally not as productive targets for fisheries. Examination across their biogeographic distribution demonstrates an interesting feature of rabbitfish fisheries: perhaps more than most families, the nature of these fisheries appears to be intrinsically linked to the dynamic biological rhythms of the rabbitfish species they exploit (Taylor et al. – Chapter 5). For instance, rabbitfish harvests tend to be temporally cyclical and linked to the timing of reproduction (harvest of adults) and subsequent recruitment (harvest of newly settled juveniles). Ultimately, while detailed fishery data remains sparse for many regions, there are signs that rabbitfishes may represent an increasingly prevalent fishery resource in many areas where more biologically vulnerable species are declining.

Top-down versus Bottom-up Processes

Piscine herbivory is widely viewed as a key process on coral reefs with the mechanical removal of algal material thought to maintain a healthy balance between corals and algae. This view has been largely driven by rapid increases in macroalgal biomass following the exclusion of herbivorous fishes from small areas of reef (e.g., Lewis 1986, Hughes et al. 2007) and dramatic changes in benthic reef communities following the large reductions in, or complete loss, of herbivores (including parrotfishes) in some locations (e.g., Hughes 1994, Rasher et al. 2013, Graham et al. 2015). While a large body of work has focussed on the top-down effects of herbivorous fishes on algal biomass (e.g., Bellwood et al. 2006, Hoey and Bellwood 2009) and benthic communities (e.g., Carpenter 1986), this top-down control of algal communities by herbivores is being increasingly questioned in favour of a bottom-up approach (e.g., Russ et al. 2015, Clements et al. 2017, Nicholson and Clements 2021). Indeed, numerous studies have related the abundance, biomass (e.g., Adam et al. 2011, Gilmour et al. 2013, Russ et al. 2015) and growth (Taylor et al. 2020) of herbivorous fish to the availability of substrata covered by turfs (both algae and cyanobacteria). Moreover, the top-down view of herbivory typically overlooks the nutritional targets of individual species, rather grouping species based on feeding substrata (often broad categories such as turfs vs macroalgae). However, recent investigation of the nutritional targets of parrotfishes has shown a high degree trophic specialisation among co-occurring species (Nicholson and Clements 2023a, 2023b, 2024). It is therefore reasonable to assume that other groups of piscine herbivores may exhibit similar trophic variation among species, and that the availability of these dietary resources may shape the distributions of herbivorous fish assemblages rather than the converse.

Feeding Ecology and Foraging Dynamics of Rabbitfishes

The acquisition of nutrients through feeding is central to an individual's fitness, and hence the availability and quality of dietary resources can strongly influence the distribution, growth and condition of individuals and populations (Krebs 1972). Despite considerable research effort that has focused on or included rabbitfishes our understanding of their nutritional ecology is limited (Clements – Chapter 4). For example, numerous studies have reported rabbitfish feeding on tethered pieces of macroalgae (e.g., Fox and Bellwood 2008, Pell et al. 2024), and while these studies may indicate what algae a particular species is capable of feeding on, it is unlikely to replicate the natural availability of all potential food resources and hence may be a poor indicator of diet. Although generally considered to be herbivorous, analyses of stomach content of rabbitfishes show that some species include substantial amounts of animal material (e.g., Hoey et al. 2013), and that stomach contents vary both among and within species (i.e., among life stages, seasons, habitats and geographic locations; Moleana et al. 2016, Martin et al. 2024, Zarco-Perello et al. 2024, reviewed in Clements – Chapter 4). It is important to note that examination of stomach content provides an indication of the material ingested, however it does not provide any insight into the relative importance of those materials in providing nutrition to rabbitfish. Unfortunately, the number of studies that have examined nutritional biomarkers on rabbitfishes is limited and have included few species. As highlighted by Clements (Chapter 4), studies comparing stomach content and/or nutritional biomarkers (e.g., stable isotopes, fatty acids) among co-occurring rabbitfishes, life stages and locations should be a priority for future research.

Ontogenetic Shifts in Habitat use

Rabbitfish are often described as being coral reef fish or reef-associated fish as adults (e.g., Woodland 1990), yet observations and an increasing number of studies indicate the vast majority of rabbitfish species appear to settle to and associate with macrophyte habitats as juveniles (primarily macroalgae, seagrass and mangroves) (Hoey et al. – Chapter 7). While the benefits of using these macrophyte habitats is unclear and warrants investigation (e.g., provision of food and/or refugia from predation), this association with macrophyte habitats may partly explain the predominantly inshore distribution of rabbitfishes across Indo-Pacific reefs (Cheal et al. 2012, Hoey et al. 2013, Pombo-Ayora et al. 2020). Indeed, the proximity of settlement, juvenile and adult habitats has been shown to influence local populations of other tropical fish taxa. For example, the abundance of adult parrotfish in reef habitats has been shown to decrease exponentially with distance from their juvenile mangrove habitat (Claydon et al. 2015), with some species experiencing local extinctions following mangrove loss (Mumby et al. 2004). Indeed, the limited availability or absence of macrophyte habitats on offshore and oceanic reefs (e.g., Wismer et al. 2009) likely contributes to the low abundance and species richness of rabbitfishes in these areas (Hoey et al. 2013). The only exceptions appear to be the sister taxa *S. argenteus* and *S. woodlandi* that are commonly observed on oceanic reefs of the Coral Sea (Hoey et al. 2020). While little is known of the settlement habitat or ecology of *S. woodlandi*, settlement stage *S. argenteus* do not appear to be reliant on macrophyte habitats with several studies reporting juveniles to be associated with shallow reef flat habitats. *Siganus argenteus* (and likely its sister taxon *S. woodlandi*) are unique among rabbitfish species in that the larvae pass through a distinctive pre-juvenile stage (Woodland 1990). This pre-juvenile stage prolongs their pelagic larval duration and hence their potential dispersal, and coupled with their lack of reliance on macrophyte habitats may contribute to their presence on offshore reefs and the broad geographic distribution of *S. argenteus*.

Over the past few years there has been an increasing research focus on quantifying the settlement and early post-settlement habitat associations of herbivorous reef fishes (e.g., Hamilton et al. 2017), in particular rabbitfishes (e.g., Priest et al 2016, Tano et al. 2017, Tang et al. 2020, Martin et al. 2024, Webber et al. 2024), and the importance of seascape configuration in shaping population size and assemblage structure (Taylor et al. 2015, Sambrook et al. 2019, 2020, Roff et al. 2019, Eggertsen et al. 2020, Bradley et al. 2024). While reductions in seagrass and mangrove habitats (Goldberg et al. 2020, Dunic et al. 2021) may be threatening the persistence of some species that rely on these habitats in their early life stages, restoration efforts on coral reefs that are engaging in macroalgal removal activities may inadvertently be threatening the ecological process they are attempting to promote—herbivory.

Despite the increased focus and understanding of habitat associations of recently-settled and juvenile rabbitfishes, this research has typically focused on a limited suite of species. There has been limited or no effort to understand the drivers of habitat use, and the timing and drivers of ontogenetic shifts in habitat are unknown. These represent key areas of future research, especially given the rapidly changing composition and configuration of tropical coastal seascapes.

Environmental Tolerances in a Changing World

Rabbitfishes, perhaps more so than any other family of tropical fishes, are tolerant of a broad range of environmental conditions, and dietary and habitat resources (Hoey et al. – Chapter 7). For example, *S. fuscescens* spans over 70° of latitude, from northern Japan to southern Australia, and its sister taxon *S. canaliculatus* is abundant within the extreme thermal environment of the southern Arabian Gulf (thermal ranges of *ca.* 15°C and > 20°C, respectively). Moreover, these species are found in a broad range of habitats (seagrass, macroalgae, mangrove, and coral reef) and appear to have flexible diets (Hoey et al. – Chapter 7). This flexibility in resource use, and in particular the ability to exploit soft sediment habitats, has likely allowed rabbitfishes to extend beyond tropic habitats more effectively than the other piscine herbivores and contribute to their success in invading novel habitats, namely the Mediterranean Sea and temperate habitats of Australia and Japan (Zarco-Perello et al. – Chapter 8). Moreover, their tolerance to a wide range of environmental conditions (temperature, salinity, dissolved oxygen, turbidity) together with their predominantly herbivorous diet and rapid growth has led to rabbitfishes being viewed as ideal targets for aquaculture. It is unclear, however, whether their flexibility in terms of environmental conditions, diet and habitat requirements will translate to future conditions, reducing their vulnerability to ongoing climate change and local anthropogenic stressors.

Increasing seawater temperature due to climate change is having a pervasive effect on tropical marine ecosystems (e.g., Hoey et al. 2016, Hughes et al. 2017), largely due to the predominance of ectothermic species in these ecosystems. Rabbitfishes (like all teleosts) are ectotherms, and as such their rates of biochemical and cellular processes are largely governed by environmental temperature (Pörtner and Farrell 2008). Despite a rapidly growing body of experimental research that has shown increasing seawater temperature to increase rates of oxygen uptake and food intake, decrease aerobic scope and activity levels, and alter habitat associations of coral reef fishes (e.g., Nilsson et al. 2009, Johansen et al. 2014, 2015, Motson and Donelson 2017, Nay et al. 2020, 2021), there has been very limited investigation on the thermal tolerances or effect of increasing temperature on the performance of rabbitfishes (see Graba-Landry 2020, LaMonica et al. 2021, Zarco-Perello et al. 2022 for exceptions). The results of these initial studies on the effect of increasing temperature on *S. doliatus* and *S. lineatus* are consistent with their use of broad thermal environments and have suggested that these two rabbitfish species may be less vulnerable to elevated temperatures than other coral reef fish (Graba-Landry 2020, LaMonica et al. 2021). However, our understanding of the physiological responses of rabbitfishes to predicted future increases in temperatures is limited. The effects of increasing temperature on the growth, survival, physiological condition, feeding rate, diet, reproduction, behaviour (e.g., behavioural thermoregulation), and population sustainability of rabbitfishes represents key areas for future research. Moreover, understanding how increasing temperature may affect the availability and nutritional quality of their dietary resources, and hence the outcomes of trophic interactions will be key to understanding how rabbitfish populations may respond to future conditions.

The Demographics of Rabbitfish Populations

A feature of rabbitfish biology that pervades as important to their evolution, ecology, expansion, fisheries, and aquaculture potential is the general notion that rabbitfishes have rapid life histories. Indeed, compared to other reef-associated species of similar body sizes, they are fast-growing, short-lived, and highly fecund, features that promote rapid turnover of populations. However, age-based biological information is available from only a small portion of extant species, and new evidence on the life spans of pair-forming species highlights a greater diversity of underlying mortality patterns than previously understood (Taylor et al. – Chapter 5).

This diversity appears to have a phylogenetic underpinning that reflects differences in the behavioural ecology of species associated with trade-offs between survival and reproduction (Brandl and Tebbett – Chapter 6). Given the increased application of population and ecosystem modelling in tropical reef sciences, and the dependency of these techniques on underlying biological data, simple assessments of age-based and reproductive life histories of a wider range of rabbitfish species represents an important area of future research.

Conclusions

In drawing up this list of emerging research directions, thoughts necessarily turn to what the future holds for populations and assemblages of rabbitfishes across their native and expanding ranges. Rabbitfishes, like other tropical fishes, are exposed to a range of threats; from the local human stressors such as fishing and declining water quality to the global effects of climate change. Collectively, these stressors are transforming tropical seascapes, the biological communities they support, and the goods and services they provide. As we write, the Great Barrier Reef (GBR) has just experienced another widespread coral bleaching event (the fifth bleaching event since 2016), and reefs in Australia's Coral Sea Marine Park has just experienced its sixth major bleaching event in the past nine years, with some regions experiencing > 16 degree heating weeks (Hoey et al. 2024). These bleaching events come on the back of numerous severe cyclones and ongoing population irruptions of the corallivorous Crown-of-Thorns starfish. With predictions of more frequent, more intense, and more diverse disturbances under ongoing climate change further loss and degradation of habitats are likely, if not inevitable. But how will the degradation of tropical seascapes impact rabbitfishes?

The apparent insensitivity of rabbitfishes to habitat change and degradation, coupled with their broad tolerances for physico-chemical conditions (Hoey et al. – Chapter 7), rapid growth, short generation times and high fecundity (Taylor et al. – Chapter 5) may allow rabbitfish populations to persist and even flourish in future tropical seascapes. Given several hundred million people globally are dependent on tropical fishes as a source of protein and micronutrients and predicted decreases in other fish taxa under climate change, it is more important than ever to further our understanding of the biology and ecology of rabbitfishes.

References

Adam, T.C., R.J. Schmitt, S.J. Holbrook, A.J. Brooks, P.J. Edmunds, R.C. Carpenter et al. 2011. Herbivory, connectivity, and ecosystem resilience: response of a coral reef to a large-scale perturbation. PLoS One 6: e23717.

Bauman, A.G., A.S. Hoey, G. Dunshea, D. Feary, J. Low and P.A. Todd. 2017. Macroalgal browsing on a heavily degraded, urbanized equatorial reef system. Sci. Repo. 7: 8352.

Bellwood, D.R. and C.H.R. Goatley. 2017. Can biological invasions save Caribbean coral reefs? Current Biology 27: R13–R14.

Bellwood, D.R., T.P. Hughes and A.S. Hoey. 2006. Sleeping functional group drives coral-reef recovery. Current Biology 16: 2434–2439.

Bradley, M., A. Dubuc, C.V. Piggott, K. Sambrook, A.S. Hoey, M. Depczynski et al. 2024. The fish–mangrove link is context dependent: Tidal regime and reef proximity determine the ecological role of tropical mangroves. Fish and Fisheries 25: 523–541.

Cheal, A., M. Emslie, I. Miller and H. Sweatman. 2012. The distribution of herbivorous fishes on the Great Barrier Reef. Mar. Biol. 159: 1143–1154.

Choat, J.H., O.S. Klanten, L. Van Herwerden, D.R. Robertson and K.D. Clements. 2012. Patterns and processes in the evolutionary history of parrotfishes (Family Labridae). Biol. J. Linn. Soc 107: 529–557.

Claydon, J.A.B., M.C. Calosso, G.A. De Leo and R.B.J. Peachey. 2015. Spatial and demographic consequences of nursery-dependence in reef fishes: an empirical and simulation study. Mar. Ecol. Prog. Ser. 525: 171 183.

Clements, K.D., D.P. German, J. Piché, A. Tribollet and J.H. Choat. 2017. Integrating ecological roles and trophic diversification on coral reefs: multiple lines of evidence identify parrotfishes as microphages. Biol. J. Linn. Soc. 120: 729–751.

Dunic, J.C., C.J. Brown, R.M. Connolly, M.P. Turschwel and I.M. Côté. 2021. Long-term declines and recovery of meadow area across the world's seagrass bioregions. Glob. Change Biol. 27: 4096–4109.

Eggertsen, L., W. Goodell, C.A. Cordeiro, T.C. Mendes, G.O. Longo, C.E. Ferreira et al. 2020. Seascape configuration leads to spatially uneven delivery of parrotfish herbivory across a Western Indian Ocean seascape. Diversity 12: 434.

Fox, R.J. and D.R. Bellwood. 2008. Remote video bioassays reveal the potential feeding impact of the rabbitfish *Siganus canaliculatus* (f: Siganidae) on an inner-shelf reef of the Great Barrier Reef. Coral Reefs 27: 605–615.

Friedman, M. and G. Carnevale. 2018. The Bolca Lagerstätten: shallow marine life in the Eocene. J. Geolog. Soc. 175: 569–579.

Froese, R. and D. Pauly. 2024. Fishbase. www.fishbase.de.

Gilmour, J.P., L.D. Smith, A.J. Heyward, A.H. Baird and M.S. Pratchett. 2013. Recovery of an isolated coral reef system following severe disturbance. Science 340: 69–71.

Goldberg, L., D. Lagomasino, N. Thomas and T. Fatoyinbo. 2020. Global declines in human-driven mangrove loss. Glob. Change Biol. 26: 5844–5855.

Graba-Landry, A. 2020. The effect of increasing temperature on algae-fish interactions on coral reefs. Doctoral dissertation, James Cook University, Townsville.

Graham, N.A., S. Jennings, M.A. MacNeil, D. Mouillot and S.K. Wilson. 2015. Predicting climate-driven regime shifts versus rebound potential in coral reefs. Nature 518: 94–97.

Hamilton, R.J., G.R. Almany, C.J. Brown, J. Pita, N.A. Peterson and J.H. Choat. 2017. Logging degrades nursery habitat for an iconic coral reef fish. Biol. Conserv. 210: 273–280.

Hoey, A.S. and D.R. Bellwood. 2008. Cross-shelf variation in the role of parrotfishes on the Great Barrier Reef. Coral Reefs 27: 37–47.

Hoey, A.S. and D.R. Bellwood. 2009. Limited functional redundancy in a high diversity system: single species dominates key ecological process on coral reefs. Ecosystems 12: 1316–1328.

Hoey, A.S., S.J. Brandl and D.R. Bellwood. 2013. Diet and cross-shelf distribution of rabbitfishes (f. Siganidae) on the northern Great Barrier Reef: implications for ecosystem function. Coral Reefs 32: 973–984.

Hoey, A.S., H. Harrison and M. Pratchett. 2020. Coral Reef Health in the Coral Sea Marine Park – Surveys 2018–2020. Report prepared for Parks Australia.

Hoey, A.S., E. Howells, J.L. Johansen, J.P.A. Hobbs, V. Messmer, D.M. McCowan et al. 2016. Recent advances in understanding the effects of climate change on coral reefs. Diversity 8: 12.

Hoey, A.S., D. Burn, J.F. Chandler, V. Huertas, B. Cresswell, G. Galbraith et al. 2024. Coral Sea Marine Park coral reef health survey 2023-24. Report prepared for Parks Australia.

Hughes, T.P. 1994. Catastrophes, phase shifts, and large-scale degradation of a Caribbean coral reef. Science 265: 1547–1551.

Hughes, T.P., M.J. Rodrigues, D.R. Bellwood, D. Ceccarelli, O. Hoegh-Guldberg, L. McCook et al. 2007. Phase shifts, herbivory, and the resilience of coral reefs to climate change. Curr. Biol. 17: 360-365.

Hughes, T.P., J.T. Kerry, M. Álvarez-Noriega, J.G. Álvarez-Romero, K.D. Anderson, A.H. Baird et al. 2017. Global warming and recurrent mass bleaching of corals. Nature 543: 373–377.

Ignatius, B. 2009. Broodstock development, breeding and larval rearing of *Siganus canaliculatus*. pp. 54–55. *In*: Madhu, K. and R. Madhu [eds.]. Recent Advances in Breeding and Larviculture of Marine Finfish and Shellfish. Central Marine Fisheries Institute, Kerala, India.

Johansen, J.L., V. Messmer, D.J. Coker, A.S. Hoey and M.S. Pratchett. 2014. Increasing ocean temperatures reduce activity patterns of a large commercially important coral reef fish. Glob. Change Biol. 20: 1067–1074.

Johansen, J.L., M.S. Pratchett, V. Messmer, D.J. Coker, A.J. Tobin and A.S. Hoey. 2015. Large predatory coral trout species unlikely to meet increasing energetic demands in a warming ocean. Sci. Rep. 5: 13830.

Krebs, C.J. 1972. Ecology: the Experimental Analysis of Distribution and Abundance. Harper and Row, New York.

Lam, T.J. 1974. Siganids: Their biology and mariculture potential. Aquaculture 3: 325–354

LaMonica, L.E., R.J. Fox and J.M. Donelson. 2021. Thermal sensitivity of juvenile rabbitfishes *Siganus doliatus* and *S. lineatus* (Siganidae): a key role for habitat? Coral Reefs 40: 1307–1320.

Lewis, S.M. 1986. The role of herbivorous fishes in the organization of a Caribbean reef community. Ecol. Monogr. 56: 183–200.

Loffler, Z., D.R. Bellwood and A.S. Hoey. 2015. Among-habitat algal selectivity by browsing herbivores on an inshore coral reef. Coral Reefs 34: 597–605.

Maina, G.W., M. Samoilys, H. Alidina and K. Osuka. 2013. Targeted fishing of the shoemaker spinefoot rabbitfish, Siganus sutor, on potential spawning aggregations in southern Kenya. Reef fish spawning aggregations in the Western Indian Ocean: Research for management. WIOMSA/SIDA/SFA/CORDIO. WIOMSA Book Series, 13.

Martin, C., E.C. McClure, K. Webber, M. Burgo and A.S. Hoey. 2024. Utilisation of macroalgal habitats by juvenile rabbitfishes. Coral Reefs, doi: 10.1007/s00338-024-02581-x.

McCord, C.L., C.M. Nash, W.J. Cooper and M.W. Westneat. 2021. Phylogeny of the damselfishes (Pomacentridae) and patterns of asymmetrical diversification in body size and feeding ecology. PLoS One 16: e0258889.

Moleana, T., L. Della Patrona, T. Meziane and Y. Letourneur. 2016. First record of *Siganus randalli* (Teleost, Siganidae) in New Caledonia, and comments on its diet. Mar. Biodiv. Rec. 9: 1–5.

Motson, K. and J.M. Donelson. 2017. Limited capacity for developmental thermal acclimation in three tropical wrasses. Coral Reefs 36: 609–621.

Mumby, P.J., A.J. Edwards, J.E. Arias-González, K.C. Lindeman, P.G. Blackwell, A. Gall et al. 2004. Mangroves enhance the biomass of coral reef fish communities in the Caribbean. Nature 427: 533–536.

Nay, T.J., J.L. Johansen, J.L. Rummer, J.F. Steffensen and A.S. Hoey. 2021. Species interactions alter the selection of thermal environment in a coral reef fish. Oecologia 196: 363–371.

Nay, T.J., J.L. Johansen, J.L. Rummer, J.F. Steffensen, M.S. Pratchett and A.S. Hoey. 2020. Habitat complexity influences selection of thermal environment in a common coral reef fish. Conserv. Physiol. 8: coaa070.

Nicholson, G.M. and K.D. Clements. 2023a. Micro-photoautotroph predation as a driver for trophic niche specialization in 12 syntopic Indo-Pacific parrotfish species. Biol. J. Linn. Soc. 139: 91–114.

Nicholson, G.M. and K.D. Clements. 2023b. Fine-scale analysis of substrata grazed by parrotfishes (Labridae: Scarini) on the outer-shelf of the Great Barrier Reef, Australia. Mar. Biol. 170: 121.

Nicholson, G.M. and K.D. Clements. 2024. A role for encrusting, endolithic sponges in the feeding of the parrotfish *Scarus rubroviolaceus*? Evidence of further trophic diversification in Indo-Pacific Scarini. Coral Reefs 43: 541–556.

Nicholson, G.M. and K.D. Clements. 2021. Ecomorphological divergence and trophic resource partitioning in 15 syntopic Indo-Pacific parrotfishes (Labridae: Scarini). Biol. J. Linn. Soc. 132: 590–611.

Nilsson, G.E., N. Crawley, I.G. Lunde and P.L. Munday. 2009. Elevated temperature reduces the respiratory scope of coral reef fishes. Glob. Change Biol. 15: 1405–1412.

Parenti, P. and J.E. Randall. 2011. Checklist of the species of the families Labridae and Scaridae: an update. Smithiana Bull. 13: 29–44.

Pell, T.J., E.C. McClure, M.J. Emslie and A.S. Hoey. 2024. Limited susceptibility of *Lobophora* to browsing fishes on inshore reefs of the Great Barrier Reef. Coral Reefs 43: 5–17.

Pombo-Ayora, L., D.J. Coker, S. Carvalho, G. Short and M.L. Berumen. 2020. Morphological and ecological trait diversity reveal sensitivity of herbivorous fish assemblages to coral reef benthic conditions. Mar. Environ. Res. 162: 105102.

Pörtner, H.O. and A.P. Farrell. 2008. Physiology and climate change. Science 322: 690–692.

Priest, M.A., A.R. Halford, K.D. Clements, E. Douglas, S.L. Abellana and J.L. McIlwain. 2016. What a difference a bay makes: natural variation in dietary resources mediates growth in a recently settled herbivorous fish. Coral Reefs 35: 1187–1199.

Rasher, D.B., A.S. Hoey and M.E. Hay. 2013. Consumer diversity interacts with prey defenses to drive ecosystem function. Ecology 94: 1347–1358.

Robinson, J., M.A. Samoilys, E. Grandcourt, D. Julie, M. Cedras and C. Gerry. 2011. The importance of targeted spawning aggregation fishing to the management of Seychelles' trap fishery. Fish. Res. 112: 96-103.

Roff, G., S. Bejarano, M. Priest, A. Marshell, I. Chollett, R.S. Steneck et al. 2019. Seascapes as drivers of herbivore assemblages in coral reef ecosystems. Ecol. Monogr. 89: e01336.

Russ, G.R., S.A. Questel, J.R. Rizzari and A.C. Alcala. 2015. The parrotfish–coral relationship: refuting the ubiquity of a prevailing paradigm. Mar. Biol. 162: 2029–2045.

Sambrook, K., A.S. Hoey, S. Andréfouët, G.S. Cumming, S. Duce and M.C. Bonin. 2019. Beyond the reef: The widespread use of non-reef habitats by coral reef fishes. Fish Fisher. 20: 903–920.

Sambrook, K., M.C. Bonin, M. Bradley, G.S. Cumming, S. Duce, S. Andréfouët et al. 2020. Broadening our horizons: seascape use by coral reef-associated fishes in Kavieng, Papua New Guinea, is common and diverse. Coral Reefs 39: 1187–1197.

Siqueira, A.C., D.R. Bellwood and P.F. Cowman. 2019. The evolution of traits and functions in herbivorous coral reef fishes through space and time. Proc. R. Soc. B 286: 20182672.

Sorenson, L., F. Santini, G. Carnevale and M.E. Alfaro. 2013. A multi-locus timetree of surgeonfishes (Acanthuridae, Percomorpha), with revised family taxonomy. Mol. Phylogen. Evol. 68: 150–160.

Tang, S., A. Graba-Landry and A.S. Hoey. 2020. Density and height of *Sargassum* influence rabbitfish (f. Siganidae) settlement on inshore reef flats of the Great Barrier Reef. Coral Reefs 39: 467–473.

Tano, S.A., M. Eggertsen, S.A. Wikström, C. Berkström, A.S. Buriyo and C. Halling. 2017. Tropical seaweed beds as important habitats for juvenile fish. Mar. Freshw. Res. 68: 1921–1934.

Taylor, B.M., S.J. Lindfield and J.H. Choat. 2015. Hierarchical and scale-dependent effects of fishing pressure and environment on the structure and size distribution of parrotfish communities. Ecography 38: 520–530.

Taylor, B.M., C.E. Benkwitt, J.H. Choat, K.D. Clements, N.A. Graham and M.G. Meekan. 2020. Synchronous biological feedbacks in parrotfishes associated with pantropical coral bleaching. Glob. Change Biol. 26: 1285–1294.

Tebbett, S.B., A.C. Siqueira and D.R. Bellwood. 2022. The functional roles of surgeonfishes on coral reefs: past, present and future. Rev. Fish Biol. Fisher. 32: 387–439.

Tyler, J.C. and A.F. Bannikov. 1997. Relationships of the fossil and Recent genera of rabbitfishes (Acanthuroidei: Siganidae). Smithson. Contrib. Paleobiol. No. 14., Smithsonian Institution Press, Washington DC.

Vergés, A., C. Doropoulos, H.A. Malcolm, M. Skye, M. Garcia-Pizá, E.M. Marzinelli et al. 2016. Long-term empirical evidence of ocean warming leading to tropicalization of fish communities, increased herbivory, and loss of kelp. Proc. Natl Acad. Sci. USA 113: 13791–13796.

Webber, K., K. Fabricius, S.K. Wilson and A.S. Hoey. 2024. The influence of *Sargassum* biomass and thallus density on the recruitment of coral reef fishes. Coral Reefs 43: 687–699.

Wismer, S., A.S. Hoey and D.R. Bellwood. 2009. Cross-shelf benthic community structure on the Great Barrier Reef: relationships between macroalgal cover and herbivore biomass. Mar. Ecol. Prog. Ser. 376: 45–54.

Woodland, D.J. 1990. Revision of the fish family Siganidae with descriptions of two new species and comments on distribution and biology. Indo-Pacific Fishes 19.

Zarco-Perello, S., N.E. Bosch, S. Bennett, M.A. Vanderklift and T. Wernberg. 2021. Persistence of tropical herbivores in temperate reefs constrains kelp resilience to cryptic habitats. J. Ecol. 109: 2081–2094.

Zarco-Perello, S., D. Fairclough, C. Dowling, J. DiBattista, R. Austin, T. Wernberg et al. 2022. Maximization of fitness by phenological and phenotypic plasticity in range expanding rabbitfish (Siganidae). J. Anim. Ecol. 91: 1666–1678.

Zarco-Perello, S., S. Martin and A.S. Hoey. 2024. Biogeographical diet variation within and between the rabbitfishes *Siganus corallinus, Siganus virgatus, Siganus doliatus* and *Siganus trispilos*. Ecol. Evol. 14: e11326.

Index

G

H

I

J

K

L

M

N

O

P

R

S

T

For Product Safety Concerns and Information please contact our
EU representative GPSR@taylorandfrancis.com Taylor & Francis
Verlag GmbH, Kaufingerstraße 24, 80331 München, Germany